DATE DUE

Vitamins and Hormones

Volume 68

Editorial Board

TADHG P. BEGLEY
ANTHONY R. MEANS
BERT W. O'MALLEY
LYNN RIDDIFORD
ARMEN H. TASHJIAN, JR.

VITAMINS AND HORMONES

ADVANCES IN RESEARCH AND APPLICATIONS

Nuclear Receptor Coregulators

Editor-in-Chief

GERALD LITWACK

Professor and Chair Emeritus
Department of Biochemistry and Molecular Pharmacology
Thomas Jefferson University Medical College
Philadelphia, Pennsylvania
Visiting Scholar
Department of Biological Chemistry
David Geffen School of Medicine at UCLA
Toluca Lake, California

VOLUME 68

ELSEVIER
ACADEMIC
PRESS

AMSTERDAM • BOSTON • HEIDELBERG • LONDON
NEW YORK • OXFORD • PARIS • SAN DIEGO
SAN FRANCISCO • SINGAPORE • SYDNEY • TOKYO

Academic Press is an imprint of Elsevier

Elsevier Academic Press
525 B Street, Suite 1900, San Diego, California 92101-4495, USA
84 Theobald's Road, London WC1X 8RR, UK

This book is printed on acid-free paper. ∞

Copyright © 2004, Elsevier Inc. All Rights Reserved.

No part of this publication may be reproduced or transmitted in any form or by any means, electronic or mechanical, including photocopy, recording, or any information storage and retrieval system, without permission in writing from the Publisher.

The appearance of the code at the bottom of the first page of a chapter in this book indicates the Publisher's consent that copies of the chapter may be made for personal or internal use of specific clients. This consent is given on the condition, however, that the copier pay the stated per copy fee through the Copyright Clearance Center, Inc. (www.copyright.com), for copying beyond that permitted by Sections 107 or 108 of the U.S. Copyright Law. This consent does not extend to other kinds of copying, such as copying for general distribution, for advertising or promotional purposes, for creating new collective works, or for resale. Copy fees for pre-2004 chapters are as shown on the title pages. If no fee code appears on the title page, the copy fee is the same as for current chapters.
0083-6729/2004 $35.00

Permissions may be sought directly from Elsevier's Science & Technology Rights Department in Oxford, UK: phone: (+44) 1865 843830, fax: (+44) 1865 853333, e-mail: permissions@elsevier.com.uk. You may also complete your request on-line via the Elsevier homepage (http://elsevier.com), by selecting "Customer Support" and then "Obtaining Permissions."

For all information on all Academic Press Publications
visit our Web site at www.academicpress.com

ISBN: 0-12-709868-2

PRINTED IN THE UNITED STATES OF AMERICA
04 05 06 07 08 9 8 7 6 5 4 3 2 1

Former Editors

ROBERT S. HARRIS
Newton, Massachusetts

JOHN A. LORRAINE
*University of Edinburgh
Edinburgh, Scotland*

PAUL L. MUNSON
*University of North Carolina
Chapel Hill, North Carolina*

JOHN GLOVER
*University of Liverpool
Liverpool, England*

GERALD D. AURBACH
*Metabolic Diseases Branch
National Institute of Diabetes
and Digestive and Kidney Diseases
National Institutes of Health
Bethesda, Maryland*

KENNETH V. THIMANN
*University of California
Santa Cruz, California*

IRA G. WOOL
*University of Chicago
Chicago, Illinois*

EGON DICZFALUSY
*Karolinska Sjukhuset
Stockholm, Sweden*

ROBERT OLSEN
*School of Medicine
State University of New York
at Stony Brook
Stony Brook, New York*

DONALD B. MCCORMICK
*Department of Biochemistry
Emory University School of Medicine
Atlanta, Georgia*

Contents

Contributors xiii
Preface xv

1

Gene Silencing by Nuclear Orphan Receptors

Ying Zhang and Maria L. Dufau

I. Introduction 2
II. COUP-TFI, II (EAR3, ARP-1), and EAR2: Inhibitors of Diverse Genes by Multiple Mechanisms 7
III. DAX-1: Silencing in the Control of Steroidogenic and Sex-Determining Target Genes 14
IV. GCNF: Negative Control during Gametogenesis and Embryonic Development 20
V. SHP: Generic Heterodimeric Partner-Inhibiting Multiple Nuclear Receptor Pathways 24
VI. TR4 and TR2: Testicular Receptor with Homologous but Not Redundant Functions 30
VII. Conclusions 34
References 35

2

STRUCTURE AND FUNCTION OF THE GLUCOCORTICOID RECEPTOR LIGAND BINDING DOMAIN

RANDY K. BLEDSOE, EUGENE L. STEWART, AND KENNETH H. PEARCE

    I. Functional Significance of the GR   51
   II. Expression, Purification, and Characterization of the GR LBD   55
  III. Fold of the Nuclear Receptor LBD and Specific Features of GR   59
  IV. Ligand Recognition by GR   64
   V. Modes of Ligand Recognition by GR Compared to Other Oxosteroid Receptors   68
  VI. Characteristics of Cofactor Association with the GR LBD   71
 VII. Role of the GR LBD in Chaperone Protein Association and Receptor Dimerization   74
VIII. Mutations in the GR LBD and Their Functional Consequences   76
  IX. Progress Toward a Selective GR Modulator   82
   X. Conclusions and Directions of Future Research   83
       References   84

3

NUCLEAR RECEPTOR RECRUITMENT OF HISTONE-MODIFYING ENZYMES TO TARGET GENE PROMOTERS

CHIH-CHENG TSAI AND JOSEPH D. FONDELL

   I. Overview of Nuclear Receptors   94
  II. Chromatin Structure   96
 III. Histone Modifications in Regulated Transcription   98
 IV. NR Coactivators with Histone-Modifying Activity   103
  V. NR Corepressors with Histone-Modifying Activity   109
 VI. Conclusion   114
     References   115

4

COREPRESSOR RECRUITMENT BY AGONIST-BOUND NUCLEAR RECEPTORS

JOHN H. WHITE, ISABELLE FERNANDES, SYLVIE MADER, AND XIANG-JIAO YANG

I. The Nuclear Receptor Superfamily 124
II. Coregulatory Proteins in Hormone-Dependent Regulation of Transcription 127
III. Histone Deacetylases in Regulation of Gene Expression 132
IV. LCoR and RIP140 Recruit the Corepressor C-Terminal Binding Protein (CtBP) 135
V. Potential Roles of LCoR and RIP140 in Hormone-Dependent Receptor Function 136
VI. Concluding Remarks 137
References 138

5

PHARMACOLOGY OF NUCLEAR RECEPTOR—COREGULATOR RECOGNITION

RAJESH S. SAVKUR, KELLI S. BRAMLETT, DAVID CLAWSON, AND THOMAS P. BURRIS

I. Introduction 146
II. Coregulators Involved in Nuclear Receptor Action 150
III. Nuclear Receptor—Coregulator Recognition 164
IV. Pharmacology of Nuclear Receptor—Coregulator Recognition 170
V. Conclusion 173
References 174

6

THYROID HORMONE RECEPTOR SUBTYPES AND THEIR INTERACTION WITH STEROID RECEPTOR COACTIVATORS

ROY E. WEISS AND HELTON E. RAMOS

    I. Introduction   186
   II. Physiology of TR Function   190
  III. Physiology of SRC Function (SRC-1, SRC-2, and SRC-3)   194
  IV. Interaction of TRs and SRC in the Pituitary   196
   V. SRC-1 and TRs in Peripheral Tissues   200
  VI. Conclusions   202
      References   203

7

COREPRESSOR REQUIREMENT AND THYROID HORMONE RECEPTOR FUNCTION DURING XENOPUS DEVELOPMENT

LAURENT M. SACHS

    I. Introduction   210
   II. Mechanism of TR Action   212
  III. Dual Role of TRs During Amphibian Development   216
  IV. Corepressor Function During Amphibian Development   219
   V. Conclusion and Perspectives   225
      References   226

8

CDC25B AS A STEROID RECEPTOR COACTIVATOR

STEVEN S. CHUA, ZHIQING MA, ELLY NGAN, AND SOPHIA Y. TSAI

    I. Introduction   233
   II. The Cdc25 Family of Proteins   234
  III. Evaluation of the Cdc25B Role in Murine Mammary Glands   236
  IV. Cdc25B: A Coactivator that Enhances Steroid Receptor-Dependent Transcription   239

V. Coactivator Function of Cdc25B in the Prostate 245
VI. Conclusions 248
   References 252

9

VITAMIN D RECEPTOR—DNA INTERACTIONS

PAUL L. SHAFFER AND DANIEL T. GEWIRTH

I. Introduction 258
II. VDR–DR3 Binding 261
III. Alternative Response Elements 266
IV. RXR–VDR Formation 269
V. Conclusions 270
   References 271

INDEX 275

CONTRIBUTORS

Numbers in parentheses indicate the pages on which the authors' contributions begin.

Randy K. Bledsoe (49) Department of Gene Expression and Protein Biochemistry, Discovery Research, GlaxoSmithKline, Research Triangle Park, North Carolina.

Kelli S. Bramlett (145) Lilly Research Laboratories, Eli Lilly and Company, Indianapolis, Indiana.

Thomas P. Burris (145) Lilly Research Laboratories, Eli Lilly and Company, Indianapolis, Indiana.

Steven S. Chua (231) Department of Molecular and Cellular Biology, Baylor College of Medicine, Houston, Texas.

David Clawson (145) Lilly Research Laboratories, Eli Lilly and Company, Indianapolis, Indiana.

Maria L. Dufau (1) Section on Molecular Endocrinology, Endocrinology, and Reproduction Research Branch, National Institutes of Health, Bethesda, Maryland.

Isabelle Fernandes (123) Department of Physiology, McGill University, McIntyre Medical Sciences Building, Montreal, Quebec, Canada.

Joseph D. Fondell (93) Department of Physiology and Biophysics, UMDNJ, Robert Wood Johnson Medical School, Piscataway, New Jersey.

Daniel T. Gewirth (257) Department of Biochemistry, Duke University Medical Center, Durham, North Carolina.

Zhiqing Ma (231) Lexicon Genetics, Woodland, Texas.

Sylvie Mader (123) Department of Medicine, McGill University, McIntyre Medical Sciences Building, Montreal, Quebec, Canada; Department of Biochemistry, Université de Montréal, Montreal, Quebec, Canada.

Elly Ngan (231) Department of Zoology, The University of Hong Kong, Hong Kong, SAR, PRC.

Kenneth H. Pearce (49) Department of Gene Expression and Protein Biochemistry, Discovery Research, GlaxoSmithKline, Research Triangle Park, North Carolina.

Helton E. Ramos (185) University of Chicago, Thyroid Study Unit, Chicago, Illinois.

Laurent M. Sachs (209) Département Régulations, Développement et Diversité Moléculaire, USM 501, Muséum National d'Histoire Naturelle, UMR-5166 CNRS, Paris cedex 05, France.

Rajesh S. Savkur (143) Lilly Research Laboratories, Eli Lilly and Company, Indianapolis, Indiana.

Paul L. Shaffer (257) Department of Biochemistry, Duke University Medical Center, Durham, North Carolina.

Eugene L. Stewart (49) Department of Computational, Analytical, and Structural Sciences, Discovery Research, GlaxoSmithKline, Research Triangle Park, North Carolina.

Chih-Cheng Tsai (93) Department of Physiology and Biophysics, UMDNJ, Robert Wood Johnson Medical School, Piscataway, New Jersey.

Sophia Y. Tsai (231) Department of Molecular and Cellular Biology, Baylor College of Medicine, Houston, Texas.

Roy E. Weiss (185) University of Chicago, Thyroid Study Unit, Chicago, Illinois.

John H. White (123) Departments of Physiology and Medicine, McGill University, McIntyre Medical Sciences Building, Montreal, Quebec, Canada.

Xiang-Jiao Yang (123) Department of Medicine, McGill University, McIntyre Medical Sciences Building, Montreal, Quebec, Canada.

Ying Zhang (1) Section on Molecular Endocrinology, Endocrinology, and Reproduction Research Branch, National Institutes of Health, Bethesda, Maryland.

Preface

This volume is a collection of manuscripts devoted to the steroid receptor class and emphasizes the roles of coactivators and corepressors in the actions of these receptors.

The book begins with a paper on "Gene Silencing by Nuclear Orphan Receptors" by Y. Zhang and M. Dufau. A structural paper entitled "Structure and Function of the Glucocorticoid Receptor Ligand Binding Domain" is offered by R. K. Bledsoe, E. L. Stewart, and K. H. Pearce. C.-C. Tsai and J. D. Fondell review "Nuclear Receptor Recruitment of Histone Modifying Enzymes to Target Gene Promoters" and J. H. White, I. Fernandes, S. Mader, and X.-J. Yang discuss "Corepressor Recruitment by Agonist-Bound Nuclear Receptors." The "Pharmacology of Nuclear Receptor-Coactivator Recognition" is the subject offered by R. S. Savkur, K. S. Bramlett, D. Clawson, and T. P. Burris. Next, R. E. Weiss and H. E. Ramos report on "Thyroid Hormone Receptor (TR) Subtypes and Their Interaction with Steroid Receptor Coactivators," while L. M. Sachs discusses "Corepressors Requirement and Thyroid Hormone Receptor Function during *Xenopus* Development." "Cdc25B as a Steroid Receptor Coactivator" is contributed by S. S. Chua, Z. Ma, E. Ngan, and S. Y. Tsai. Finally, P. Shaffer and D. T. Gewirth provide: "Vitamin D Receptor-DNA Interactions."

The Publisher intends Vitamins and Hormones to become available electronically in the near future. This should add to the size of the audience for this book Serial and increase its availability.

Gerry Litwack
Toluca Lake, California
9th January, 2004

1

Gene Silencing by Nuclear Orphan Receptors

Ying Zhang and Maria L. Dufau

Section on Molecular Endocrinology, Endocrinology, and Reproduction Research Branch, National Institutes of Health Bethesda, Maryland 20892

I. Introduction
II. COUP-TFI, II (EAR3, ARP-1), and EAR2: Inhibitors of Diverse Genes by Multiple Mechanisms
III. DAX-1: Silencing in the Control of Steroidogenic and Sex-Determining Target Genes
IV. GCNF: Negative Control during Gametogenesis and Embryonic Development
V. SHP: Generic Heterodimeric Partner-Inhibiting Multiple Nuclear Receptor Pathways
VI. TR4 and TR2: Testicular Receptor with Homologous but Not Redundant Functions
VII. Conclusions
References

Nuclear orphan receptors represent a large and diverse subgroup in the nuclear receptor superfamily. Although putative ligands for these orphan members remain to be identified, some of these receptors possess intrinsic activating, inhibitory, or dual regulatory functions in development, differentiation, homeostasis, and reproduction. In particular, gene-silencing events elicited by chicken ovalbumin upstream promoter-transcription factors (COUP-TFs); dosage-sensitive sex reversal-adrenal hypoplasia congenita critical region on the X chromosome, gene 1 (DAX-1); germ cell nuclear factor (GCNF); short heterodimer partner (SHP); and testicular receptors 2 and 4 (TR2 and TR4) are among the best characterized. These orphan receptors are critical in controlling basal activities or hormonal responsiveness of numerous target genes. They employ multiple and distinct mechanisms to mediate target gene repression. Complex cross-talk exists between these orphan receptors at their cognate DNA binding elements and an array of steroid/nonsteroid hormone receptors, other transcriptional activators, coactivators and corepressors, histone modification enzyme complexes, and components of basal transcriptional components. Therefore, perturbation induced by these orphan receptors at multiple levels, including DNA binding activities, receptor homo- or heterodimerization, recruitment of cofactor proteins, communication with general transcriptional machinery, and changes at histone acetylation status and chromatin structures, may contribute to silencing of target gene expression in a specific promoter or cell-type context. Moreover, the findings derived from gene-targeting studies have demonstrated the significance of these orphan receptors' function in physiologic settings. Thus, COUP-TFs, DAX-1, GCNF, SHP, and TR2 and 4 are known to be required for multiple physiologic and biologic functions, including neurogenesis and development of the heart and vascular system steroidogenesis and sex determination, gametogenesis and embryonic development, and cholesterol/lipid homeostasis. © 2004 Elsevier Inc.

I. INTRODUCTION

The largest superfamily of eukaryotic transcription factors comprises nuclear receptors and contains more than 150 proteins, including steroid/nonsteroid nuclear hormone receptors and orphan receptors (Mangelsdorf et al., 1995; Perlmann and Evans, 1997). The nuclear hormone receptors are ligand-inducible transcripton factors that have critical roles in the control of development, differentiation, homeostasis, and reproduction. These diverse actions are initiated by their specific binding to small lipophilic molecules such as steroid and thyroid hormones, retinoids, and vitamin D_3 (Harvey and Williams, 2002; Haussler et al., 1997; Ruberte, 1994). The majority of nuclear receptors are orphan receptors for which no ligand was initially identified (Giguere, 1999). During recent years, the identification of novel ligands for a number of these receptors, which include the retinoid X receptor (RXR) (Mangelsdorf et al., 1992), peroxisome proliferator-activated receptor (PPAR) (Devchand et al., 1999), farnesoid X receptor (FXR), (Niesor et al., 2001; Tu et al., 2000), liver X receptor (LXR) (Peet et al., 1998), and pregnane X receptor (PXR) (Honkakoski et al., 2003), has led to the characterization of their functional roles and an understanding of their significance in metabolic regulation (Fitzgerald et al., 2002). At this stage, it is not known whether the numerous unliganded orphan receptors exert their diverse actions in a ligand-independent manner or are regulated by specific ligands that are yet to be identified. Since nuclear orphan receptors occur in nearly all species examined and outnumber the steroid/nonsteroid hormone receptors, they probably represent an ancient and complex class of regulatory proteins that participate in multiple physiologic functions. Many nuclear orphan receptors exert constitutively active regulation on numerous target genes through their intrinsic activation/repression activities or through multiple modes of cross-talk with nuclear hormone receptors. This chapter focuses on the silencing mechanisms of nuclear orphan receptor-mediated regulation of gene expression. To facilitate the understanding of down-regulation in the context of a nuclear receptor superfamily, a brief account of the structural organization of nuclear receptors in relationship with their function is presented.

Nuclear receptors are modular proteins and consist of five characteristic domains (Fig. 1A). These include an N-terminal region (A/B), a DNA binding domain (C), a hinge region (domain D), and C-terminal ligand binding domains (E and F) (Evans, 1988). The A/B domain is highly variable among the members of the nuclear receptor family. It possesses a ligand-independent transcriptional activation function, which is referred to as AF-1. The AF-1 mediated gene activation often displays promoter-dependent and cell-type-dependent specificities, indicating that this domain may be responsible for interactions with cell-specific coregulatory proteins (Bastien et al., 2000; Rochette-Egly et al., 1997). Domain C, the DNA binding domain (DBD), is highly conserved and is a signature motif of the nuclear receptor superfamily. The DBD targets receptors to specific DNA sequences, known as hormone response elements (HREs), in the promoter region of target genes. Two zinc finger motifs in this domain serve as an interaction surface to contact a hexameric DNA core sequence containing nucleotides AGGTCA (Freedman and Towers, 1991; Umesono and Evans, 1989). The unique actions of nuclear receptors are highly specific and dependent on DNA-receptor interactions and on receptor-receptor dimerization. Accordingly, they usually bind as homodimers or heterodimers to two copies of the response elements that can be arranged as inverted (palindromic), everted, or direct repeats. Receptors for estrogen or glucocorticoid steroid hormones recognize palindromic HREs as homodimers (Evans, 1988). In contrast, most nonsteroid receptors heterodimerize with RXR at various direct-repeat motifs, and the spacing between the two repeats often dictates the binding specificities of different heterodimers (Mangelsdorf and Evans, 1995). Such heterodimers have been shown to function as dynamic transcription factors in which the two partners influence each other's ability to interact with ligands and cofactors. The DBDs also serve to stabilize receptor

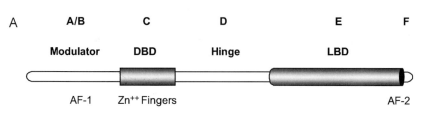
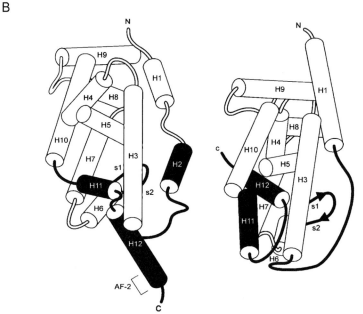

FIGURE 1. Schematic structure of a nuclear receptor and the ligand binding domain (LBD) in absence or presence of a ligand. (A) A typical nuclear receptor with its five characteristic domains. The N-terminal A/B domain is a variable among nuclear receptors family members and harbors a ligand-independent activation function (AF-1). The domain C represents the highly conserved DNA binding domain (DBD), which is the signature motif of the nuclear receptors. The hinge domain D as a nonconserved linker region connects the DBD to the C-terminal E and F domains where E is the LBD binding domain (LBD) responsible for multiple functions of the receptor. The ligand-inducible activation function (AF-2) is located at the C-terminal of the LBD. (B, Left) Crystal structure of LBD of the unligand human RXRα receptor. (B, Right) Crystal structure of LBD of the RARγ receptor bound to all-trans retinoic acid. There are 12 α-helices (numbered H1 to H12) that are indicated by cylinders, where the orientation and position of H12 (containing AF-2) undergoes significant changes after ligand binding (Bourguet et al., 1995; Renaud et al., 1995). Reprinted with permission from *Nature*.

dimerization on the DNA template and have a fundamental role in DNA sequence recognition (Rastinejad et al., 1995). Moreover, several nuclear orphan receptors, including steroidogenic factor-1 (SF-1), nerve growth factor inducible-B (NGFI-B) nuclear receptor, and retinoid-related orphan receptor (ROR) bind to their DNA targets as monomers; the binding requires an A/T-rich sequence 5' adjacent to the core motif (Giguere et al., 1995; Meinke and Sigler, 1999; Wilson et al., 1993).

Domain D, the hinge region that links the DBD and ligand binding domain (LBD), is not well conserved and varies significantly in length. This domain has been proposed to participate in DNA rotation and corepressor interaction (McBroom et al., 1995). The C-terminal region of nuclear receptors is a multifunctional domain and mediates ligand binding, receptor dimerization, and transcriptional activation or repression. Crystallographic studies of unliganded versus liganded nuclear receptors have provided significant insights into the structural and functional relationships of ligand-induced gene activation (Bourguet et al., 1995; Jacobs et al., 2003; Renaud et al., 1995; Rochel et al., 2000) (Fig. 1B). The LBD, domain E, is composed of 12 alpha–helices (H1–H12) that are packed together in a sandwich-like manner, in which H12 contains an AF-2 ligand-dependent activation function. Nuclear receptors function in concert with transcriptional coregulatory proteins, which include basal transcriptional machinery components, chromatin-modifying complexes, corepressors, and coactivators (Jaskelioff and Peterson, 2003; Jepsen and Rosenfeld, 2002; Xu et al., 1999). The AF-2 domain is critical for interaction with these various factors. Although certain receptors act as potent transcription repressors in the absence of ligands by binding to corepressors, ligand binding induces several major conformational changes in this region (Brzozowski et al., 1997; Egea et al., 2000; Wagner et al., 1995). In particular, reorientation of H12 upon agonist binding is a prerequisite for dissociation of corepressors and recruitment of coactivators. Conversely, antagonistic ligands are thought to induce a different conformation of the LBD, which prevents the association of coactivators. Therefore, the precise conformation of the LBD and the position of H12 that a nuclear receptor holds upon a specific gene promoter largely determines the transcriptional properties of a nuclear receptor.

It has become evident that the functions of nuclear receptors can be modulated by several different mechanisms: (1) DNA binding specificity and affinity; (2) nature of the bound ligands (agonists versus antagonists); (3) covalent modifications, usually by phosphorylation, in response to specific stimuli and environmental cues; and (4) protein–protein interactions occurring between nuclear receptors themselves and among multiple cofactor proteins. When controlling expression of a specific target gene, individual mechanisms may function cooperatively to achieve precise regulation. For example, unliganded hormone receptors (e.g., TR, RAR) are coupled to a silenced chromatin state due to recruitment of histone deacetylase complexes by corepressor proteins (Urnov and Wolffe, 2001). Conversely, ligand-bound receptors may bind more than one coactivator protein, some of which, such as the cAMP response element binding protein (CREB) binding protein (CBP/p300) and the CBP association factor (p/CAF), harbor intrinsic histone acetyltransferase activity. The resultant histone hyperacetylation induces a competent open chromatin configuration, which in turn facilitates the interplay of nuclear receptors with other transcription factors and components of the basal transcription machinery. Thus, it is conceivable that perturbation of any function by nuclear orphan receptors will have a negative impact on the transcriptional control of a target gene. Moreover, the process of orphan nuclear receptor-mediated

TABLE I. Subgroups of Mammalian Nuclear Receptors

Class 0	Class I	Class II	Class III	Class IV	Class V	Class VI
SHP (o*)	TR α, β	RXR α, β, γ (o)	GR	NGFI-B α, β, γ (o*)	SF-1/ FTZ-F1 α, β (o*)	GCNF (o*)
DAX-1 (o*)	RAR α, β, γ	COUP-TFs (I, II) & EAR2 (o*)	AR			
	VDR	HNF4 α, β, γ (o)	PR			
	PPAR α, β, γ (o)	TLX (o*)	ER α, β			
	PXR (o)	PNR (o*)	ERR (o) α, β, γ			
	CAR & MB67 (o)	TR2 & TR4 (o*)				
	LXR (o)					
	FXR (o)					
	RevErb α, β (o*)					
	RZR/ROR α, β, γ (o)					

O: nuclear orphan receptors; O*: Orphan receptors without identified ligand; TR: thyroid hormone receptor; RAR: retinoic acid receptor; VDR: vitamin D receptor; PPAR: peroxisome proliferator-activated receptor; PXR: pregnane X receptor; CAR/MB67: constitutive androstane receptor (previously MB67); LXR: liver X receptor; FXR: farnesoid X receptor; RevErb: reverse ErbA; RZR/ROR: retinoid Z receptor/retinoid acid-related orphan receptor; RXR: retinoid X receptor; COUP-TF: chicken ovalbumin upstream promoter-transcription factor; HNF4: hepatocyte nuclear factor 4; TLX: tailes-related receptor; PNR: photoreceptor-specific nuclear receptor; TR2/TR4: testicular receptors 2 and 4; GR: glucocorticoid receptor; AR: androgen receptor; PR: progesterone receptor; ER: estrogen receptor; ERR: estrogen-related receptor; NGFI-B: nerve growth factor-induced nuclear receptor B; SF-1/FTZ-FI: steroidogenic factor-1/ fushi tarazu-factor 1; GCNF: germ cell nuclear factor; SHP: small heterodimeric partner; DAX-1: dosage-sensitive sex reversal-adrenal hypoplasia congenita critical region on the X chromosome, gene 1; α and β: alpha and beta, the subtypes of receptors.

down-regulation is often diverse and complex because multiple modulation mechanisms can be employed simultaneously (Giguere, 1999).

The nuclear orphan receptor subfamily is increasing rapidly and, based on molecular phylogeny studies, can be classified into seven groups (0 to VI) (Laudet, 1997). Some of the identified orphan receptors have been well characterized, but the functions of other candidates remain to be elucidated, particularly in physiologic settings (Table I). Both activation and repression of target gene expression by orphan nuclear receptors are critical for appropriate maintenance of key biologic functions. Furthermore, some orphan receptors exert dual regulatory mechanisms depending on the specific target genes that they recognize. This chapter reviews recent progresses in the silencing mechanisms of gene transcription by nuclear orphan receptors chicken ovabulmin upstream promoter-transcription factors (COUP-TFs), dosage-sensitive sex reversal-adrenal hypoplasia congenita critical region on the X chromosome, gene 1 (DAX-1), germ cell nuclear factor (GNCF), short heterodimer partner (SHP), and testicular receptors 2 and 4 (TR2/TR4). These orphan receptors have been well characterized, and they represent four categories within the seven defined groups of the nuclear receptor subfamily. The repression of gene transcription by these orphan receptors contributes significantly to the control of a wide range of physiologic functions, including development, differentiation, and reproduction. Furthermore, in addition to their intrinsic activities, these orphan receptors interfere with steroid and nonsteroid hormonal signaling pathways via interactions with hormone receptors. These orphan receptors' silencing mechanisms are diverse and complex, which is consistent with the receptors' multiple functions in regulating distinct sets of target genes.

II. COUP-TFI, II (EAR3, ARP-1), AND EAR2: INHIBITORS OF DIVERSE GENES BY MULTIPLE MECHANISMS

Among the rapidly growing numbers of nuclear orphan receptors that have been identified, chicken ovalbumin upstream promoter-transcription factor (COUP-TF) is one of the most extensively studied and best characterized transcription factors (Cooney et al., 2001; Tsai and Tsai, 1997; Zhou et al., 2000). Human COUP-TFI was initially identified as an activator protein critical for expression of the chicken ovalbumin gene (Ing et al., 1992; Pastorcic et al., 1986; Wang et al., 1989). This receptor was also cloned simultaneously by another group based on its homology to the human erbA gene and termed EAR3 (Miyajima et al., 1988). COUP-TFI AI-regulatory protein-1 (ARP-1) was subsequently isolated by its high homology to COUP-TFI and characterized as a factor that regulates the apolipoprotein AI (ApoAI) gene expression (Ladias and Karathanasis,

1991). COUP-TF homologues have been cloned in multiple species, including humans, mice, rats, chickens, and *Drosophila* (Lu *et al.*, 1994; Qiu *et al.*, 1994; Tsai and Tsai, 1997). Sequence analyses revealed that COUP-TFs are typical nuclear receptors and function in the absence of a known ligand (Wang *et al.*, 1989). COUP-TFs are highly conserved across the species (identities greater than 90%) (Fig. 2). Such a strikingly high similarity suggests that the function of COUP-TFs is conserved during evolution. In addition, EAR2, which was cloned at the same time as the EAR3/COUP-TFI gene, is the closest orphan nuclear receptor to COUP-TFs within the nuclear receptor family (Miyajima *et al.*, 1988). It shares 86% identity with COUP-TFs at its DBD and negatively regulates the same subsets of target genes. For this reason, EAR2 has often been regarded as a subtype of COUP-TFs' family members.

Although COUP-TFs act predominantly as transcription repressors, they also exert positive regulation on several target genes. One the other hand, EAR2 only exerts negative regulatory actions. COUP-TFs bind preferentially to direct repeats (DR) of the AGGTCA core motif, and they can also recognize inverted (palindromic) or everted repeats of the core sequence with reduced binding affinity (Cooney *et al.*, 1992). Compared to other nuclear receptors that recognize DR domains with restricted spacing between the two half-sites, COUP-TFs exhibit substantial binding flexibility where they bind as homodimers to a variety of DR elements with different lengths of spacers. COUP-TFs' binding affinity is highest for the DR1 element and decreases in the order of DR6>DR4>DR8>DR0>DR11 for other DR motifs (Cooney *et al.*, 1992; Kadowaki *et al.*, 1992). Consistent with this, characteristic COUP-TFs bind to different DR or palindromic response elements that are recognized by the following nuclear receptors:

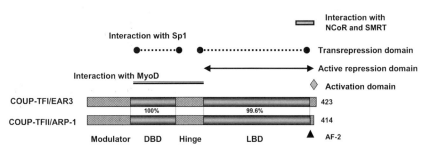

FIGURE 2. Functional domains of COUP-TFs. COUP-TFI/EAR3 and COUP-TFII/ARP-1 are highly homologous. The LBD of COUP-TFs harbors multiple functions, including active repression (solid line with everted arrow heads), transrepression (dashed dots), and the interaction surface for corepressors NCoR and SMRT (filled rectangular box). A region containing the most C-terminal 15 amino acids also harbors an activation function (filled diamond). COUP-TFI interacts with Sp1 through its DBD, while its interaction with MyoD requires the DBD, the hinge region, and a small stretch of sequences at the N-terminal region of the LBD.

RAR, RXR, TR (Cooney *et al.*, 1992), peroxisome proliferator-activated receptor (PPAR) (Marcus *et al.*, 1996; Nishiyama *et al.*, 1998), hepatocyte nuclear factor 4 (HNF4) (Mietus-Snyder *et al.*, 1992; Yanai *et al.*, 1999), and estrogen receptor (ER) (Liu *et al.*, 1993). COUP-TFs consequently antagonize the hormone-induced target gene transactivation. In addition, COUP-TFs inhibit basal promoter activities for a number of genes. They repress the genes for apolipoproteins (Ladias *et al.*, 1992), c-mos proto-oncogene (Lin *et al.*, 1999), and Oct4 (Schoorlemmer *et al.*, 1994) via DR1 domains. COUP-TFs also repress rat insulin 2 (Hwung *et al.*, 1988), arrestin (Lu *et al.*, 1994), and HIV-LTR genes (Cooney *et al.*, 1991) by binding to a DR6, DR7, or DR9 element, respectively. COUP-TFs and EAR2 cause potent repression of the luteinizing hormone receptor (LHR) gene (Zhang and Dufau, 2000, 2001), the oxytocin gene (Chu and Zingg, 1997), and a neurotransmitter gene, GRIK5 (Chew *et al.*, 1999). They also inhibit the acyl-CoA dehydrogenase gene promoter at everted repeats separated by 8 and 14 nucleotides (Carter *et al.*, 1994). Moreover, gene-targeting studies have demonstrated that COUP-TFs play critical roles in neurogenesis, angiogenesis, and heart development. COUP-TFI null mice die shortly after birth from starvation and dehydration (Qiu *et al.*, 1997). Death is caused by defects in the development of the peripheral nervous system, where swallowing and suckling functions are compromised. Disruption of the COUP-TFII gene is lethal at the embryonic stage because of defects in angiogenesis, vascular remodeling, and heart development (Pereira *et al.*, 1999).

COUP-TFs induce gene silencing through several mechanisms, including competition for binding to DNA cis-elements, quenching through titrating out RXR, and active repression and transrepression (Leng *et al.*, 1996) (Figs. 3 and 4). In the presence of their cognate hormones, vitamin D_3 receptor (VDR) RAR, or TR activate target gene expression through binding to DR3, DR4, or DR5 hormone response elements, respectively. The promiscuous binding of COUP-TFs to DR domains thus provides a molecular mechanism in which their competitive occupancy of the same DNA elements inhibits the gene transactivation induced by these hormone receptors (Ben-Shushan *et al.*, 1995; Cooney *et al.*, 1992; Stephanou *et al.*, 1996; Tran *et al.*, 1992) (Fig. 3, IA). In addition, binding of EAR2 to the retinoic acid response element (RARE) motif suppresses both basal and RA-induced mouse renin gene promoter activity (Liu *et al.*, 2003). Repression by COUP-TFs of the PPAR-induced gene activation and HNF4 function in the control of many liver-specific genes is attributed to a similar mechanism. In the regulation of the estrogen responsive lactoferrin gene, COUP-TFs down-regulate the estrogen receptor activation through an overlapping COUP/ER binding module (Liu *et al.*, 1993). Their occupancy of this composite site blocks the binding of ER, which is followed by the loss of estrogen response of these target genes. This mechanism also applies to

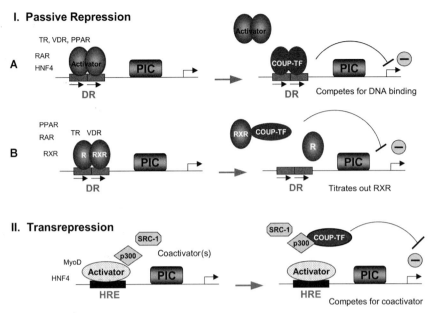

FIGURE 3. Models for passive and transrepression of target gene expression by COUP-TFs. (I) The COUP-TFs-elicited passive gene repression is achieved by competing with nuclear receptors (RAR, TR, VDR, PPAR, and HNF4) for occupancy of the same DR elements (A) or by titrating out RXR via protein–protein interaction between COUP-TF and RXR (B). Disruption of the formation of active heterodimer between RAR, TR, VDR, or PPAR and RXR or of RXR homodimer results in gene silencing (B). (II) Transrepression by COUP-TFs is independent of their binding to the target gene promoter regions, whereas the repression is achieved through competition of COUP-TFs with the transcription activator proteins (e.g., MyoD, HNF4) for the same sets of coactivators (e.g., p300 or SRC-1) via protein–protein interactions. PIC, preinitiation complex.

COUP-TFs' repressed SF-1 target gene activation, during which the SF-1 site is embedded in a COUP-TF binding element (Wehrenberg *et al.*, 1994; Xing *et al.*, 2002).

The notion that heterodimeric association of VDR, RAR, TR, or PPAR with RXR is a prerequisite for these hormone receptors to achieve high-efficiency binding to their cognate target sites has identified RXR as a generic partner critical in these hormone signaling pathways (Zhang *et al.*, 1992). COUP-TFs, in addition to their homodimeric binding activity, can heterodimerize with RXR in the presence of DNA binding elements. Such formation of an inactive COUP-TF:RXR dimer titrates out RXR for binding to RAR or other hormone receptors (e.g., VDR, TR) and thereby alleviates the hormone-induced gene expression (Kliewer *et al.*, 1992; Widom *et al.*, 1992) (Fig. 3, IB). Overexpression of RXR relieved the squelching effect of COUP-TFs, confirming that quenching of the RXR receptor resulted in the inhibition. Moreover, the genes that are activated by

Active Repression

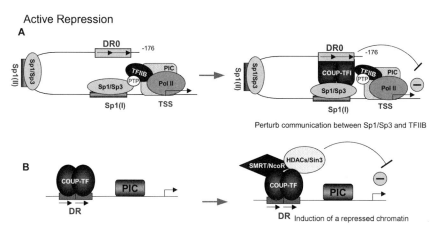

FIGURE 4. Models for active silencing of target gene expression by COUP-TFs. (A) Repression of human LHR gene expression by COUP-TFI/EAR3 from cross-talk among Sp1/Sp3, COUP-TFI/EAR3, and TFIIB. COUP-TFI/EAR3 bound to the DR0 motif interacts with Sp1/Sp3 bound to the Sp1(I) site. Such interaction prevents the association of TFIIB to the Sp1 (I) site without affecting the recruitment of TFIIB to the hLHR gene core promoter region. Anchoring of TFIIB at the Sp1(I) site is independent of prior binding of COUP-TFI/EAR3 to the DR0 element, and the interaction of TFIIB with Sp1/Sp3 is indirectly bridged by a currently unidentified protein(s), putative tethering protein (PTP). The COUP-TFI/EAR3-reduced association of TFIIB to the Sp1/Sp3–DNA complex may induce a nonproductive or less productive form of preinitiation complex (PIC), where the recruitment of RNA Pol II to the hLHR promoter is decreased in JAR cells. This occurs when the hLHR gene expression is subjected to a repressed state by COUP-TFI/EAR3. TSS stands for the transcriptional start site (B) Alternative mechanism of active repression by COUP-TFs depends on specific binding of COUP-TFs to their cognate response elements in the target gene promoters. Corepressor proteins SMRT/NCoR that are recruited to COUP-TFs through protein–protein interaction act as bridging molecules to exert a negative impact on the PIC complex. In addition HDAC/Sin3 complexes may associate with COUP-TFs by direct interaction with COUP-TFs or by association with NCoR/SMRT. The resultant compressed/closed chromatin structure induced by histone hypoacetylation contributes to the target gene silencing (Zhang and Dufaul (2003a). Reprinted with permission from *Mol. Cell. Biol.*)

RXR:RXR homodimer in presence of 9-cis retinoid acid are suppressed by COUP-TFs through their association with RXR. Therefore, competition for hormone response elements and for RXR are dual parallel mechanisms for COUP-TFs to suppress the response of the target genes to vitamin D_3 T3, retinoid acid, or PPARs.

COUP-TFs also possess intrinsic repression activity, which causes active repression of several target genes (Fig. 4B). The active repression domain is located in LBD (Achatz *et al.*, 1997), which interacts with corepressor proteins nuclear receptor corepressor (NCoR) and its variant RIP13deltal as well as silencing mediator for retinoid acid and thyroid hormone receptors (SMART) that potentiate COUP-TFs' silencing effect (Bailey *et al.*, 1997; Shibata *et al.*, 1997). Such active repression by COUP-TFs also involves participation of histone-modification complexes with interactions between COUP-TFs and histone deacetylase (HDAC) (Smirnov et al., 2000) or between NCoR or SMRT and HDAC1/mSin3 corepressor complex (Alland et al., 1997; Nagy et al., 1997). Specific inhibition of HDAC activity by trichostatin A (TSA) abolishes the silencing effect, which further indicates that the condensed chromatin structure caused by histone hypoacetylation may provide an appropriate environment, favoring or stabilizing the recruitment of corepressor proteins by COUP-TFs. In addition, two COUP-TF interacting proteins CIP1 and CIP2 have been isolated through yeast two-hybrid screening of an adenoma cDNA library (Kobayashi et al., 2002). Coexpression of CIP1 or CIP2 with COUP-TFs enhanced the COUP-TFs' mediated repression of several steroidogenic genes, including CYP17 and CYP11B2 (Kobayashi et al., 2002). Predominant expression of CIP proteins in steroidogenic tissues such as testes, ovaries, and adrenal glands, where COUP-TFs are coexpressed, indicates that CIP1 and CIP2 function as corepressor proteins for COUP-TFs in the control of steroidogenesis. Moreover, COUP-TFII and EAR2 heterodimer displayed high but distinct binding specificity compared to the respective homodimers of each receptor (Avram et al., 1999). The fact that EAR2 transcript expression overlaps precisely with COUP-TFII expression in several mouse tissues and embryonic carcinoma cell lines suggests that heterodimeric interaction between COUP-TFII and EAR2 may define an alternative mechanism for the function of these orphan receptors.

COUP-TFI and EAR2 cause marked repression of the LHR gene promoter activity in a dose-dependent manner. The repression is mediated by the binding of these receptors to an imperfect DR0 motif in the LHR gene promoter, while mutation of the DR0 domain abolishes the silencing effect (Zhang and Dufau, 2000, 2001). The finding that the inhibition is independent of changes in histone acetylation levels suggests a mechanism other than involvement of corepressor/HDAC activity in the negative modulation of the LHR gene expression (Zhang and Dufau, 2002, 2003b). Recent studies have revealed a novel mechanism for active repression of the LHR gene by COUP-TFs caused by cross-talk among COUP-TFI, Sp1/Sp3, and basal transcription factor TFIIB (Zhang and Dufau, 2003a) (Fig. 4A). The repression depends on a proximal Sp1/Sp3 binding site, designated as Sp1(I) site, which is one of the two essential Sp1 sites in the control of the basal promoter activity. Both Sp1 and Sp3 are required for the inhibition, and mutation of the Sp1(I) site to disrupt the Sp1/Sp3 binding abolishes the COUP-TFI silencing effect. The functional cooperation between the DR0 and the Sp1 domain is supported by mutual recruitment of COUP-TFI and Sp1/Sp3 bound to their cognate sites. COUP-TFI interacts with Sp1 through its DBD but also requires its N-terminal region, and deletion of these two domains greatly compromises the repression. Furthermore, TFIIB interacts with Sp1(I)-bound Sp1/Sp3 in addition to its association with COUP-TFI and the

TATA-less human LHR (hLHR) gene core promoter region. Such interaction is indirect, relying on adaptor protein(s) present in the nuclear extracts of human placental choriocarcinoma JAR cells. COUP-TFI specifically decreases the association of TFIIB to the Sp1(I) site without interfering with its interaction with the core promoter. The C-terminal region of COUP-TFI, which does not participate in its interaction with Sp1, is required for its inhibitory function and may affect the association of TFIIB with Sp1. The COUP-TFI-elicited disassociation of TFIIB with Sp1 is also reflected in the reduced recruitment of RNA PoLII to the hLHR gene promoter. Overexpression of TFIIB counteracts the inhibition by COUP-TFI and activates hLHR gene transcription in an Sp1(I) site-dependent manner. Taken together, these findings indicate that TFIIB is a key component in the regulatory control of COUP-TFI and Sp1/Sp3 on the initiation complex, where repression of the hLHR gene transcription by COUP-TFI results from its perturbation of communication between Sp1/Sp3 at the Sp1(I) site and the basal transcription initiator complex. Thus, the control of the TATA-less hLHR gene by active repression employs a novel mechanism that is different from those operative for most of the TATA-box genes.

COUP-TFs transrepress the expression of target genes by a mechanism that is independent of their cognate DNA elements (Fig. 3, II). In the case of transactivation of the human apo-B gene by HNF4, HNF3α, or C/EBP, COUP-TFII markedly antagonizes the individual transactivator's function without binding to the apo-B gene promoter (Achatz *et al.*, 1997). The domain that harbors such a transrepression function is located in the COUP-TFII DNA binding region and also in a segment spanning amino acid residues 193 to 399 but does not include its repressive domain. Tethering of COUP-TFs to TR or RAR via protein–protein interactions between their respective LBDs also results in potent transrepression of TR- or RAR-activated gene expression (Berrodin *et al.*, 1992; Butler and Parker, 1995). In addition, COUP-TFII inhibits myogenesis through repression of MyoD-dependent transcription in the absence of its binding site (Bailey *et al.*, 1998). Such repression was mediated through association of COUP-TFII's DBD and the hinge region with the N-terminal domain of MyoD. Overexpression of coactivator p300 relieves the inhibition by COUP-TFII, indicating cross-talk between COUP-TFII, MyoD, and p300. COUP-TFII, MyoD, and p300 interact in a competitive manner. Increasing the amount of COUP-TFII reduces the association of MyoD with p300 and elicits an inhibitory regulation of MyoD-induced myogenic differentiation. Moreover, EAR2 directly interacts with and decreases thyroid hormone receptor TRβ1 binding activity to its target genes. EAR2 represses both the 3,3,5-triiodo-L-thyronine (T3) induced and T3-independent TRβ1 gene activation in a cell-specific manner. Such repression is reversed by the coactivator SRC-1, indicating that the balance of corepressor (e.g., EAR2) and coactivator actions has an important role in the control of the TR-mediated responses (Zhu *et al.*, 2000).

Up-regulation of target gene expression by COUP-TFs is also mediated through several mechanisms. One involves protein–protein interactions of COUP-TFs with other transcription factors, whereas the binding of COUP-TFs to the response elements are either disposable or absent. COUP-TFI exhibits strong activation of HIV long terminal repeat and NGFI-A genes through a Sp1 binding site and up-regulates vHNF gene promoter activity through an octomer protein-binding site element (Pipaon et al., 1999; Power and Cereghini, 1996; Rohr et al., 1997). Interaction of COUP-TFI with Sp1 or Oct1 is dependent on COUP-TFI's DBD, while the C-terminal 15 amino acids are also crucial for the transactivating function. The extreme C-terminal region of COUP-TFI is responsible for the recruitment of a coactivator. Steroid receptor coactivator-1 (SRC-1), glucocorticoid receptor interacting protein 1 (GRIP1), and p300 interact with COUP-TF and potentiate its activation of HIV and NGFI-A genes. In addition, COUP-TFs act as an accessory factor for induction of phosphoenolpyruvate carboxykinase (PEPCK) gene in response to glucocorticoids (Hall et al., 1995; Scott et al., 1996). Although the effect of glucocorticoids is mediated by a glucocorticoid receptor (GR) binding to two glucocorticoid response elements (GRE), the maximal induction depends on two COUP-TF binding sites that flank the GRE domains. Although COUP-TFs binding to these sites has no effect on the basal promoter activity, they work synergistically with GR to elicit marked transactivation. In addition, COUP-TFI is required for induction of the RARβ gene in the presence of retinoic acid, which contributes to the RA-induced growth inhibition and apoptosis in cancer cells (Lin et al., 2000). In this case, specific binding of COUP-TFI to a DR8 motif in the RARβ promoter enhanced the RARα:RXR transactivation by increasing recruitment of coactivator, cyclic AMP response element-binding (CREB) protein to RARα. Taken together, multiple mechanisms have been derived for COUP-TFs-regulated gene expression, in which positive or negative regulation is primarily determined by the specific target genes recognized. Such diversity and complexity is thus consistent with the multiple functions of COUP-TFs in the control of different target genes.

III. DAX-1: SILENCING IN THE CONTROL OF STEROIDOGENIC AND SEX-DETERMINING TARGET GENES

DAX-1 (dosage-sensitive sex reversal-adrenal hypoplasia congenita critical region on the X chromosome, gene 1) is a member of the nuclear receptor superfamily with major action in development of steroidogenic and reproductive tissues (Lalli and Sassone-Corsi, 2003; Meeks et al., 2003). DAX-1 mutations cause the X-linked form of adrenal hypoplasia congenita (AHC) that is invariably associated with hypogonadotropic hypogonadism

(HHG) (Muscatelli *et al.*, 1994; Tabarin, 2001). Duplication of the DAX-1 gene locus on chromosome Xp21 causes male to female sex reversal in XY individuals with a normal SRY (male sex determinant) gene (Bardoni *et al.*, 1994). Consistent with its critical role in adrenal and gonadal development and differentiation, expression of the DAX-1 gene is predominantly observed in the adrenal cortex, testicle, ovary, hypothalamus, and anterior pituitary (Guo *et al.*, 1995; Swain *et al.*, 1996). The DAX-1 gene consists of two exons separated by an intron that encodes a protein of 470 amino acids (Burris *et al.*, 1996). DAX-1 is an atypical nuclear receptor because it does not contain a canonical zinc finger DBD characteristic of other members of the nuclear receptor superfamily (Fig. 5). Instead, the N-terminal of the DAX-1 gene is composed of three- and -a-half repeats of 65 to 67 amino acids. The C-terminal region of the DAX-1 gene harbors a putative LBD that resembles the ligand binding region of other nuclear receptors. In particular, its helix 12 (H12) region contains a conserved $\Phi\Phi X E \Phi\Phi$ motif (Φ: a hydrophobic amino acid; X: a nonconserved amino acid) that in other nuclear receptors mediates ligand-inducible transcriptional activation. However, no ligand for this orphan receptor has yet been identified.

The unique N-terminal of DAX-1 confers novel binding activity by which this orphan receptor recognizes its target genes (Fig. 5). DAX-1 binds to a DNA hairpin structure in the promoter of the steroidogenic acute regulatory protein (StAR) gene and potently represses its transcription

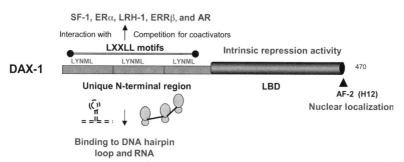

FIGURE 5. Structure and function relationship in DAX-1-mediated regulation of target gene expression. The N-terminal region of DAX-1 contains three repeats of leucine-rich motifs as indicated by LYNML. DAX-1 binds to DNA hairpin loop structure and RNA in the polyribosome complexes through the leucine-repeat motifs. This region also mediates its interaction with SF-1, ERα, LRH-1, ERRβ, and AR receptors, in which competitive binding with coactivators for AF-2 domain of the respective nuclear receptor causes target gene silencing. The C-terminal LBD region of DAX-1 harbors intrinsic repressive activity and also participates cooperatively in the RNA binding activity. Nuclear corepressors NCoR and Alien interact with this region to potentiate the DAX-1's inhibitory function. H12 is critical for the nuclear localization of DAX-1.

(Zazopoulos *et al.*, 1997). The StAR gene encodes a protein that mediates transport of cholesterol from the outer to the inner mitochondrial membrane, where it is converted to pregnenolone by the cytochrome P450 cholesterol-side chain cleavage enzyme) (CYP11A1) during steroid biosynthesis (Stocco, 2001). Consequently, DAX-1-blocked StAR gene transcription is followed by a marked reduction of steroidogenesis. The silencing effect of DAX-1 on StAR gene transcription is attributed to an intrinsic repression function located in the DAX-1 C-terminal LBD (Ito *et al.*, 1997). Such DAX-1-mediated inhibition is both cell-type and promoter specific. Furthermore, DAX-1 is also an efficient RNA binding protein and associates with mRNA in polyribosome complexes (Lalli *et al.*, 2000). In this case, the N-terminal repeat sequences and the C-terminal domain of DAX-1 are both important since the two domains function cooperatively to achieve competent RNA binding. Moreover, the DNA/RNA binding activities of DAX-1 are compromised by mutations present in patients with hypoplasia, suggesting that DAX-1 exerts its regulation on target gene expression at both transcriptional and posttranscriptional levels under certain physiologic conditions.

The observation that the spatial and temporal expression of DAX-1 is closely related to that of another orphan receptor, SF-1, suggested that these receptors might work in concert to modulate endocrine functions (Ikeda *et al.*, 1996, 2001; Swain *et al.*, 1996). SF-1 has been directly implicated in the regulation of Müllerian-inhibiting substance (MIS) gene expression and also in the transcriptional regulation of steroid hydroxylase and aromatase genes (Giuili *et al.*, 1997; Parker, 1998). Mice harboring a SF-1 null mutation display adrenal and gonadal agenesis, loss of pituitary gonadotropins, and altered structural properties of the ventromedial hypothalamus (Ingraham *et al.*, 1994; Luo *et al.*, 1994). The close resemblance of the phenotype by SF-1 disruption to that of the DAX-1 mutation-induced human hypoplasia congenital symptoms supports a functional correlation between these two orphan receptors in adrenal and reproductive development and function (Achermann *et al.*, 2001b; Beuschlein *et al.*, 2002). DAX-1 strongly represses the SF-1-mediated reporter gene activation through its direct interaction with SF-1; such inhibition does not affect the SF-1 binding activity to its response element (Ito *et al.*, 1997). Although the DAX-1 N-terminal region is responsible for its interaction with SF-1, its inhibitory function is attributed to its C-terminal LBD. To date, more than 60 different DAX-1 mutations have been found in patients with X-link AHC disease (Phelan and McCabe, 2001). Most are frameshift or nonsense mutations that result in premature truncation/deletion of the DAX-1 protein. Relatively few missense mutations have also been detected, and these mutations appear to cluster in certain regions of the C-terminal of DAX-1. Hence, the loss of DAX-1-mediated negative regulation of the SF-1 gene due to mutations (deletion or point mutations) at the C-terminal inhibitory domain may contribute to AHC disease (Lalli et al., 1997). Consistent with the inhibitory role of DAX-1 in SF-1-modulated target gene expression, the expression of the aromatase CYP19 gene, which is under control of SF-1, is significantly elevated in DAX-1 deficient male mice (Wang et al., 2001). So far, the target genes that are negatively regulated by DAX-1 in this manner also include genes for CYP11A1, 3 beta-hydroxysteroid dehydrogenase (3β-HSD) (Hu et al., 2001; Lalli et al., 1998), CYP17 (Hanley et al., 2001), the relaxin-like factor (Koskimies et al., 2002), inhibin alpha (Achermann et al., 2001a), high-density lipoprotein (HDL) receptor (Lopez et al., 2001), and others. It is has been proposed that silencing of expression of differentiation genes (e.g., StAR, steroid hydroxylases) by DAX-1 in the adrenal cortex may be a prerequisite for proliferation of the definitive adrenocortical zone in the critical postnatal period for humans In the absence of repressive function of DAX-1, abnormal early expression of steroidogenic genes occurs in the definitive zone, and consequently cell proliferation, differentiation, and zonation are disrupted. Adrenal hypoplasia would then follow the physiologic regression of the fetal zone (Lalli, 2003).

DAX-1 can also interfere with SF-1's interactions with other transcriptional factors. It is known that expression of the MIS gene is cooperatively regulated by actions of several transcriptional factors, including Sox9, SF-1, Wilms' tumor 1 (WT1) GATA binding protein 4 (GATA-4), and DAX-1 (Arango et al., 1999; Nachtigal et al., 1998; Shimamura et al., 1997). Appropriate spatial-temporal expression of the MIS gene is critical for mammalian sex differentiation because the MIS hormone produced by Sertoli cells mediates regression of the Müllerian ducts in males during testes differentiation (MacLaughlin and Donahoe, 2002). WT1 or GATA-4 synergize with SF-1 to promote MIS gene expression in Sertoli cells. In contrast, DAX-1 causes strong repression of the SF-1-induced gene activation by disruption of synergism between SF-1 and WT1 or GATA-4 (Nachtigal et al., 1998; Tremblay and Viger, 2001). Such inhibitory effect is mediated by direct protein–protein interaction of DAX-1 with the DNA-bound SF-1. Moreover, targeted disruption of the DAX-1 gene in mice reveals that DAX-1 is also essential for the maintenance of spermatogenesis in males (Yu et al., 1998). The knockout animals display progressive degeneration of the testicular germinal epithelium, which causes sterility in males. In contrast, the loss of DAX-1 function in females does not affect ovarian development or fertility, indicating that DAX-1 is not an ovarian determining gene. Transgenic studies further demonstrate that XY mice carrying extra copies of the DAX-1 gene antagonizes Sry action in mammalian sex determination, and overexpression of DAX-1 causes sex reversal (Swain et al., 1998). These findings confirm that DAX-1 is responsible for dosage-sensitive sex (DSS) reversal as initially identified in human DSS reversal patients.

The corepressors NCoR and Alien but not SMRT (Crawford *et al.*, 1998) participate in the silencing mechanism of DAX-1 (Fig. 5). Recruitment of such corepressor proteins by DAX-1 suggests a novel regulatory mechanism by which corepressors can be tethered to a transactivator protein such as SF-1, in addition to the widely-held view that corepressor proteins usually associate with transcriptional repressors. It has been shown that natural AHC mutations of DAX-1 markedly diminish the recruitment function of DAX-1 for both NCoR and Alien, significantly impairing the silencing effect of DAX-1 (Altincicek *et al.*, 2000). The abolishment of corepressor binding activities of DAX-1 due to its mutations in patients with AHC may contribute to the pathogenesis of this disease. More recent findings have shed further light on the molecular mechanism of the DAX-1 and SF-1 interactions. The N-terminal domain of DAX-1 contains repeated leucine-rich LXXLL-related sequences, a motif that usually occurs in nuclear receptor coactivator proteins and mediates their interaction with nuclear hormone receptors (Suzuki *et al.*, 2003). This leucine-rich motif determines the interactions of DAX-1 with SF-1 and several other nuclear receptors, including estrogen receptor α (ERα), liver receptor homologue-1 (LRH-1) estrogen-related receptor 2 (ERR2), and fly fushi tarazu factor 1. Thus, it is possible that the N-terminal LXXLL-repeated motifs of DAX-1 serve as an anchor site to tether DAX-1 to SF-1 and that the C-terminal intrinsic repression domain inhibits SF-1's transactivating function through recruitment of corepressor proteins (NCoR and Alien).

On the other hand, the silencing effect of DAX-1 on SF-1 action is not limited to SF-1 activated genes but also extends to gene(s) potentially inhibited by SF-1. In the case of the human CYP11B2 gene, which encodes an enzyme that is critical for aldosterone production in the adrenal gland, coexpression of DAX-1 and SF-1 overcomes SF-1-repressed gene transcription in transient transfection studies (Bassett *et al.*, 2003). These results demonstrate the difference in the control of the human CYP11B2 gene expression from other genes that are up-regulated by SF-1 during steroidogenesis.

DAX-1 has also been found to participate in hormonal signaling pathways through cross-talk with nuclear hormone receptors. DAX-1 interacts with the ligand-activated estrogen receptors alpha and beta (ERα and ERβ) through DAX-1 N-terminal LXXLL-related motifs and significantly reduces ER activation (Zhang *et al.*, 2000). Since DAX-1 shares a similar interaction domain with coactivator proteins in the recognition of nuclear hormone receptors, by occupying the LBD of ERs, DAX-1 may block or mask recruitment of coactivators to ERs and, therefore, have a negative impact on ER-induced gene transcription. It is also possible that subsequent recruitment of some corepressors by DAX-1 upon binding to ERs may contribute to the down-regulation. Furthermore, identification of the androgen receptor (AR) as a novel target of DAX-1-mediated repression provides further insights into the critical role of DAX-1 in male reproduction *in vivo* because DAX-1 (−/−) mice have defects in spermatogenesis (Holter *et al.*, 2002; Yu *et al.*, 1998). Studies showed that DAX-1 also interacts with AR and markedly inhibits ligand-dependent AR transactivation activities. Direct physical interaction was observed between these two receptors, and the interaction domains are mapped to the N-terminal of DAX-1 and the ligand binding AF-2 activation domain of AR. In addition, besides its intrinsic silencing function, DAX-1 exerts a novel mechanism in regulation of AR action by tethering AR to cytoplasm. Therefore, the altered cellular localization of AR causes an impaired AR function. The physiologic relevance of DAX-1-mediated repression of AR is supported by evidence that expression of DAX-1 is also observed in the human prostate, an important target site of androgen action. In particular, expression of DAX-1 was detected mainly in epithelial cells, where high levels of AR and ERβ were expressed.

In addition to the observation that the DAX-1 AHC mutants exhibited impaired activities in corepressor recruitment and silencing effect, the subcellular localization of the DAX-1 AHC mutant proteins was invariably shifted to the cytoplasm, independent of an intact nuclear localization signal (NLS) present at DAX-1's N-terminal (Lehmann *et al.*, 2002). The altered localization is also evident for the GAL4 DBD with its own NLS fused to the mutant DAX-1 C-terminal region, indicating that the nuclear-to-cytoplasmic shift is solely attributed to the mutations of the DAX-1 C-terminal region. The cytoplasmic localization of DAX-1 AHC mutants inversely correlates with their transcription repressive activities, and the H12 region is identified as critical for maintaining DAX-1 as a nuclear protein. Consistent with these findings, a DAX-1 mutant found in a patient with late adult onset of adrenal insufficiency and incomplete HHG exhibited least shifting effect on DAX-1 localization. Recent studies further show that the nuclear-to-cytoplasmic shift of the DAX-1 AHC mutants is likely caused by protein misfolding because the DAX-1 mutants are much more sensitive to proteolysis when compared to the wild-type DAX-1 (Lehmann *et al.*, 2003). It is therefore suggested that the altered conformation present in the C-terminal mutant DAX-1 may induce formation of a novel interacting surface for putative cytoplasmic anchoring site(s), consequently causing the shift of DAX-1 from nucleus to cytoplasm.

Taken together, the current understandings have revealed that DAX-1, an atypical nuclear orphan receptor, functions as a potent repressor in adrenal and reproductive axis development and function through perturbation of targeted gene expression modulated by StAR, SF-1, ER, and AR. DAX-1 also possesses unique DNA/RNA binding activities. More than one mechanism exists for DAX-1-mediated gene silencing, in which combined effects of active repression, recruitment of a corepressor, mask of coactivator binding, and alteration of a hormone receptor's localization may work together to achieve a precisely programmed biologic function. These various studies, together with the findings that all DAX-1 AHC mutants display cytoplasmic-oriented relocalization, have provided a molecular basis for DAX-1 defect-induced adrenal and gonadal disorders characterized by naturally occurring DAX-1 mutations that abolish or significantly impair DAX-1 function as a transcriptional repressor.

IV. GCNF: NEGATIVE CONTROL DURING GAMETOGENESIS AND EMBRYONIC DEVELOPMENT

Germ cell nuclear factor (GCNF), which is also known as retinoid acid receptor-related testis-associated receptor (RTR) and neuronal cell nuclear factor (NCNF), is an orphan member of the nuclear receptor superfamily; no ligand has been identified for its action. The GCNF gene was initially isolated from a mouse cDNA library and later was identified in human and *Xenopus laevis* (Chen *et al.*, 1994; Hirose *et al.*, 1995b; Joos *et al.*, 1996; Susens and Borgmeyer, 1996). The mouse GCNF (mGCNF) gene is composed of 11 exons, 2 that encode a protein of 495 amino acids. The human GCNF (hGCNF) contains 476 amino acids, and its gene has been located on chromosome 9 at locus q33–34 (Agoulnik *et al.*, 1998). Sequence alignment of GCNF genes isolated from different species has demonstrated that the genes are highly homologous and evolutionary conserved (Greschik and Schule, 1998). In addition, GCNF genes from humans and *Xenopus laevis* harbor deletions at their N-termini when compared to that of mouse GCNF, but it is yet unclear if these differences exert any species-specific impact on the function of GCNF. Furthermore, GCNF contains typical structural characteristics of the nuclear receptor superfamily, including a DNA binding domain (DBD) and a putative ligand binding domain (LBD) (Fig. 6). Because of the lack of the AF-2 transcriptional activation function in H12 of GCNF's LBD, this orphan receptor was proposed to function as a transcriptional repressor (Susens *et al.*, 1997).

Sequence analyses show that GCNF does not display substantial homology to any other nuclear receptors and that it is the only member of group VI of the nuclear receptor family (Giguere, 1999). The closest homologue of GCNF is the retinoid X receptor (RXR), with which it shares 32 to 34% overall amino acid identity and 61% identity at the DBD. Temporal and tissue-specific expression of GCNF has been intensively studied in humans, rodents, and *Xenopus laevis*. In addition, its binding properties were investigated and the response elements have been identified. Significant evidence about the physiologic role of GCNF during early embryonic development has been derived from gene-targeting studies using

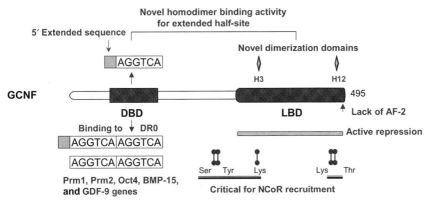

FIGURE 6. DNA binding and inhibitory properties of GCNF. GCNF binds as homodimer to AGGTCA direct-repeat motifs with zero spacing (DR0) with or without a 5′ extension sequence (filled square box). GCNF also recognizes an extended AGGTCA half-site as homodimer, where dimerization involves H3 and H12 as novel association domains. GCNF lacks AF-2 domain but possesses intrinsic repression activity in its C-terminal LBD region (hatched region). Recruitment of corepressor NCoR potentiates the GCNF-induced gene silencing, and the amino acids represented by vertical bars are critical for GCNF to interact with NCoR. GCNF inhibits expression of Prm1, Prm2, Oct4, BMP-15, and GDF-9 genes through binding to DR0 motifs within these gene promoters.

animal models; the studies reveal that GCNF is essential for embryonic survival and development (Hummelke and Cooney, 2001). Other studies have demonstrated that GCNF silences transcription of genes for protamines 1 and 2 as well as Oct4. GCNF also transrepresses the gene transcription modulated by the estrogen related-receptor alpha 1 (ERRα1).

The name of GCNF originated from the evidence that GCNF expression is predominant in germ cells of testes and ovaries of adult mammals. The highest GCNF expression in the mouse is observed in round spermatids at stages VII and VIII and less in spermatocytes (Zhang et al., 1998). Moreover, mouse GCNF (mGCNF) mRNA is not detected in hypogonadal mouse testes where the development of spermatogenic cells stops after the first prophase of meiosis (Mason et al., 1986). In contrast to the mouse, GCNF expression in human testes is predominant in spermatocytes but less in round spermatids (Agoulnik et al., 1998). The cell-specific expression of GCNF in spermatogenic cells of adult male tests indicates that it may play a regulatory role during terminal differentiation of spermatogenic cells, in particular in spermatids prior to the initiation of nuclear elongation and condensation. GCNF expression also has been observed in growing ovarian follicles but not in primordial oocytes, and its expression persists in the ovulatory follicles (Katz et al., 1997; Lan et al., 2003a). GCNF is also present in ovulated oocytes and preimplantation embryos, indicating that

GCNF may play a maternal role in zygotic development prior to implantation (Lan et al., 2003a). Recent studies have shown that mice with an oocyte-specific GCNF gene knockout display impaired fertility and prolonged diestrus resulting from reduced steroid hormone levels (testosterone, estradiol, and progesterone). In addition, abnormal double-oocyte follicles are observed in $GCNF^{-/-}$ female mice. Mechanistically, disruption of GCNF expression abrogates the GCNF-mediated silencing of BMP-15 and GDP-9 gene transcription at diestrus. The aberrant steroidogenesis with reduction of StAR, 3β-HSD, and 17α-hydroxylase gene expression in the somatic cells of the ovary result from overexpression of the BMP-15 and GDP-9 genes in oocytes at diestrus. The fact that BMP-15 and GDF-9 are important components of the paracrine signaling pathway in the ovary indicates that GCNF affects female fertility by regulating paracrine communication between oocyte and somatic steroidogenic cells via repression of the BMP-15 and GDF-9 gene expression (Lan et al., 2003b).

Expression of GCNF is widely detected in mouse embryos after gastrulation and later mainly exists in the developing nervous system (Bauer et al., 1997; Susens et al., 1997). Strong expression of GCNF is also shown in embryonic carcinoma cells (Lei et al., 1997). Mouse embryos harboring germ line-targeted mutation of GCNF died around 10.5 days postcoitum (E10.5) because of cardiovascular complication (Lan et al., 2002). Prior to death, significant developmental defects were observed in $GCNF^{-/-}$ mice; these defects included posterior truncation, ectopic tail-bud formation, unclosed neural tube, compromised somitogenesis, and other abnormalities. In addition, studies of *Xenopus* embryos indicate that the xGCNF gene is expressed between the gastrula and mid-neurula stages. Depletion of embryonic xGCNF expression shows that xGCNF function is required for morphogenetic cell movement during neurulation (Barreto et al., 2003).

Characterization of DNA binding activity of GCNF has revealed unique properties for this orphan receptor as a sequence-specific DNA binding protein. GCNF binds specifically and with high affinity as homodimer to a direct-repeat (DR) motif of AGGTCA half-site with zero spacing (DR0) (Cooney et al., 1998; Yan et al., 1997). GCNF can also bind strongly as a homodimer to a consensus SF-1 response element, thus displaying a novel dimeric binding property for an extended half-site (Greschik et al., 1999). Monomeric binding of GCNF to the SF-1 site with markedly lower activity has also been observed. Homodimeric binding of GCNF to an extended half-site requires a novel dimerization function present in its DBD and the participation of helix 3 (H3) and H12 of its putative LBD. Identification of H3 and H12 as critical for dimerization is unique for GCNF since dimerization of other receptors depends on either helices 9 and 10 or helices 5 through 7. Moreover, GCNF does not bind direct-repeat sequences separated by one to six nucleotides (DR1–6) (Cooney et al., 1998). In contrast to other nuclear orphan receptors, such as COUP-TFs that display flexibility in recognition of DR motifs with varied spacing lengths (Tsai and Tsai, 1997), GCNF is the first nuclear receptor identified that solely binds to DR0 motifs. Moreover, unlike the thyroid hormone receptor (TR) or retinoid acid receptor (RAR) that heterodimerize with RXR to achieve high affinity bindings to DR elements, no heterodimerization is observed between GCNF and RXR (Borgmeyer, 1997).

GCNF represses activities of heterologous and natural gene promoters (protamines 1 and 2 and Oct4) through binding to DR0 elements and harbors its intrinsic silencing function at the LBD (Cooney et al., 1998; Greschik et al., 1999; Lan et al., 2002; Yan et al., 1997). GCNF specifically binds to one of the two DR0 motifs in the Prm1 promoter and to the DR0 element in the Prm2 promoter (Fuhrmann et al., 2001). Antibody supershift assays show that endogenous GCNF isolated from mouse testes nuclear extracts and elutriated round spermatid nuclear extracts bind to both Prm1 and Prm2 promoters. A reporter gene bearing the protamine gene DR motif is potently repressed by GCNF. Thus, the findings demonstrate protamines as target genes silenced by GCNF.

Silenced target gene transcription by GCNF for appropriate maintenance of physiologic functions is so far best reflected in its repression of the Oct4 gene transcription (Fuhrmann et al., 2001). The POU domain transcription factor Oct4 plays an essential role in the maintenance of embryonic stem cell potency and the establishment of the germ cell lineage. Oct4 is down-regulated during gastrulation when embryonic stem cells differentiate, and its expression is subsequently confined to the germ cell lineage. Consistent with this finding, expression of Oct4 is significant in embryonic stem cells, carcinoma cells, and germ cells, whereas its expression is rapidly decreased in cells undergoing retinoid acid (RA)-induced differentiation. The requirement of GCNF for embryo viability and its dynamic expression profiles during embryogenesis imply that target genes critical for embryonic development (e.g. Oct4) may be subject to modulation by GCNF. In this regard, GCNF expression in both P19 embryonic carcinoma cells and mouse embryos inversely correlates with Oct4 expression. While Oct4 expression is shut off during gastrulation, GCNF expression is significantly up-regulated. Furthermore, dosage-dependent marked repression of Oct4 gene transcription by GCNF is observed in differentiating P19 embryonic carcinoma cells. The repression is mediated through a DR0 element in the Oct4 gene promoter. Mutation of each individual half-site to disrupt the DR0 element abolishes the GCNF-mediated silencing effect, confirming that specific homodimeric binding of GCNF to the Oct4 DR0 motif is a prerequisite for the repression. In addition, GCNF interacts with corepressor proteins NCoR and SMRT but not Alien in yeast and mammalian two-hybrid systems and in vitro pull-down assays, indicating that repression of Oct4 gene by GCNF is mediated through recruitment of corepressors via direct protein–protein interaction. Studies in GCNF null mice have revealed that repression of Oct4 activity by GCNF is critical for confining Oct4 expression to the germ line during pluripotent stem cell differentiation. In wild-type animals, Oct4 expression is restricted to primordial germ cells in the posterior of embryos at E8.5 to E8.75. In contrast, markedly altered expression of Oct4 is observed in the GCNF deficient embryos, in which suppression of Oct4 expression in certain differentiating somatic cells is lost compared to the wild type. The Oct4 expression is no longer restricted to the germ cell lineage, while additional expression of Oct4 is detected in the putative hindbrain regions.

Transrepression of target gene expression by GCNF has been observed in ERRα1-mediated gene activation (Yan and Jetten, 2000). Members of ERR subfamily and GCNF recognize similar response elements and display overlapped expression patterns. GCNF represses the ERRα1-induced gene transactivation through an ERRα1 response element. No protein–protein interaction is detected between GCNF and ERRα1, and repression is mediated at least in part through competitive binding of GCNF to the same ERRα1 response element. This also involves the participation of the corepressor, NCoR, which directly interacts with GCNF. Studies of the interaction between GCNF and NCoR have revealed some unique properties of GCNF. The hinge region and H3 and H12 of GNCF's LBD are required for the interaction between GCNF and NCoR, for which the residues Ser^{246}–Tyr^{247} in the hinge domain, Lys^{318} in H3, and Lys^{489}–Thr^{490} in H12 are critical. Taken together, these findings support the important role of GCNF in embryonic development, neurogenesis, and reproduction through down-regulation of the expression of several target genes that are actively involved in these physiologic functions.

V. SHP: GENERIC HETERODIMERIC PARTNER-INHIBITING MULTIPLE NUCLEAR RECEPTOR PATHWAYS

Short heterodimer partner (SHP) was initially cloned in an effort to isolate interacting protein(s) for the mouse constitutive adrostane receptor (CAR) in yeast two-hybrid screening (Seol et al., 1996). As indicated by its name, SHP has been recognized as being able to interact and heterodimerize with a number of nuclear receptors, including MB67 (the human homologue of mouse CAR), thyroid hormone receptor (TR), retinoid acid receptor (RAR), retinoid X receptor (RXR), estrogen receptor (ER), glucocorticoid receptor (GR), and hepatocyte nuclear factor 4 (HNF4) (Borgius et al., 2002; Gobinet et al., 2001; Johansson et al., 1999; Klinge et al., 2001; Lee et al., 2000; Seol et al., 1996, 1998). The interaction of SHP with TR, RAR, or RXR is enhanced in the presence of their cognate ligand

(Seol et al., 1996). The ability of SHP to down-regulate signaling pathways modulated by these multiple nuclear receptors indicates that the SHP-mediated repression represents a generic mechanism in the control of divergent physiologic functions. Several distinct mechanisms have been identified for SHP-induced gene silencing, including inhibition of DNA binding, competition with cofactor(s) for binding to nuclear receptors, and active repression through SHP's intrinsic inhibitory activity. Furthermore, a critical physiologic requirement of SHP activity was demonstrated by studies using SHP-deficient animal models. In addition, studies of the SHP gene promoter activity and its regulatory mechanism have provided further insights for the understanding of SHP function. This is of particular importance under the context of differential expression of a single protein because SHP may trigger complex changes in signal cascades due to its capacity to cross-talk with a wide range of nuclear receptors.

SHP is an atypical orphan member in the nuclear receptor superfamily (Fig. 7). It contains a putative LBD but does not have a conventional DBD that is conserved among the nuclear receptors (Seol et al., 1996). No DNA binding activity has been recorded for SHP to date, and there is no ligand yet identified for its function. The SHP gene encodes a small nuclear protein that is composed of 273 amino acids in humans and 260 amino acids in mice. Studies of its genomic structure reveal that the SHP gene consists of two exons separated by a single intron, and the human SHP (hSHP) gene is located at chromosome 1p36.1 (Lee et al., 1998b). The hSHP gene contains a potential TATA box at its proximal 5' flanking region, and the

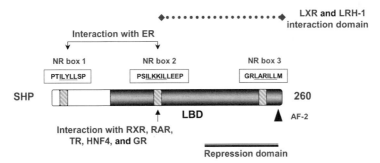

FIGURE 7. Function domains of SHP. SHP gene encodes a small protein containing 260 amino acids. The SHP protein contains a conserved LBD but lacks the conventional DBD domain. Three leucine-rich nuclear receptor boxes (NR boxes 1, 2, and 3) are located within its N-terminal, central, and C-terminal regions, where the amino acid sequences of each NR box with the LXXLL motif underlined are shown. SHP interacts with estrogen receptor (ERα or ERβ) in NR boxes 1 and 2, while its interaction with RXR, RAR, TR, HNF4, and GR depends on the NR box 2. SHP's C-terminal LBD region harbors the domain for interaction with LXR and LRH-1 receptors (dotted line). The intrinsic repression domain of SHP is also located at the C-terminal (solid double lines).

transcription start site is mapped to 32 nucleotides downstream of the TATA sequence. The expression of the SHP gene has been observed in various tissues, with high expression shown in the liver, heart, brain, adrenal gland, small intestine, and pancreas (Lee et al., 1998b). The wide expression pattern of the SHP gene, thus supports the physiologic relevance of this orphan receptor for interaction with numerous different nuclear receptors under distinct biologic settings.

Functionally, SHP inhibits binding of RAR:RXR heterodimer to its response element and represses retinoid acid-induced gene transactivation (Seol et al., 1996). The interaction of SHP with RAR or RXR may result in formation of novel heterodimers of SHP with RAR or RXR that are incapable of binding to the DNA. Similarly, SHP also exerts potent repression of the CAR-mediated or liganded TR-mediated gene activation by protein–protein interaction. Moreover, SHP interacts with RXR through its central domain and therein differs from most other nonsteroid hormone receptors that use their C-terminal LBDs to dimerize with RXR (Seol et al., 1997). Furthermore, SHP, C-terminal domain harbors intrinsic repression activity, and this active inhibitory domain as well as its central RXR-interacting sequences are both required for SHP to achieve the maximal inhibition of the RXR-induced transactivation. The intrinsic C-terminal inhibitory region of SHP, however, does not interact with corepressor NCoR (Seol et al., 1997). This indicates that a distinct mechanism is employed in SHP-induced gene silencing since interaction and recruitment of NCoR is usually required for repression incurred by several nuclear receptors (i.e., unliganded TR or RAR/RXR and some orphan receptors) (McKenna et al., 1999).

SHP participates in estrogen signaling pathways via its direct interaction with ERα and ERβ (Johansson et al., 1999; Seol et al., 1998). As observed for nonsteroid hormone receptors, the interaction of SHP with ERs is significantly increased by estradiol (E2) but decreased by 4-OH tamoxifen. This suggests that agonist binding promotes formation of SHP-interacting surfaces in ERs. Consistent with this notion, SHP interacts with the ligand binding AF-2 domain of ERα to prevent its recruitment of the coactivator protein transcriptional intermediary factor 2 (TIF2) (Johansson et al., 1999). Similarly, SHP dose-dependently displaces the association of ERα/AF-2 domain with another coactivator: receptor-interacting protein 140 (RIP140) (Johansson et al., 1999). In agreement with the above studies, mutation of the ERα/AF-2 domain, or binding of an antagonist that is known to cause a distinct allosteric arrangement of AF-2, abolishes SHP binding to ERα. Consequently, SHP causes marked repression of E2-activated ERα transactivation and antagonizes TIF2-potentiated ERα-modulated target gene expression. SHP uses its N-terminal LXXLL-related (L represents leucine) motifs to bind to estrogen receptors, while its C-terminal LBD is not required for such interaction (Johansson et al., 2000). Similar leucine-rich motifs, also called nuclear receptor (NR) boxes, are usually identified in the coactivator proteins as critical for their binding to the liganded AF-2 domains of nuclear receptors (McKenna et al., 1999). Cases in which SHP elicited marked inhibition of hepatocyte nuclear factor 4 (HNF4) and glucocorticoid receptor (GR)-induced gene activation, SHP also competed with coactivator proteins to bind HNF4 or ligand-bound GR (Borgius et al., 2002; Lee et al., 2000). Taken together, this evidence has revealed that competition with coactivators to occupy the same AF-2 domain through the leucine-rich motif is likely a generic mechanism in the SHP-mediated target gene inhibition for this class of receptors. The critical involvement of the SHP C-terminal repressive domain supports a two-step mode for SHP's action, with initial displacement of a coactivator via its N-terminal LXXLL receptor-interaction motif. This is followed by a direct inhibitory effect exerted by SHP's C-terminal repression region (Lee et al., 2000). In addition, coexpression of SHP causes an intranuclear redistribution of GR molecules to cytoplasmic sites; however, such relocalization does not occur in the presence of inhibition-deficient SHP mutants (Borgius et al., 2002). Thus, the tethering effect of SHP may represent a novel mechanistic aspect in SHP-modulated gene repression.

A direct interaction of SHP with peroxisome proliferator-activated receptors (PPARs) alpha and gamma and the subsequent regulation of PPAR-targeted genes, however, have shown several major differences. Whereas SHP binding to TR/RAR/RXR or ER/GR depends on their cognate ligand, its interaction with PPARα or PPARγ is not affected by absence or presence of peroxisome proliferator ligands (i.e., Wy-14643) (Nishizawa et al., 2002). PPARγ interacts with SHP through its DBD and the hinge region, rather than the ligand binding AF-2 domain; SHP strongly augments, rather than represses the PPARγ-mediated gene activation at both basal and ligand-inducible conditions. The SHP-interacting domain within PPARγ overlaps with its NCoR-interacting domain, and disassociation of NCoR binding from PPARγ has been shown to contribute to the SHP-mediated PPARγ activation. On the other hand, SHP up- or down-regulates the genes involved in the peroxisomal β-oxidation pathways through heterodimerization with PPARα (Kassam et al., 2001). The dual regulatory mechanisms of SHP are likely attributed to the different PPAR receptor subtypes and different promoter contexts SHP recognizes.

Critical physiologic contribution of SHP has been demonstrated in the coordinate regulation of bile acid synthesis pathways by SHP and several other nuclear receptors (Fig. 8). The catabolism of cholesterol into bile acids in the liver is essential for cholesterol homeostasis, and this process is tightly controlled through regulation of the activity of cholesterol 7α-hydroxylase (CYP7A1), the rate-limiting enzyme of this metabolic cascade (Hoffmann, 1994). While the promoter activity of CYP7A1 is positively regulated by cholesterol derivative oxysterol, it is markedly repressed by bile acid.

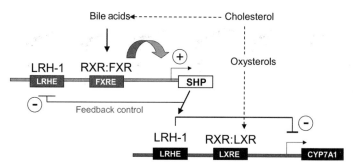

FIGURE 8. Control of bile acid synthesis by a regulatory cascade of nuclear receptors SHP, FXR, LRH-1, and LXR. Elevated cholesterol levels cause up-regulation of the CYP7A1 gene expression by binding of its oxysterol metabolites to RXR:LXR receptor heterodimmer in the CYP7A1 gene promoter. This is followed by increased catabolism of cholesterol and synthesis of bile acids as the end product. Accumulated bile acids in turn induce the transcription of the SHP gene by activation of the FXR receptor that binds as RXR:FXR heterodimer to an FXRE element in the SHP gene promoter. The SHP protein interacts with LRH-1 receptor and results in formation of an inactive SHP:LRH-1 heterodimer. This process represses the expression of the CYP7A1 gene and induces feedback control of the SHP gene due to the presence of LRH-1 binding sites in both gene promoters.

Mechanistically, the oxysterol receptor liver X receptor (LXR) heterodimerizes with RXR to activate the CYP7A1 gene transcription when binding to a cholesterol response element in the CYP7A1 gene promoter (Schoonjans and Auwerx, 2002; Stroup et al., 1997). This ligand-dependent LXR:RXR transactivation requires simultaneous binding of liver receptor homologue-1 (LRH-1) as an auxiliary competence factor to a LRH-1 element of the CYP7A1 promoter (Crestani et al., 1998; Gupta et al., 2002; Tu et al., 2000).

In contrast to the direct feed-forward effect of oxysterols through the promoter-bound RXR:LXR dimer (Peet et al., 1998), the feedback regulation of CYP7A1 activity by bile acid is indirect and is mediated by combined efforts of SHP and farnesoid X receptor (FXR), LHR-1, and RXR (Goodwin et al., 2000; Lu et al., 2000; Makishima et al., 1999). Bile acid initially binds to its specific receptor FXR that dimerizes with RXR to recognize an FXR response element in the SHP gene promoter. The significantly induced SHP gene expression by bile acid-bound RXR:FXR, subsequently causes marked repression of CYP7A1 gene promoter activity due to interaction of SHP with LRH-1. By forming a heterodimer with LRH-1, SHP is shown to inhibit both basal and oxysterol–RXR:LXR transactivated CYP7A1 gene transcription.

Mice lacking the SHP gene displayed mild defects in bile acid homeostasis, whereas $SHP^{-/-}$ or $FXR^{-/-}$ null mice showed complete abrogation of SHP-mediated repression of CYP7A1 gene expression (Kerr et al., 2002; Wang et al., 2002). The down-regulation of CYP7A1 gene expression by bile acid feeding is retained in the SHP null mice, indicating the existence of redundant or compensatory inhibitory pathway(s). Moreover, SHP exerts autoregulatory control of its own expression since its promoter contains a consensus LRH-1 binding site. This feedback regulation may provide a mechanism for attenuating the SHP-mediated gene silencing effect. Therefore, the suppression of CYP7A1 gene transcription by SHP illustrates a complex regulatory network, which requires concerted participation of FXR, SHP, LHR-1, and RXR (Fig. 8).

SHP has also been shown to negatively regulate human ATP-binding cassette transporter 1 (ABCA1) gene transcription through perturbation of LXRα:RXR-mediated gene transactivation in the presence of bile acid (Brendel et al., 2002). The activation of SHP expression by FXR:RXR, followed by its interaction with LXRα, accounts for the repression of human ABCA1 (hABCA1) gene expression by bile acid. The bile acid-induced repression of the sodium taurocholate cotransforting polypeptide (ntcp) gene, which is the principal hepatic bile acid transporter, is also attributed to SHP via its interaction with an RAR/RXR element of the ntcp promoter (Denson et al., 2001). Taken together, these findings demonstrate that formation of inactive heterodimers of SHP with other nuclear receptors may be a common mechanism in the control of genes involved in bile acid metabolism.

The activation of SHP gene promoter activity by FXR and its autorepression through LRH-1 in the presence of elevated level of bile acid reveals that well-balanced SHP gene expression is important for accommodating a particular biologic function. Also, SHP gene transcription is activated by nuclear receptors SF-1 and fetoprotein transcription factor (FTF) (Lee et al., 1999c). At the molecular level, SF-1, FTF, and LHR-1 recognize the same five SF-1 binding elements identified in the SHP gene promoter because FTF and LHR-1 receptors are liver-specific close homologues of SF-1. Moreover, estrogen-related receptor gamma (ERRγ), but not alpha (ERRα) or beta (ERRβ), was recently reported to strongly increase the SHP gene transcription by binding to one of the five SF-1 domains (Sanyal et al., 2002). This is consistent with the previous observation that the ERR isoforms α, β, and γ recognize both the estrogen response element and the SF-1 binding site. The lack of activation of the SHP gene by ERRα and ERRβ likely results from the association of these two receptors with putative corepressor protein(s). Although direct interaction of SHP with all three ERR subtypes has been demonstrated, SHP itself represses solely the ERRγ-mediated activation of the SHP gene transcription. Therefore, cross-talk between SHP and ERRs has identified another novel autoregulatory loop in the control of SHP gene promoter activity. Because SHP is able to interact with a variety of nuclear receptors, a precise regulation of the SHP gene is particularly essential for its differential contribution in multiple divergent signaling pathways.

VI. TR4 AND TR2: TESTICULAR RECEPTOR WITH HOMOLOGOUS BUT NOT REDUNDANT FUNCTIONS

Testicular receptors 4 (also termed TAK1) and 2 (TR4 and TR2) are two evolutionarily conserved members within the nuclear receptor superfamily. Although they display structural features characteristic of the nuclear receptors, they are orphan receptors without a known ligand. The overlapping expression patterns of TR4 and TR2, as well as their functional participation in common signaling pathways, indicate that TR4 and TR2 may participate in modulation of the same sets of target genes. Both receptors exert dual regulatory mechanisms and achieve a positive or a negative effect depending on the specific target genes they recognize. In TR4/TR2-mediated gene silencing events, active repression and transrepression of multiple hormone-regulated pathways are observed. Several mechanisms are operative in the inhibitory process, including competition with nuclear hormone receptors for the same cis-elements, impairment of hormone receptor binding activities, competition for coactivator proteins, and modification of chromatin structures. In addition, differences have also been noted between TR4 and TR2 despite their striking similarities, which indicates that these two orphan receptors may not be physiologically redundant.

The initial cloning of the human TR4 gene from the prostate and testes and the TR2 gene from the prostate was followed by identification of their homologues in different species, including rats, mice and *Drosophila* (Chang and Kokontis, 1988; Chang *et al.*, 1994; Hirose *et al.*, 1994; Lee *et al.*, 2002; Sanyal *et al.*, 2003). In humans and rodents, several variants of TR4 and TR2 gene transcripts are generated by alternative splicing (Chang and Kokontis, 1988; Chang *et al.*, 1989; Yoshikawa *et al.*, 1996a,b). Human full-length TR4 and TR2 are composed of 615 and 603 amino acids, respectively, and display high homology at their amino acid sequences, with 82% and 65% identities at their respective DBDs and LBDs (Fig. 9). At the genomic level, the human TR4 and TR2 genes are located at chromosome 3p24 and 12q22, respectively (Lee *et al.*, 1995; Lin *et al.*, 1998). These receptors are expressed in multiple tissues of adult mammals, and their abundance varies significantly (Chang and Kokontis 1988; Chang *et al.*, 1989; Hirose *et al.*, 1994). Expression is highest in the prostate and testes, and high levels of expression are present in the adrenal gland, spleen, and thyroid gland. During embryonic development, TR4 and TR2 are predominantly expressed in the central nervous system during proliferation and differentiation in embryonic neurogenesis (Lee *et al.*, 1996; Young *et al.*, 1998). In addition, the expression of TR4 and TR2 is influenced by a number of factors (i.e., retinoid, p53 and others), which may contribute to a dynamic regulatory network of their function (Inui *et al.*, 1999; Lee and Wei, 1999; Lin and Chang, 1996).

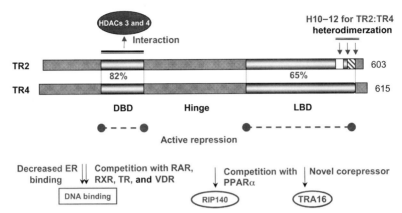

FIGURE 9. TR2 and TR4: Functional domains and repressive mechanisms. Human TR2 and TR4 are highly homologous. Active repression of target gene expression by TR2 and TR4 is mediated by the DBD and LBD (dotted lines). The DBD also serves as an interface for TR2 to associate with classes I and II histone deacetylases (HDACs 3 and 4). Heterodimerization between TR2:TR4 is dependent on the region of helices 10 to 12 (H10–12) of TR2. TR2 or TR4 represses the hormone-induced gene activation by competing for binding with RAR, RXR, TR, or VDR receptors, decreasing the binding activity of ER through protein–protein interaction, or titrating out of coactivator RIP140 from PPARα. In addition, TRA16 functions as corepressor for TR4-induced gene transactivation.

The binding properties of TR4 and TR2 have been intensively investigated since the original cloning of these two receptor genes. TR4 binds effectively to synthetic or natural DNA response elements that are composed of direct repeats (DR) of core motif PuGGTCA separated by zero to five nucleotides (DR0–5) (Hirose et al., 1995a). In contrast, TR4 binds poorly to palindromic or inverted repeat sequences. Similar binding kinetics are observed for TR2; it prefers to bind to direct-repeat motifs with various lengths of spacers. Binding affinities follow the order of DR1>DR2>DR4=DR5=DR6>DR3 (Lin et al., 1995). TR4 and TR2 usually bind to a DR element as homodimers. Active repression by TR4 or TR2 through various DR elements has been shown in regulation of many viral and cellular genes. Both TR4 and TR2 exert marked repression of early and late SV40 promoters via binding to a DR2-type response element, and TR2 suppresses the promoter activities of human erythropoietin (EPO) and hepatitis B virus genes (Lee and Chang, 1995; Lee et al., 1996; Yu and Mertz, 1997). Heterodimeric binding of TR4 and TR2 to a DR5-containing reporter gene elicits much stronger repression when compared to the action by either homodimer alone (Lee et al., 1998a). In this case, protein–protein interaction between TR4 and TR2 involves helix 10 (H10) of TR2. In addition, a 540-kDa protein complex containing TR4:TR2 heterodimer as its core component represses embryonic and fetal globin gene transcription in definitive erythroid cells (Tanabe *et al.*, 2002). Based on the coexistence of TR4 and TR2 at certain physiologic stages, heterodimerization of these two receptors may provide a more effective inhibitory mechanism in the control of specific target gene expression. Although initial studies did not detect binding of TR4 or TR2 to a single half-site of the core element, specific monomeric binding of TR4 to an AGGTCA element results in suppression of human steroid 21-hydroxylase gene expression (Lee *et al.*, 2001).

The notion that TR4 and TR2 play significant roles in down-regulation of hormonal signaling pathways is derived from evidence that the various DR elements recognized by TR4 and TR2 also serve as consensus binding sites for nonsteroid nuclear hormone receptors. TR4 and TR2 significantly repress RA-transactivated cellular retinal-binding protein II (CRBPII) and RARβ gene transcription via potent binding of TR4 or TR2 to the DR1 or DR5 motif (Chinpaisal *et al.*, 1997; Lin *et al.*, 1995). TR4 and TR2 do not interact directly with RXR or RAR, and the repression is attributed to competitive occupancy of the same response elements by TR4 or TR2 homodimer or putative TR4:TR2 heterodimer and RARα:RXRα heterodimer. TR2 represses T_3-activated reporter gene activity by competitive binding with TRα:RXRα heterodimer to a DR4 element (Hirose *et al.*, 1995a). This is also the case for the TR4-mediated repression of vitamin D_3-induced 25-hydroxyvitamin D_3 24-hydroxylase gene activity at a DR3 motif (Lee *et al.*, 1999b). In addition, silencing of PPARα-modulated transactivation of rat enoyl-CoA hydratase and peroxisomal fatty acyl-CoA oxidase gene expression by TR4 is caused by competition of TR4 with PPARα:RARα heterodimer (Yan *et al.*, 1998). A mechanism involving titration out of RXR from PPARα:RXRα is excluded since TR4 does not interact with PPARα or RXRα. Furthermore, the PPAR ligand 8(S)-hydroxy-eicosate-traenoic acid strongly promotes the recruitment of coactivator protein RIP140 to PPARα, while it decreases PPARα association with corepressor SMRT. TR4 can directly interact with RIP140, but not with SMRT, and competes with PPARα for binding to RIP140 (Yan *et al.*, 1998). Thus, at least two cooperative mechanisms participate in TR4-induced down-regulation of PPAR responsive genes: competition with PPARα:RXRα heterodimer for binding to the PPAR response element and competition for binding to coactivators, such as RIP140.

In addition, TR4 and TR2 display differential regulatory function in the PPARα signaling pathways. In human HaCaT keratinocytes, TR4, but not TR2, causes marked inhibition of the PPARα-activated reporter gene activity in the presence of Wy-14643, a peroxisome proliferator ligand (Inui *et al.*, 2003). In addition, the endogenous expression of TR4 protein is induced by Wy-14643, whereas TR2 protein expression is decreased under the same condition. Hence, it can be proposed that selective up-regulation of

TR4 but not TR2 expression, followed by subsequent increase of TR4's repressive activity, may be required for fine-tuning of target genes in response to the agonist. Furthermore, TR2 is found to directly interact with histone deacetylases 3 and 4 (HDAC3 and HDAC4) (Franco et al., 2001). Functionally, the TR2-repressed RARβ gene activation by RA is significantly compromised by trichostatin A (TSA), a specific histone deacetylase inhibitor, which suggests that HDACs may work cooperatively with TR2 in the inhibition of gene expression.

Repression of target gene expression by TR4 and TR2 via mechanism(s) distinct from direct competitive DNA binding has also emerged. Although TR2 or TR4 does not bind palindromic repeat motifs that include the binding sites for estrogen receptors (ERα and ERβ), it shows that TR4 and TR2 repress the ER-transactivated gene expression by direct protein–protein interaction of TR4 or TR2 with ER (Shyr et al., 2002a,b). Such interaction prevents the formation of an active ER homodimer and abrogates ER binding to the estrogen response element. Functionally, TR4 markedly represses two estrogen responsive genes, cyclin D1 and pS2, in human mammary gland carcinoma cells (MCF-7), and the ER-mediated cell growth and proliferation is suppressed in cells harboring stable transfected TR4. Similar findings have been also observed for TR2. Also, cross-talk of TR4 with other nuclear receptors has been noted. Interaction of TR4 with androgen receptor (AR) is found to trigger bidirectional mutual repression on TR4- and AR-modulated target gene expression (Lee et al., 1999a). While AR represses TR4-activated expression of ciliary neurotrophic factor (CNTF) receptor gene, TR4 exhibits potent repression of AR-activated promoter activities of mammary tumor virus (MMTV) and prostate-specific antigen (PSA) genes in the presence of AR ligand dihydrotestosterone (DHT). The repression by TR4 is shown to be AR-specific since it does not affect the glucocorticoid receptor- or progesterone receptor-induced transactivation of the MMTV gene. Convergence of TR4- and AR-regulated pathways may have significant physiologic relevance because TR4 is most abundantly expressed in the prostate, the major target site of androgen action. TR4 interacts with hepatocyte nuclear receptor 4 (HNF4) to repress HNF4-transactivated hepatitis B virus (HBV) core promoter activity through an interaction that is mediated by a DNA-bound form of TR4 at a DR1 motif (Lin et al., 2003). Since TR4 does not affect the HNF4 binding activity to the HBV promoter, a different mechanism other than cross-talk between TR4 and ER/AR is operative in this case.

In addition to the potent repression of multiple target gene expression by TR4 and TR2, these two orphan receptors also exert positive transcriptional control of several genes. TR4 and TR2 activate reporter gene activity through a synthetic DR4 thyroid hormone response element (DR4-TR) (Chang and Pan 1998; Lee et al., 1997). By binding competitively to this motif, the TR2-induced activation antagonizes the inhibitory effect of the reporter gene activity by the unliganded TRα receptor. In the presence of T$_3$, while an activating form of ligand-bound TRα is induced, TR2 exerts an additive effect on T$_3$-activated gene expression. Moreover, TR4 up-regulates the α-myosin heavy chain and rat S14 genes through a DR4-like element, the human luteinizing hormone receptor gene through an imperfect DR0 element, as well as the human CNTF (ciliary neutrotrophic factor) receptor gene expression via a DR1 domain (Lee et al., 1997; Young et al., 1997; Zhang and Dufau, 2000). CNTF elicits a feed-forward regulatory effect by increasing the expression of both TR4 and TR2 proteins, resulting in enhanced TR4/TR2 activation of the CNTF receptor gene. The two orphan receptors are coexpressed with the CNTF receptor in many developing neural structures, indicating their active participation in neurogenesis.

Although the mechanism of TR4/TR2-induced gene activation is poorly understood at this stage, current evidence suggests that these orphan receptors may adapt distinct conformations depending on the specific promoter context. Gel mobility shift assays of TR4 to a DR3/VDR element as a repressor and to a DR4/TR element as an activator reveal the formation of different antibody supershift complexes of TR4 when it binds to these two different motifs (Lee et al., 1999b). Moreover, proteolytic analyses of the DNA-bound TR4 protein have revealed distinct peptide digestion patterns, indicating the different conformations of TR4 receptor on these elements. It is possible that different allosteric structures of TR4 may favor its selective interaction with corepressors or coactivators, thereby eliciting opposite effects through a DR3/VDR or a DR4/TR element. Furthermore, a novel 16-kD TR4-interacting protein, TRA16, has been recently identified through yeast two-hybrid screening and has been shown to act as a repressor of TR4-mediated gene activation (Yang et al., 2003). Mechanistically, TRA16 decreases the binding activity of TR4, blocks formation of TR4 homodimer, and consequently abrogates TR4-induced gene activation. In summary, the orphan nuclear receptors TR4 and TR2, through recognition of various DR motifs located in a number of viral and cellular genes, play an important role in the control of multiple biologic functions.

VII. CONCLUSIONS

Identification of the orphan receptors COUP-TFs, DAX-1, GCNF, SHP, and TR4 and 2, as well as the multiple processes by which they negatively regulate their target genes, has revealed the molecular mechanisms by which these receptors operate in the inhibitory control of cell functions. These findings and studies in knockout animals (COUP-TFs, DAX-1, GCNF, and SHP) and in patients with naturally occurring mutations (DAX-1) have provided direct evidence linking the orphan receptors' inhibitory actions to physiologic states and pathologic conditions.

Despite the large amount of knowledge that has been accumulated in this field of research, much remains to be elucidated. The manner in which multiple mechanisms are coordinated in a tissue- or cell-specific manner to achieve concise regulatory control has yet to be explained. Further studies by tissue- and cell-specific knockouts of certain orphan receptors to avoid embryonic lethality will clarify additional specific functions and provide insights into novel regulatory pathways. The possible involvement of specific unidentified corepressor or coadaptor proteins still needs to be determined, and the relevance of orphan receptor-mediated repression with epigenetic control of gene expression through histone modification and remodeling and DNA methylation events requires detailed analysis. Because most evidence is based on studies of TATA-box genes, which differ in their regulation from TATA-less genes, further study of the silencing mechanisms operative for the latter could provide new insights into their functional control by orphan receptors. Moreover, identification of the putative ligands for these orphan receptors could lead to the discovery of novel therapeutic pathways and targets for drug development. Finally, characterization of the regulatory cascades governing their own expression will delineate the complex networks of cross-talk between COUP-TFs, DAX-1, GCNF, SHP, or TR2 and 4 and other transcription factors involved in repression of target gene expression.

REFERENCES

Achatz, G., Holzl, B., Speckmayer, R., Hauser, C., Sandhofer, F., and Paulweber, B. (1997). Functional domains of the human orphan receptor ARP-1/COUP-TFII involved in active repression and transrepression. *Mol. Cell. Biol.* **17,** 4914–4932.

Achermann, J. C., Ito, M., Silverman, B. L., Habiby, R. L., Pang, S., Rosler, A., and Jameson, J. L. (2001a). Missense mutations cluster within the carboxyl-terminal region of DAX-1 and impair transcriptional repression. *J. Clin. Endocrinol. Metab.* **86,** 3171–3175.

Achermann, J. C., Meeks, J. J., and Jameson, J. L. (2001b). Phenotypic spectrum of mutations in DAX-1 and SF-1. *Mol. Cell. Endocrinol.* **185,** 17–25.

Agoulnik, I. Y., Cho, Y., Niederberger, C., Kieback, D. G., and Cooney, A. J. (1998). Cloning, expression analysis, and chromosomal localization of the human nuclear receptor gene GCNF. *FEBS Lett.* **424,** 73–78.

Alland, L., Muhle, R., Hou, H., Potes, J., Chin, L., Schreiber-Agus, N., and DePinpo, R. A. (1997). Role for NCoR and histone deacetylase in Sin3-mediated transcriptional repression. *Nature* **387,** 49–55.

Altincicek, B., Tenbaum, S. P., Dressel, U., Thormeyer, D., Renkawiz, R., and Baniahma (2000). Interaction of the corepressor Alien with DAX-1 is abrogated by mutations of DAX-1 involved in adrenal hypoplasia congenita. *J. Biol. Chem.* **27,** 7662–7667.

Arango, N. A., Lovell-Badge, R., and Behringer, R. R. (1999). Targeted mutagenesis of the endogenous mouse Mis gene promoter: *In vivo* definition of genetic pathways of vertebrate sexual development. *Cell* **99,** 409–419.

Avram, D., Ishmael, J. E., Nevrivy, D. J., Peterson, V. J., Lee, S. H., Dowell, P., and Leid, M. (1999). Heterodimeric interactions between chicken ovalbumin upstream promoter-transcription factor family members ARP-1 and EAR2. *J. Biol. Chem.* **274,** 14331–14336.

Bailey, P., Sartorelli, V., Hamamori, Y., and Muscat, G. E. (1998). The orphan nuclear receptor, COUP-TF II, inhibits myogenesis by post-transcriptional regulation of MyoD function: COUP-TF II directly interacts with p300 and MyoD. *Nucleic Acids Res.* **26**, 5501–5510.

Bailey, P. J., Dowhan, D. H., Franke, K., Burke, L. J., Doalnes, M., and Muscat, G. E. (1997). Transcriptional repression by COUP-TF II is dependent on the C-terminal domain and involves the NCoR variant, RIP13delta1. *J. Steroid Biochem. Mol. Biol.* **63**, 165–174.

Bardoni, B., Zanaria, E., Guioli, S., Floridia, G., Worley, K. C., Tonini, G., Ferrante, E., and Chiumello, G. (1994). A dosage-sensitive locus at chromosome Xp21 is involved in male to female sex reversal. *Nat. Genet.* **7**, 497–501.

Barreto, G., Reintsch, W., Kaufmann, C., and Dreyer, C. (2003). The function of *Xenopus* germ cell nuclear factor (xGCNF) in morphogenetic movements during neurulation. *Dev. Biol.* **257**, 329–342.

Bassett, M. H. White, P. C., and Rainey, W. E. (2003). Controlling the capacity to produce aldosterone: Regulation of aldosterone synthase transcription through novel mechanisms 85th Endocrine Society Annual Meeting (abstract), p. 66.

Bastien, J., Adam-Stitah, S., Riedl, T., Egly, J. M., Chambon, P., and Rochette-Egly, C. (2000). TFIIH interacts with the retinoic acid receptor gamma and phosphorylates its AF-1-activating domain through cdk7. *J. Biol. Chem.* **275**, 21896–21904.

Bauer, U. M., Schneider-Hirsch, S., Reinhardt, S., Pauly, T., Maus, A., Wang, F., Heiermann, R., Rentrop, M., and Maelicke, A. (1997). Neuronal cell nuclear factor—A nuclear receptor possibly involved in the control of neurogenesis and neuronal differentiation. *Eur. J. Biochem.* **249**, 826–887.

Ben-Shushan, E., Sharir, H., Pikarsky, E., and Bergman, Y. (1995). A dynamic balance between ARP-1/COUP-TFII, EAR-3/COUP-TFI, and retinoic acid receptor: Retinoid X receptor heterodimers regulates Oct3/4 expression in embryonal carcinoma cells. *Mol. Cell. Biol.* **15**, 1034–1048.

Berrodin, T. J., Marks, M. S., Ozato, K., Linney, E., and Lazar, M. A. (1992). Heterodimerization among thyroid hormone receptor, retinoic acid receptor, retinoid X receptor, chicken ovalbumin upstream promoter-transcription factor, and an endogenous liver protein. *Mol. Endocrinol.* **6**, 1468–1478.

Beuschlein, F., Keegan, C. E., Bavers, D. L., Mutch, C., Hutz, J. E., Shah, S., Ulrich-Lai, Y. M., Engeland, W. C., Jeff, B., Jameson, J. L., and Hammer, G. D. (2002). SF-1, DAX-1, and acd: Molecular determinants of adrenocortical growth and steroidogenesis. *Endocr. Res.* **28**, 597–607.

Borgius, L. J., Steffensen, K. R., Gustafsson, J. A., and Treuter, E. (2002). Glucocorticoid signaling is perturbed by the atypical orphan receptor and corepressor SHP. *J. Biol. Chem.* **277**, 49761–49776.

Borgmeyer, U. (1997). Dimeric binding of the mouse germ cell nuclear factor. *Eur. J. Biochem.* **244**, 120–127.

Bourguet, W., Ruff, M., Chambon, P., Gronemeyer, H., and Moras, D. (1995). Crystal structure of the ligand binding domain of the human nuclear receptor RXR alpha. *Nature* **375**, 377–382.

Brendel, C., Schoonjans, K., Botrugno, O. A., Treuter, E., and Auwerx, J. (2002). The small heterodimer partner interacts with the liver X receptor alpha and represses its transcriptional activity. *Mol. Endocrinol.* **16**, 2065–2076.

Brzozowski, A. M., Pike, A. C., Dauter, Z., Hubbard, R. E., Bonn, T., Engstrom, O., Ohman, L., Greene, G., Gustafsson, J. A., and Carlquist, M. (1997). Molecular basis of agonism and antagonism in the oestrogen receptor. *Nature* **389**, 753–758.

Burris, T. P., Guo, W., and McCabe, E. R. (1996). The gene responsible for adrenal hypoplasia congenita, DAX-1, encodes a nuclear hormone receptor that defines a new class within the superfamily. *Recent Prog. Horm. Res.* **51**, 241–259.

Butler, A. J., and Parker, M. G. (1995). COUP-TF II homodimers are formed in preference to heterodimers with RXR alpha or TR beta in intact cells. *Nucleic Acids Res.* **23,** 4143–4150.
Carter, M. E., Gulick, T., Moore, D. D., and Kelly, D. P. (1994). A pleiotropic element in the medium-chain acyl coenzyme A dehydrogenase gene promoter mediates transcriptional regulation by multiple nuclear receptor transcription factors and defines novel receptor-DNA binding motifs. *Mol. Cell. Biol.* **14,** 4360–4372.
Chang, C., Da Silva, S. L., Ideta, R., Lee, Y., Yeh, S., and Burbach, I. P. (1994). Human and rat TR4 orphan receptors specify a subclass of the steroid receptor superfamily. *Proc. Natl. Acad. Sci. USA* **91,** 6040–6044.
Chang, C., and Kokontis, J. (1988). Identification of a new member of the steroid receptor superfamily by cloning and sequence analysis. *Biochem. Biophys. Res. Commun.* **55,** 971–977.
Chang, C., Kokontis, J., Acakpo-Satchivi, L., Liao, S., Takeda, H., and Chang, Y. (1989). Molecular cloning of new human TR2 receptors: A class of steroid receptor with multiple ligand binding domains. *Biochem. Biophys. Res. Commun.* **165,** 735–741.
Chang, C., and Pan, H. J. (1998). Thyroid hormone direct repeat 4 response element is a positive regulatory element for the human TR2 orphan receptor, a member of steroid receptor superfamily. *Mol. Cell. Biochem.* **189,** 195–200.
Chen, F., Cooney, A. J., Wang, Y., Wang, F., Law, S. W., and O'Malley, B. W. (1994). Cloning of a novel orphan receptor (GCNF) expressed during germ cell development. *Mol. Endocrinol.* **8,** 1434–1444.
Chew, L. J., Huang, F., Boutin, J. M., and Gallo, V. (1999). Identification of nuclear orphan receptors as regulators of expression of a neurotransmitter receptor gene. *J. Biol. Chem.* **274,** 29366–29375.
Chinpaisal, C., Chang, L., Hu, X., Lee, C. H., Wen, W. N., and Wei, L. N. (1997). The orphan nuclear receptor TR2 suppresses a DR4 hormone response element of the mouse CRABP-I gene promoter. *Biochemistry* **36,** 14088–14095.
Chu, K., and Zingg, H. H. (1997). The nuclear orphan receptors COUP-TFII and EAR2 act as silencers of the human oxytocin gene promoter. *J. Mol. Endocrinol.* **19,** 163–172.
Cooney, A. J., Hummelke, G. C., Herman, T. *et al.* (1998). Germ cell nuclear factor is a response element-specific repressor of transcription. *Biochem. Biophys. Res. Commun.* **245,** 94–100.
Cooney, A. J., Lee, C. T., Lin, S. C., Tsai, S. Y., and Tsai, M. J. (2001). Physiological function of the orphans GCNF and COUP-TF. *Trends Endocrinol. Metab.* **12,** 247–251.
Cooney, A. J., Tsai, S. Y., O'Malley, B. W., and Tsai, M. J. (1991). Chicken ovalbumin upstream promoter-transcription factor binds to a negative regulatory region in the human immunodeficiency virus type 1 long terminal repeat. *J. Virol.* **65,** 2853–2860.
Cooney, A. J., Tsai, S. Y., O'Malley, B. W., and Tsai, M. J. (1992). Chicken ovalbumin upstream promoter-transcription factor (COUP-TF) dimers bind to different GGTCA response elements, allowing COUP-TF to repress hormonal induction of the vitamin D_3, thyroid hormone, and retinoic acid receptors. *Mol. Cell. Biol.* **12,** 4153–4163.
Crawford, P. A., Dorn, C., Sadovsky, Y., and Milbrandt, J. (1998). Nuclear receptor DAX-1 recruits nuclear receptor corepressor NCoR to steroidogenic factor-1. *Mol. Cell. Biol.* **18,** 2949–2956.
Crestani, M., Sadeghpour, A., Stroup, D., Galli, G., and Chiang, J. Y. (1998). Transcriptional activation of the cholesterol 7 alpha-hydroxylase gene (CYP7A1) by nuclear hormone receptors. *J. Lipid Res.* **39,** 2192–2200.
Denson, L. A., Sturm, E., Echevarria, W., Zimmerman, T. L., Makishima, M., Mangelsdorf, D. J., and Karpen, S. J. (2001). The orphan nuclear receptor, SHP, mediates bile acid-induced inhibition of the rat bile acid transporter, ntcp. *Gastroenterology* **121,** 140–147.
Devchand, P. R., Ijpenberg, A., Devesvergne, B., and Wahli, W. (1999). PPARs: Nuclear receptors for fatty acids, eicosanoids, and xenobiotics. *Adv. Exp. Med. Biol.* **469,** 231–236.

Egea, P. F., Mitschler, A., Rochel, N., Ruff, M., Chambon, and Moras, D. (2000). Crystal structure of the human RXR alpha ligand binding domain bound to its natural ligand: 9-cis retinoic acid. *EMBO J.* **19**, 2592–2601.

Evans, R. M. (1988). The steroid and thyroid hormone receptor superfamily. *Science* **240**, 889–895.

Fitzgerald, M. L., Moore, K. J., and Freeman, M. W. (2002). Nuclear hormone receptors and cholesterol trafficking: The orphans find a new home. *J. Mol. Med.* **80**, 271–281.

Franco, P. J., Farooqui, M., Seto, E., and Wei, L. N. (2001). The orphan nuclear receptor TR2 interacts directly with both class I and class II histone deacetylases. *Mol. Endocrinol.* **15**, 1318–1328.

Freedman, L. P., and Towers, T. L. (1991). DNA binding properties of the vitamin D_3 receptor zinc finger region. *Mol. Endocrinol.* **5**, 1815–1826.

Fuhrmann, G., Chung, A. C., Jackson, K. J., Hummelke, G., Baniahmad, A., Sutter, J., Sylvester, Scholer, H. R., and Cooney, A. J. (2001). Mouse germline restriction of Oct4 expression by germ cell nuclear factor. *Dev. Cell.* **1**, 377–387.

Giguere, V. (1999). Orphan nuclear receptors: From gene to function. *Endocr. Rev.* **20**, 689–725.

Giguere, V., McBroom, L. D., and Flock, G. (1995). Determinants of target gene specificity for ROR alpha 1: Monomeric DNA binding by an orphan nuclear receptor. *Mol. Cell. Biol.* **15**, 2517–2526.

Giuili, G., Shen, W. H., and Ingraham, H. A. (1997). The nuclear receptor SF-1 mediates sexually dimorphic expression of Müllerian inhibiting substance, *in vivo*. *Development* **124**, 1799–1807.

Gobinet, J., Auzou, G., Nicolas, J. C., Sultan, C., and Jalaguier, S. (2001). Characterization of the interaction between androgen receptor and a new transcriptional inhibitor, SHP. *Biochemistry* **40**, 15369–15377.

Goodwin, B., Jones, S. A., Price, R. R., Watson, M. A., Mckee, D. D., Moore, L. B., Galardi, C., Wilson, J. G., Lewis, M. C., Roth, M. E., Maloney, P. R., Willson, T. M., and Kliewer, S. A. (2000). A regulatory cascade of the nuclear receptors FXR, SHP-1, and LRH-1 represses bile acid biosynthesis. *Mol. Cell* **6**, 517–526.

Greschik, H., and Schule, R. (1998). Germ cell nuclear factor: An orphan receptor with unexpected properties. *J. Mol. Med.* **76**, 800–810.

Greschik, H., Wurtz, J. M., Hublitz, P., Kohler, F., Moras, D., and Schule, R. (1999). Characterization of the DNA-binding and dimerization properties of the nuclear orphan receptor germ cell nuclear factor. *Mol. Cell. Biol.* **19**, 690–703.

Guo, W., Burris, T. P., and McCabe, E. R. (1995). Expression of DAX-1, the gene responsible for X-linked adrenal hypoplasia congenita and hypogonadotropic hypogonadism, in the hypothalamic–pituitary–adrenal/gonadal axis. *Biochem. Mol. Med.* **56**, 8–13.

Gupta, S., Pandak, W. M., and Hylemon, P. B. (2002). LXR alpha is the dominant regulator of CYP7A1 transcription. *Biochem. Biophys. Res. Commun.* **293**, 338–343.

Hall, R. K., Sladek, F. M., and Granner, D. K. (1995). The orphan receptors COUP-TF and HNF4 serve as accessory factors required for induction of phosphoenolpyruvate carboxykinase gene transcription by glucocorticoids. *Proc. Natl. Acad. Sci. USA* **92**, 412–416.

Hanley, N. A., Rainey, W. E., Wilson, D. I., Ball, S. G., and Parker, K. L. (2001). Expression profiles of SF-1, DAX-1, and CYP17 in the human fetal adrenal gland: Potential interactions in gene regulation. *Mol. Endocrinol.* **15**, 57–68.

Harvey, C. B., and Williams, G. R. (2002). Mechanism of thyroid hormone action. *Thyroid* **12**, 441–444.

Haussler, M. R., Haussler, C. A., Jurutka, P. W., Thompson, P. D., Hsieh, J. C., Remus, L. S., Selznick, S., and Whitfield, G. K. (1997). The vitamin D hormone and its nuclear receptor: Molecular actions and disease states. *J. Endocrinol.* **154**, S57–S73.

Hirose, T., Apfel, R., Pfahl, M. *et al.* (1995a). The orphan receptor TAK1 acts as a repressor of RAR-, RXR-and T3R-mediated signaling pathways. *Biochem. Biophys. Res. Commun.* **211,** 83–91.

Hirose, T., Fujimoto, W., Tamaai, T., Kim, K. H., Matsuura, H., and Jatten, A. M. (1994). TAK1: Molecular cloning and characterization of a new member of the nuclear receptor superfamily. *Mol. Endocrinol.* **8,** 1667–1680.

Hirose, T., O'Brien, D. A., and Jetten, A. M. (1995b). RTR: A new member of the nuclear receptor superfamily that is highly expressed in murine testis. *Gene* **152,** 247–251.

Hoffmann, A. F. (1994). Bile acids. *In* "The Liver: Biology and Pathobiology" (I. M. Arias, J. L. Boyer, N. Fausto, W. B. Jakoby, D. A. Schachter, and D. A. Shafritz, Eds.), pp. 677–718. Raven Press, New York.

Holter, E., Kotaja, N., Makela, S., Strauss, L., Kietz, S., Janne, O. A., Gustafsson, J. A., Palvimo, J. J., and Treuter, E. (2002). Inhibition of androgen receptor (AR) function by the reproductive orphan nuclear receptor DAX-1. *Mol. Endocrinol.* **16,** 515–528.

Honkakoski, P., Sueyoshi, T., and Negishi, M. (2003). Drug-activated nuclear receptors CAR and PXR. *Ann. Med.* **35,** 172–182.

Hu, M. C., Hsu, N. C., Pai, C. I., Wang, C. K., and Chung, B. (2001). Functions of the upstream and proximal steroidogenic factor-1 (SF-1)-binding sites in the CYP11A1 promoter in basal transcription and hormonal response. *Mol. Endocrinol.* **15,** 812–818.

Hummelke, G. C., and Cooney, A. J. (2001). Germ cell nuclear factor is a transcriptional repressor essential for embryonic development. *Front. Biosci.* **6,** D1186–D1191.

Hwung, Y. P., Wang, L. H., Tsai, S. Y., and Tsai, M. J. (1988). Differential binding of the chicken ovalbumin upstream promoter (COUP) transcription factor to two different promoters. *J. Biol. Chem.* **26,** 13470–13474.

Ikeda, Y., Swain, A., Weber, T. J., Hentges, K. E., Zanaria, E., Lalli, E., Tamai, K. T., Sassone-Corsi, P., Lovell-Badge, R., Camerino, G., and Parker, K. L. (1996). Steroidogenic factor-1 and DAX-1 colocalize in multiple cell lineages: Potential links in endocrine development. *Mol. Endocrinol.* **10,** 1261–1272.

Ikeda, Y., Takeda, Y., Shikayama, T., Mukai, T., Hisano, S., and Morohashi, K. I. (2001). Comparative localization of DAX-1 and Ad4BP/SF-1 during development of the hypothalamic–pituitary–gonadal axis suggests their closely related and distinct functions. *Dev. Dyn.* **220,** 363–376.

Ing, N. H., Beekman, J. M., Tsai, S. Y., Tsai, M. J., and O'Malley, B. W. (1992). Members of the steroid hormone receptor superfamily interact with TFIIB (S300-II). *J. Biol. Chem.* **26,** 17617–17623.

Ingraham, H. A., Lala, D. S., Ikeda, Y., Luo, X., Shen, W. H., Nachtigal, M. W., Nilson, J. H., and Parker, K. L. (1994). The nuclear receptor steroidogenic factor-1 acts at multiple levels of the reproductive axis. *Genes Dev.* **8,** 2302–2312.

Inui, S., Lee, Y. F., Chang, E., Shyr, C. R., and Chang, C. (2003). Differential and bidirectional regulation between TR2/TR4 orphan nuclear receptors and a specific ligand mediated-peroxisome proliferator-activated receptor alpha in human HaCaT keratinocytes. *J. Dermatol. Sci.* **31,** 65–71.

Inui, S., Lee, Y. F., Haake, A. R., Goldsmith, L. A., and Chang, C. (1999). Induction of TR4 orphan receptor by retinoic acid in human HaCaT keratinocytes. *J. Invest. Dermatol.* **112,** 426–431.

Ito, M., Yu, R., and Jameson, J. L. (1997). DAX-1 inhibits SF-1-mediated transactivation via a carboxy-terminal domain that is deleted in adrenal hypoplasia congenita. *Mol. Cell. Biol.* **1,** 1476–1483.

Jacobs, M. N., Dickins, M., and Lewis, D. F. (2003). Homology modelling of the nuclear receptors: Human oestrogen receptorbeta (hERbeta), the human pregnan X receptor (PXR), the Ah receptor (AhR) and the constitutive androstane receptor (CAR) ligand binding domains from the human estrogen receptor alpha (hERα) crystal structure, and the human peroxisome proliferator-activated receptor alpha (PPARα) ligand binding domain from the human PPARgamma crystal structure. *J. Steroid Biochem. Mol. Biol.* **84,** 117–132.

Jaskelioff, M., and Peterson, C. L. (2003). Chromatin and transcription: Histones continue to make their marks. *Nat. Cell. Biol.* **5,** 395–399.

Jepsen, K., and Rosenfeld, M. G. (2002). Biological roles and mechanistic actions of corepressor complexes. *J. Cell. Sci.* **15,** 689–698.

Johansson, L., Bavner, A., Thomsen, J. S., Farnegardh, M., Gustafsson, J. A., and Treuter, E. (2000). The orphan nuclear receptor SHP utilizes conserved LXXLL-related motifs for interactions with ligand-activated estrogen receptors. *Mol. Cell. Biol.* **20,** 1124–1133.

Johansson, L., Thomsen, J. S., Damdimopoulos, A. E., Spyrou, G., Gustafsson, J. A., and Treuter, E. (1999). The orphan nuclear receptor SHP inhibits agonist-dependent transcriptional activity of estrogen receptors ER alpha and ER beta. *J. Biol. Chem.* **274,** 345–353.

Joos, T. O., David, R., and Dreyer, C. (1996). xGCNF, a nuclear orphan receptor is expressed during neurulation in *Xenopus laevis*. *Mech. Dev.* **60,** 45–57.

Kadowaki, Y., Toyoshima, K., and Yamamoto, T. (1992). EAR3/COUP-TF binds most tightly to a response element with tandem repeat separated by one nucleotide. *Biochem. Biophys. Res. Commun.* **183,** 492–498.

Kassam, A., Capone, J. P., and Rachubinski, R. A. (2001). The short heterodimer partner receptor differentially modulates peroxisome proliferator-activated receptor alpha-mediated transcription from the peroxisome proliferator-response elements of the genes encoding the peroxisomal beta-oxidation enzymes acyl-CoA oxidase and hydratase-dehydrogenase. *Mol. Cell. Endocrinol.* **76,** 49–56.

Katz, D., Niederberger, C., Slaughter, G. R., and Cooney, A. J. (1997). Characterization of germ cell-specific expression of the orphan nuclear receptor, germ cell nuclear factor. *Endocrinology* **138,** 4364–4372.

Kerr, T. A., Saeki, S., Schneider, M., Schaefer, K., Berdy, S., Redder, T., Shan, B., Russell, D. W., and Schwarz, M. (2002). Loss of nuclear receptor SHP impairs but does not eliminate negative feedback regulation of bile acid synthesis. *Dev. Cell.* **2,** 713–720.

Kliewer, S. A., Umesono, K., Heyman, R. A., Mangelsdorf, D. J., Dyck, J. A., and Evans, R. M. (1992). Retinoid X receptor–COUP-TF interactions modulate retinoic acid signaling. *Proc. Natl. Acad. Sci. USA* **89,** 1448–1452.

Klinge, C. M., Jernigan, S. C., Risinger, K. E., Lee, J. E., Tyulmenkov, V. V., Falkner, K. C., and Prough, R. A. (2001). Short heterodimer partner (SHP) orphan nuclear receptor inhibits the transcriptional activity of aryl hydrocarbon receptor (AHR)/AHR nuclear translocator (ARNT). *Arch. Biochem. Biophys.* **390,** 64–70.

Kobayashi, S., Shibata, H., Kurihara, I., Saito, I., and Saruta, T. (2002). CIP-1 is a novel corepressor for nuclear receptor COUP-TF: A potential negative regulator in adrenal steroidogenesis. *Endocr. Res.* **28,** 579.

Koskimies, P., Levallet, J., Sipila, P., Huhtaniemi, I., and Poutanen, M. (2002). Murine relaxin-like factor promoter: Functional characterization and regulation by transcription factors steroidogenic factor-1 and DAX-1. *Endocrinology* **143,** 909–919.

Ladias, J. A., Hadzopoulou-Cladaras, M., Kardassis, D., Cardot, P., Cheng, J., Zannis, V., and Cladaras (1992). Transcriptional regulation of human apolipoprotein genes ApoB, ApoCIII, and ApoAII by members of the steroid hormone receptor superfamily HNF4, ARP-1, EAR2, and EAR3. *J. Biol. Chem.* **267,** 15849–15860.

Ladias, J. A., and Karathanasis, S. K. (1991). Regulation of the apolipoprotein AI gene by ARP-1, a novel member of the steroid receptor superfamily. *Science* **251,** 561–565.

Lalli, E., Bardoni, B., Zazopoulos, E., Wurtz, J. M., Strom, T., and Sassone-Corsi, P. (1997). A transcriptional silencing domain in DAX-1 whose mutation causes adrenal hypoplasia congenita. *Mol. Endocrinol.* **11,** 1950–1960.

Lalli, E., Melner, M. H., Stocco, D. M., and Sassone-Corsi, P. (1998). DAX-1 blocks steroid production at multiple levels. *Endocrinology* **39,** 4237–4243.
Lalli, E., Ohe, K., Hindelang, C., and Sassone-Corsi, P. (2000). Orphan receptor DAX-1 is a shutting RNA binding protein associated with polyribosomes via mRNA. *Mol. Cell. Biol.* **20,** 4910–4921.
Lalli, E., and Sassone-Corsi, P. (2003). DAX-1, an unusual orphan receptor at the crossroads of steroidogenic function and sexual differentiation. *Mol. Endocrinol.* **17,** 1445–1453.
Lan, Z. J., Chung, A. C., Xu, X., DeMayo, F. J., and Cooney, A. J. (2002). The embryonic function of germ cell nuclear factor is dependent on the DNA binding domain. *J. Biol. Chem.* **277,** 50660–50667.
Lan, Z. J., Gu, P., Xu, X., and Cooney, A. J. (2003a). Expression of the orphan nuclear receptor, germ cell nuclear factor, in mouse gonads and preimplantation embryos. *Biol. Reprod.* **68,** 282–289.
Lan, Z. J., Gu, P., Xu, X., Jackson, K. J., DeMayo, F. J., O'Malley, B. W., and Cooney, A. J. (2003b). GCNF-dependent repression of BMP-15 and GDF-9 mediates gamete regulation of female fertility. *EMBO J.* **22,** 4070–4081.
Laudet, V. (1997). Evolution of the nuclear receptor superfamily: Early diversification from an ancestral orphan receptor. *J. Mol. Endocrinol.* **19,** 207–226.
Lee, C. H., Chinpaisal, C., and Wei, L. N. (1998a). A novel nuclear receptor heterodimerization pathway mediated by orphan receptors TR2 and TR4. *J. Biol. Chem.* **273,** 25209–25215.
Lee, C. H., Copeland, N. G., Gilbert, D. J., Jenkins, N. A., and Wei, L. N. (1995). Genomic structure, promoter identification, and chromosomal mapping of a mouse nuclear orphan receptor expressed in embryos and adult testes. *Genomics* **30,** 46–52.
Lee, C. H., and Wei, L. N. (1999). Characterization of an inverted repeat with a zero spacer (IRO)-type retinoic acid response element from the mouse nuclear orphan receptor TR2–11 gene. *Biochemistry* **38,** 8820–8825.
Lee, H. J., and Chang, C. (1995). Identification of human TR2 orphan receptor response element in the transcriptional initiation site of the simian virus 40 major late promoter. *J. Biol. Chem.* **270,** 5434–5440.
Lee, H. J., Lee, Y. F., and Chang, C. (2001). TR4 orphan receptor represses the human steroid 21-hydroxylase gene expression through the monomeric AGGTCA motif. *Biochem. Biophys. Res. Commun.* **285,** 1361–1368.
Lee, H. J., Young, W. J., Shih, C. Y., and Chang, C. (1996). Suppression of the human erythropoietin gene expression by the TR2 orphan receptor, a member of the steroid receptor superfamily. *J. Biol. Chem.* **271,** 10405–10412.
Lee, H. K., Lee, Y. K., Park, S. H., Kim, Y. S., Lee, J. W., Kwon, H. B., Soh, J., Moore, D. D., and Choi, H. S. (1998b). Structure and expression of the orphan nuclear receptor SHP gene. *J. Biol. Chem.* **273,** 14398–14402.
Lee, Y. F., Lee, H. J., and Chang, C. (2002). Recent advances in the TR2 and TR4 orphan receptors of the nuclear receptor superfamily. *J. Steroid Biochem. Mol. Biol.* **81,** 291–308.
Lee, Y. F., Pan, H. J., Burbach, J. P., Morkin, E., and Chang, C. (1997). Identification of direct-repeat 4 as a positive regulatory element for the human TR4 orphan receptor. A modulator for the thyroid hormone target genes. *J. Biol. Chem.* **272,** 12215–12220.
Lee, Y. F., Shyr, C. R., Thin, T. H., Lin, W. J., and Chang, C. (1999a). Convergence of two repressors through heterodimer formation of androgen receptor and testicular orphan receptor-4: A unique signaling pathway in the steroid receptor superfamily. *Proc. Natl. Acad. Sci. USA* **96,** 14724–14729.
Lee, Y. F., Young, W. J., Lin, W. J., Shyr, C. R., and Chang, C. (1999b). Differential regulation of direct-repeat 3 vitamin D_3 and direct-repeat 4 thyroid hormone signaling pathways by the human TR4 orphan receptor. *J. Biol. Chem.* **274,** 16198–16205.

Lee, Y. K., Parker, K. L., Choi, H. S., and Moore, D. D. (1999c). Activation of the promoter of the orphan receptor SHP by orphan receptors that bind DNA as monomers. *J. Biol. Chem.* **274,** 20869–20873.

Lee, Y. K., Dell, H., Dowhan, D. H., Hadzopoulou-Cladaras, M., and Moore, D. D. (2000). The orphan nuclear receptor SHP inhibits hepatocyte nuclear factor 4 and retinoid X receptor transactivation: Two mechanisms for repression. *Mol. Cell. Biol.* **20,** 187–195.

Lehmann, S. G., Lalli, E., and Sassone-Corsi, P. (2002). X-linked adrenal hypoplasia congenita is caused by abnormal nuclear localization of the DAX-1 protein. *Proc. Natl. Acad. Sci. USA* **99,** 8225–8230.

Lehmann, S. G., Wurtz, J. M., Renaud, J. P., Sassone-Corsi, P., and Lalli, E. (2003). Structure–function analysis reveals the molecular determinants of the impaired biological function of DAX-1 mutants in AHC patients. *Hum. Mol. Genet.* **12,** 1063–1072.

Lei, W., Hirose, T., Zhang, L. X., Adachi, H., Spinella, M. J., Dmitrovsky, E., and Jetten, A. M. (1997). Cloning of the human orphan receptor germ cell nuclear factor/retinoid receptor-related testis-associated receptor and its differential regulation during embryonal carcinoma cell differentiation. *J. Mol. Endocrinol.* **18,** 167–176.

Leng, X., Cooney, A. J., Tsai, S. Y., and Tsai, M. J. (1996). Molecular mechanisms of COUP-TF-mediated transcriptional repression: Evidence for transrepression and active repression. *Mol. Cell. Biol.* **16,** 2332–2340.

Lin, B., Chen, G. Q., Xiao, D., Kolluri, S. K., Cao, X., Su, H., and Zhang, X. K. (2000). Orphan receptor COUP-TF is required for induction of retinoic acid receptor beta, growth inhibition, and apoptosis by retinoic acid in cancer cells. *Mol. Cell. Biol.* **20,** 957–970.

Lin, D. L., and Chang, C. (1996). p53 is a Mediator for radiation-repressed human TR2 orphan receptor expression in MCF-7 cells, a new pathway from tumor suppressor to member of the steroid receptor superfamily. *J. Biol. Chem.* **27,** 14649–14652.

Lin, D. L., Wu, S. Q., and Chang, C. (1998). The genomic structure and chromosomal location of the human TR2 orphan receptor, a member of the steroid receptor superfamily. *Endocrine* **8,** 123–134.

Lin, H. B., Jurk, M., Gulick, T., and Cooper, G. M. (1999). Identification of COUP-TF as a transcriptional repressor of the c-mos proto-oncogene. *J. Biol. Chem.* **274,** 36796–36800.

Lin, T. M., Young, W. J., and Chang, C. (1995). Multiple functions of the TR2-11 orphan receptor in modulating activation of two key cis-acting elements involved in the retinoic acid signal transduction system. *J. Biol. Chem.* **270,** 30121–30128.

Lin, W. J., Li, J., Lee, Y. F., Lee, Y. F., Yeh, S. D., Altuwaijri, S., Ou, J. H., and Chang, C. (2003). Suppression of hepatitis B virus core promoter by the nuclear orphan receptor TR4. *J. Biol. Chem.* **278,** 9353–9360.

Liu, X., Huang, X., and Sigmund, C. D. (2003). Identification of a nuclear orphan receptor (EAR2) as a negative regulator of renin gene transcription. *Circ. Res.* **92,** 1033–1040.

Liu, Y., Yang, N., and Teng, C. T. (1993). COUP-TF acts as a competitive repressor for estrogen receptor-mediated activation of the mouse lactoferrin gene. *Mol. Cell. Biol.* **13,** 1836–1846.

Lopez, D., Shea-Eaton, W., and Sanchez, M. D. (2001). DAX-1 represses the high-density lipoprotein receptor through interaction with positive regulators sterol regulatory element-binding protein-1a and steroidogenic factor-1. *Endocrinology* **42,** 5097–5106.

Lu, T. T., Makishima, M., Repa, J. J., Schoonjans, K., Kerr, T. A., Auwerx, J., and Mangelsdorf, D. J. (2000). Molecular basis for feedback regulation of bile acid synthesis by nuclear receptors. *Mol. Cell* **6,** 507–515.

Lu, X. P., Salbert, G., and Pfahl, M. (1994). An evolutionary conserved COUP-TF binding element in a neural-specific gene and COUP-TF expression patterns support a major role for COUP-TF in neural development. *Mol. Endocrinol.* **8,** 1774–1788.

Luo, X., Ikeda, Y., and Parker, K. L. (1994). A cell-specific nuclear receptor is essential for adrenal and gonadal development and sexual differentiation. *Cell* **77,** 481–490.

MacLaughlin, D. T., and Donahoe, P. K. (2002). Müllerian inhibiting substance: An update. *Adv. Exp. Med. Biol.* **511,** 25–38.
Makishima, M., Okamoto, A. Y., Repa, J. J., Tu, H., Learned, R., Luk, A., Hull, M. V., Lustig, K. D., Mangelsdorf, D. J., and Shan, B. (1999). Identification of a nuclear receptor for bile acids. *Science* **284,** 1362–1365.
Mangelsdorf, D. J., Borgmeyer, U., Heyman, R. A., Zhou, J. Y., Ong, E. S., Oro, A. E., Kakizuka, A., and Evans, R. M. (1992). Characterization of three RXR genes that mediate the action of 9-cis retinoic acid. *Genes Dev.* **6,** 329–344.
Mangelsdorf, D. J., and Evans, R. M. (1995). The RXR heterodimers and orphan receptors. *Cell* **83,** 841–850.
Mangelsdorf, D. J., Thummel, C., Beato, M., Herrlich, P., Schutz, Umesono, K., Blumberg Kastner, P., Mark, M., and Chamborn, P. (1995). The nuclear receptor superfamily: The second decade. *Cell* **83,** 835–839.
Marcus, S. L., Capone, J. P., and Rachubinski, R. A. (1996). Identification of COUP-TFII as a peroxisome proliferator response element binding factor using genetic selection in yeast: COUP-TFII activates transcription in yeast but antagonizes PPAR signaling in mammalian cells. *Mol. Cell. Endocrinol.* **120,** 31–39.
Mason, A. J., Hayflick, J. S., Zoeller, R. T., Young, W. S., III, Phillips, H. S., Nikolics, K., and Seeburg, P. H. (1986). A deletion truncating the gonadotropin-releasing hormone gene is responsible for hypogonadism in the hpg mouse. *Science* **234,** 1366–1371.
McBroom, L. D., Flock, G., and Giguere, V. (1995). The nonconserved hinge region and distinct amino-terminal domains of the ROR alpha orphan nuclear receptor isoforms are required for proper DNA bending and ROR alpha–DNA interactions. *Mol. Cell. Biol.* **15,** 796–808.
McKenna, N. J., Lanz, R. B., and O'Malley, B. W. (1999). Nuclear receptor coregulators: Cellular and molecular biology. *Endocr. Rev.* **20,** 321–344.
Meeks, J. J., Crawford, S. E., Russell, T. A., Morohashi, K., Weiss, J., and Jameson, J. L. (2003). DAX-1 regulates testis cord organization during gonadal differentiation. *Development* **130,** 1029–1036.
Meinke, G., and Sigler, P. B. (1999). DNA-binding mechanism of the monomeric orphan nuclear receptor NGFI-B. *Nat. Struct. Biol.* **6,** 471–477.
Mietus-Snyder, M., Sladek, F. M., Ginsburg, G. S., Kuo, C. F., Ladias, Darnell, J. E., and Karathanasis, S. K. (1992). Antagonism between apolipoprotein AI regulatory protein-1, EAR3/COUP-TF, and hepatocyte nuclear factor 4 modulates apolipoprotein CIII gene expression in liver and intestinal cells. *Mol. Cell. Biol.* **12,** 1708–1718.
Miyajima, N., Kadowaki, Y., Fukushige, S., Shimizu, S., Semba, K., Yamanashi, Y., Matsubana, K., Toyshima, K., and Yamamoto, T. (1988). Identification of two novel members of erbA superfamily by molecular cloning: The gene products of the two are highly related to each other. *Nucleic Acids Res.* **16,** 11057–11074.
Muscatelli, F., Strom, T. M., Walker, A. P., Zanaria, E., Recan, D., Meindl, A., Bardoni, B., Guioli, S., Zehetner, G., and Rabl, W. (1994). Mutations in the DAX-1 gene give rise to both X-linked adrenal hypoplasia congenita and hypogonadotropic hypogonadism. *Nature* **372,** 672–676.
Nachtigal, M. W., Hirokawa, Y., Enyeart-van Houten, D. L., Flanagan, J. N., Hammer, G. D., and Ingraham, H. A. (1998). Wilms' tumor 1 and DAX-1 modulate the orphan nuclear receptor SF-1 in sex-specific gene expression. *Cell* **93,** 445–454.
Nagy, L., Kao, H. Y., Chakravarti, D., Lin, R. J., Hassig, C. A., Ayer, D. E., Schreiber, S. L., and Evans, R. M. (1997). Nuclear receptor repression mediated by a complex containing SMRT, mSin3A, and histone deacetylase. *Cell* **89,** 373–380.
Niesor, E. J., Flach, J., Lopes-Antoni, I., Perez, A., and Bentzen, C. L. (2001). The nuclear receptors FXR and LXR alpha: Potential targets for the development of drugs affecting lipid metabolism and neoplastic diseases. *Curr. Pharm. Des.* **7,** 231–259.

Nishiyama, C., Hi, R., Osada, S., and Osumi, T. (1998). Functional interactions between nuclear receptors recognizing a common sequence element the direct-repeat motif spaced by one nucleotide (DR-1). *J. Biochem.* **123**, 1174–1179.

Nishizawa, H., Yamagata, K., Shimomura, I., Takahashi, M., Kuriyama, H., Kishida, K., Hotta, K., Nagaretani, H., Maeda, N., Matsuda, M., Kihara, S., Nakamura, T., Nishigori, H., Tomura, H., Moore, D. D., Takeda, J., Funahashi, T., and Matsuzawa (2002). Small heterodimer partner, an orphan nuclear receptor, augments peroxisome proliferator-activated receptor gamma transactivation. *J. Biol. Chem.* **277**, 1586–1592.

Parker, K. L. (1998). The roles of steroidogenic factor-1 in endocrine development and function. *Mol. Cell. Endocrinol.* **140**, 59–63.

Pastorcic, M., Wang, H., Elbrecht, A., Tsai, S. Y., Tsai, M. J., and O'Malley, B. W. (1986). Control of transcription initiation *in vitro* requires binding of a transcription factor to the distal promoter of the ovalbumin gene. *Mol. Cell. Biol.* **6**, 2784–2791.

Peet, D. J., Janowski, B. A., and Mangelsdorf, D. J. (1998). The LXRs: A new class of oxysterol receptors. *Curr. Opin. Genet. Dev.* **8**, 571–575.

Pereira, F. A., Qiu, Y., Zhou, G., Tsai, M. J., and Tsai, S. Y. (1999). The orphan nuclear receptor COUP-TFII is required for angiogenesis and heart development. *Genes Dev.* **13**, 1037–1049.

Perlmann, T., and Evans, R. M. (1997). Nuclear receptors in Sicily: All in the famiglia. *Cell* **90**, 391–397.

Phelan, J. K., and McCabe, E. R. (2001). Mutations in NR0B-1 (DAX-1) and NR5A-1 (SF-1) responsible for adrenal hypoplasia congenita. *Hum. Mutat.* **18**, 472–487.

Pipaon, C., Tsai, S. Y., and Tsai, M. J. (1999). COUP-TF upregulates NGFI-A gene expression through an Sp1 binding site. *Mol. Cell. Biol.* **19**, 2734–2745.

Power, S. C., and Cereghini, S. (1996). Positive regulation of the vHNF1 promoter by the orphan receptors COUP-TF1/EAR3 and COUP-TFII/ARP-1. *Mol. Cell. Biol.* **16**, 778–791.

Qiu, Y., Cooney, A. J., Kuratani, S., DeMayo, F. J., Tsai, S. Y., and Tsai, M. J. (1994). Spatiotemporal expression patterns of chicken ovalbumin upstream promoter-transcription factors in the developing mouse central nervous system: Evidence for a role in segmental patterning of the diencephalon. *Proc. Natl. Acad. Sci. USA* **91**, 4451–4455.

Qiu, Y., Pereira, F. A., DeMayo, F. J., Lydon, J. P., Tsai, S. Y., and Tsai, M. J. (1997). Null mutation of mCOUP-TFI results in defects in morphogenesis of the glossopharyngeal ganglion, axonal projection, and arborization. *Genes Dev.* **11**, 1925–1937.

Rastinejad, F., Perlmann, T., Evans, R. M., and Sigler, P. B. (1995). Structural determinants of nuclear receptor assembly on DNA direct repeats. *Nature* **375**, 203–211.

Renaud, J. P., Rochel, N., Ruff, M., Vivat, V., Chambon, P., Gronemeyer, H., and Moras, D. (1995). Crystal structure of the RAR gamma ligand binding domain bound to all-trans retinoid acid. *Nature* **378**, 681–689.

Rochel, N., Wurtz, J. M., Mitschler, A., Klaholz, B., and Moras, D. (2000). The crystal structure of the nuclear receptor for vitamin D bound to its natural ligand. *Mol. Cell* **5**, 173–179.

Rochette-Egly, C., Adam, S., Rossignol, M., Egly, J. M., and Chambon, P. (1997). Stimulation of RAR alpha activation function AF-1 through binding to the general transcription factor TFIIH and phosphorylation by CDK7. *Cell* **90**, 97–107.

Rohr, O., Aunis, D., and Schaeffer, E. (1997). COUP-TF and Sp1 interact and cooperate in the transcriptional activation of the human immunodeficiency virus type 1 long terminal repeat in human microglial cells. *J. Biol. Chem.* **272**, 31149–31155.

Ruberte, E. (1994). Nuclear retinoic acid receptors and regulation of gene expression. *Arch. Toxicol.* **16**(Suppl.), 105–111.

Sanyal, S., Handschin, C., Podvinec, M., Song, K. H., Kim, H. J., Seo, Y. W., Kim, S. A., Kwon, H. B., Lee, K., Kim, W. S., Meyer, U. A., and Choi, H. S. (2003). Molecular cloning

and characterization of chicken orphan nuclear receptor cTR2. *Gen. Comp. Endocrinol.* **132**, 474–484.
Sanyal, S., Kim, J. Y., Kim, H. J., Takeda, J., Lee, Y. K., Moore, D. D., and Choi, H. S. (2002). Differential regulation of the orphan nuclear receptor small heterodimer partner (SHP) gene promoter by orphan nuclear receptor ERR isoforms. *J. Biol. Chem.* **277**, 1739–1748.
Schoonjans, K., and Auwerx, J. (2002). A sharper image of SHP. *Nat. Med.* **8**, 789–791.
Schoorlemmer, J., van Puijenbroek, A., and van Den Eijnden, M. (1994). Characterization of a negative retinoic acid response element in the murine Oct4 promoter. *Mol. Cell. Biol.* **14**, 1122–1136.
Scott, D. K., Mitchell, J. A., and Granner, D. K. (1996). The orphan receptor COUP-TF binds to a third glucocorticoid accessory factor element within the phosphoenolpyruvate carboxykinase gene promoter. *J. Biol. Chem.* **271**, 31909–31914.
Seol, W., Choi, H. S., and Moore, D. D. (1996). An orphan nuclear hormone receptor that lacks a DNA binding domain and heterodimerizes with other receptors. *Science* **272**, 1336–1339.
Seol, W., Chung, M., and Moore, D. D. (1997). Novel receptor interaction and repression domains in the orphan receptor SHP. *Mol. Cell. Biol.* **17**, 7126–7131.
Seol, W., Hanstein, B., Brown, M., and Moore, D. D. (1998). Inhibition of estrogen receptor action by the orphan receptor SHP (short heterodimer partner). *Mol. Endocrinol.* **12**, 1551–1557.
Shibata, H., Nawaz, Z., Tsai, S. Y., O'Malley, B. W., and Tsai, M. J. (1997). Gene silencing by chicken ovalbumin upstream promoter-transcription factor I (COUP-TFI) is mediated by transcriptional corepressors, nuclear receptor corepressor (NCoR), and silencing mediator for retinoic acid receptor and thyroid hormone receptor (SMRT). *Mol. Endocrinol.* **11**, 714–724.
Shimamura, R., Fraizer, G. C., Trapman, J., Lau, Y. C., and Saunder, G. F. (1997). The Wilms' tumor gene (WT) can regulate genes involved in sex determination and differentiation: SRY, Müllerian-inhibiting substance, and the androgen receptor. *Clin. Cancer Res.* **3**, 2571–2580.
Shyr, C. R., Hu, Y. C., Kim, E., and Chang, C. (2002a). Modulation of estrogen receptor-mediated transactivation by orphan receptor TR4 in MCF-7 cells. *J. Biol. Chem.* **277**, 14622–14628.
Shyr, C. R., Hu, Y. C., Kim, E. *et al.* (2002b). Modulation of estrogen receptor-mediated transactivation by orphan receptor TR4 in MCF-7 cells. *J. Biol. Chem.* **277**, 14622–14628.
Smirnov, D. A., Hou, S., and Ricciardi, R. P. (2000). Association of histone deacetylase with COUP-TF in tumorigenic Ad12-transformed cells and its potential role in shut-off of MHC class I transcription. *Virology* **268**, 319–328.
Stephanou, A., Shah, M., Richardson, B., and Handwerger, S. (1996). The ARP-1 orphan receptor represses steroid-mediated stimulation of human placental lactogen gene expression. *J. Mol. Endocrinol.* **16**, 221–227.
Stocco, D. M. (2001). Tracking the role of a star in the sky of the new millennium. *Mol. Endocrinol.* **15**, 1245–1254.
Stroup, D., Crestani, M., and Chiang, J. Y. (1997). Identification of a bile acid response element in the cholesterol 7 alpha-hydroxylase gene (CYP7A). *Am. J. Physiol.* **273**, G508–G517.
Susens, U., Aguiluz, J. B., Evans, R. M., and Borgmeyer, U. (1997). The germ cell nuclear factor mGCNF is expressed in the developing nervous system. *Dev. Neurosci.* **19**, 410–420.
Susens, U., and Borgmeyer, U. (1996). Characterization of the human germ cell nuclear factor gene. *Biochim. Biophys. Acta* **1309**, 179–182.
Suzuki, T., Kasahara, M., Yoshioka, H., Morohashi, K., and Umesono, K. (2003). LXXLL-related motifs in DAX-1 have target specificity for the orphan nuclear receptors Ad4BP/SF-1 and LRH-1. *Mol. Cell. Biol.* **23**, 238–249.
Swain, A., Narvaez, V., Burgoyne, P., Camerino, G., and Lovell-Badge, R. (1998). DAX-1 antagonizes Sry action in mammalian sex determination. *Nature* **391**, 761–767.

Swain, A., Zanaria, E., Hacker, A., Lovell-Badge, R., and Camerino, G. (1996). Mouse DAX-1 expression is consistent with a role in sex determination as well as in adrenal and hypothalamus function. *Nat. Genet.* **12,** 404–409.

Tabarin, A. (2001). Congenital adrenal hypoplasia and DAX-1 gene mutations. *Ann. Endocrinol.* **62,** 202–206.

Tanabe, O., Katsuoka, F., Campbell, A. D., Song, W., Yamamoto, M., Tanimoto, K., and Engel, J. D. (2002). An embryonic/fetal beta-type globin gene repressor contains a nuclear receptor TR2:TR4 heterodimer. *EMBO J.* **21,** 3434–3442.

Tran, P., Zhang, X. K., Salbert, G., Hermann, T., Lehmann, J., and Pfabl, M. (1992). COUP orphan receptors are negative regulators of retinoic acid response pathways. *Mol. Cell. Biol.* **12,** 4666–4676.

Tremblay, J. J., and Viger, R. S. (2001). Nuclear receptor DAX-1 represses the transcriptional cooperation between GATA-4 and SF-1 in Sertoli cells. *Biol. Reprod.* **64,** 1191–1199.

Tsai, S. Y., and Tsai, M. J. (1997). Chick ovalbumin upstream promoter-transcription factors (COUP-TFs): Coming of age. *Endocr. Rev.* **18,** 229–240.

Tu, H., Okamoto, A. Y., and Shan, B. (2000). FXR, a bile acid receptor and biological sensor. *Trends Cardiovasc. Med.* **10,** 30–35.

Umesono, K., and Evans, R. M. (1989). Determinants of target gene specificity for steroid/thyroid hormone receptors. *Cell* **57,** 1139–1146.

Urnov, F. D., and Wolffe, A. P. (2001). A necessary good: Nuclear hormone receptors and their chromatin templates. *Mol. Endocrinol.* **15,** 1–16.

Wagner, R. L., Apriletti, J. W., McGrath, M. E., West, B. S., Baxter, J. D., and Fletterick, R. J. (1995). A structural role for hormone in the thyroid hormone receptor. *Nature* **378,** 690–697.

Wang, L., Lee, Y. K., Bundman, D., Han, Y., Thevananther, S., Kim, C. S., Chua, S. S., Wei, P., Heyman, R. A., Karin, M., and Moore, D. D. (2002). Redundant pathways for negative feedback regulation of bile acid production. *Dev. Cell* **2,** 721–731.

Wang, L. H., Tsai, S. Y., Cook, R. G., Beattie, W. G., Tsai, M. J., and O'Malley, B. W. (1989). COUP transcription factor is a member of the steroid receptor superfamily. *Nature* **340,** 163–166.

Wang, Z. J., Jeffs, B., Ito, M., Achermann, J. C., Yu, R. N., Hales, D. B., and Jameson, J. L. (2001). Aromatase (Cyp19) expression is up-regulated by targeted disruption of DAX-1. *Proc. Natl. Acad. Sci. USA* **98,** 7988–7993.

Wehrenberg, U., Ivell, R., Jansen, M., von Goedecke, S., and Walther, N. (1994). Two orphan receptors binding to a common site are involved in the regulation of the oxytocin gene in the bovine ovary. *Proc. Natl. Acad. Sci. USA* **91,** 1440–1444.

Widom, R. L., Rhee, M., and Karathanasis, S. K. (1992). Repression by ARP-1 sensitizes apolipoprotein AI gene responsiveness to RXP alpha and retinoic acid. *Mol. Cell. Biol.* **12,** 3380–3389.

Wilson, T. E., Fahrner, T. J., and Milbrandt, J. (1993). The orphan receptors NGFI-B and steroidogenic factor-1 establish monomer binding as a third paradigm of nuclear receptor-DNA interaction. *Mol. Cell. Biol.* **13,** 5794–5804.

Xing, W., Danilovich, N., and Sairam, M. R. (2002). Orphan receptor chicken ovalbumin upstream promoter-transcription factors inhibit steroidogenic factor-1, upstream stimulatory factor, and activator protein-1 activation of ovine follicle-stimulating hormone receptor expression via composite cis-elements. *Biol. Reprod.* **66,** 1656–1666.

Xu, L., Glass, C. K., and Rosenfeld, M. G. (1999). Coactivator and corepressor complexes in nuclear receptor function. *Curr. Opin. Genet. Dev.* **9,** 140–147.

Yan, Z., and Jetten, A. M. (2000). Characterization of the repressor function of the nuclear orphan receptor retinoid receptor-related testis-associated receptor/germ cell nuclear factor. *J. Biol. Chem.* **275,** 35077–35085.

Yan, Z. H., Karam, W. G., Staudinger, J. L., Medvedev, A., Ghanayem, B. I., and Jetten, A. M. (1998). Regulation of peroxisome proliferator-activated receptor alpha-induced transactivation by the nuclear orphan receptor TAK1/TR4. *J. Biol. Chem.* **273**, 10948–10957.

Yan, Z. H., Medvedev, A., Hirose, T., Gotoh, H., and Jetten, A. M. (1997). Characterization of the response element and DNA binding properties of the nuclear orphan receptor germ cell nuclear factor/retinoid receptor-related testis-associated receptor. *J. Biol. Chem.* **272**, 10565–10572.

Yanai, K., Hirota, K., Taniguchi-Yanai, K., Shigematsu, Y., Shimamoto, Y., Saito, T., Chowdhury, S., Takiguchi, M., Arakawa, M., Nibu, Y., Sugiyama, F., Yagami, K., and Fukamizu, A. (1999). Regulated expression of human angiotensinogen gene by hepatocyte nuclear factor 4 and chicken ovalbumin upstream promoter-transcription factor. *J. Biol. Chem.* **274**, 34605–34612.

Yang, Y., Wang, X., Dong, T., Kim, E., Lin, W. J., and Chang, C. (2003). Identification of a novel testicular orphan receptor 4 (TR4)-associated protein as repressor for the selective suppression of TR4-mediated transactivation. *J. Biol. Chem.* **278**, 7709–7717.

Yoshikawa, T., DuPont, B. R., Leach, R. J., and Detera-Wadleigh, S. D. (1996a). New variants of the human and rat nuclear hormone receptor, TR4: Expression and chromosomal localization of the human gene. *Genomics* **35**, 361–366.

Yoshikawa, T., Makino, S., Gao, X. M., Xing, G. Q., Chuang, D. M., and Detera-Wadleigh, S. D. (1996b). Splice variants of rat TR4 orphan receptor: Differential expression of novel sequences in the 5′-untranslated region and C-terminal domain. *Endocrinology* **137**, 1562–1571.

Young, W. J., Lee, Y. F., Smith, S. M., and Chang, C. (1998). A bidirectional regulation between the TR2/TR4 orphan receptors and the ciliary neurotrophic factor (CNTF) signaling pathway. *J. Biol. Chem.* **273**, 20877–20885.

Young, W. J., Smith, S. M., and Chang, C. (1997). Induction of the intronic enhancer of the human ciliary neurotrophic factor receptor (CNTFRalpha) gene by the TR4 orphan receptor. A member of steroid receptor superfamily. *J. Biol. Chem.* **272**, 3109–3116.

Yu, R. N., Ito, M., Saunders, T. L., Camper, S. A., and Jameson, J. L. (1998). Role of Ahch in gonadal development and gametogenesis. *Nat. Genet.* **20**, 353–357.

Yu, X., and Mertz, J. (1997). Differential regulation of the pre-C and pregenomic promoters of human hepatitis B virus by members of the nuclear receptor superfamily. *J. Virol.* **71**, 9366–9374.

Zazopoulos, E., Lalli, E., Stocco, D. M., and Sassone-Corsi, P. (1997). DNA binding and transcriptional repression by DAX-1 blocks steroidogenesis. *Nature* **390**, 311–315.

Zhang, H., Thomsen, J. S., Johansson, L., Gustafsson, J. A., and Treuter, E. (2000). DAX-1 functions as an LXXLL-containing corepressor for activated estrogen receptors. *J. Biol. Chem.* **275**, 39855–39859.

Zhang, X. K., Hoffmann, B., Tran, P. B., Graupner, G., and Pfahl, M. (1992). Retinoid X receptor is an auxiliary protein for thyroid hormone and retinoic acid receptors. *Nature* **355**, 441–446.

Zhang, Y., and Dufau, M. L. (2000). Nuclear orphan receptors regulate transcription of the gene for the human luteinizing hormone receptor. *J. Biol. Chem.* **275**, 2763–2770.

Zhang, Y., and Dufau, M. L. (2001). EAR2 and EAR3/COUP-TFI regulate transcription of the rat LH receptor. *Mol. Endocrinol.* **15**, 1891–1905.

Zhang, Y., and Dufau, M. L. (2002). Silencing of transcription of the human luteinizing hormone receptor gene by histone deacetylase–mSin3A complex. *J. Biol. Chem.* **277**, 33431–33438.

Zhang, Y., and Dufau, M. L. (2003a). Repression of luteinizing hormone receptor gene promoter by cross-talk among EAR3/COUP-TFI, Sp1/Sp3, and TFIIB. *Mol. Cell. Biol.* **23**, 6958–6972.

Zhang, Y., and Dufau, M. L. (2003b). Dual mechanisms of regulation of transcription of luteinizing hormone receptor gene by nuclear orphan receptors and histone deacetylase complexes. *J. Steroid Biochem. Mol. Bio.* **85,** 401–414.

Zhang, Y. L., Akmal, K. M., Tsuruta, J. K., Shang, Q., Hirose, T., Jetten, A. M., Kim, K. H., and O'Brien, D. A. (1998). Expression of germ cell nuclear factor (GCNF/RTR) during spermatogenesis. *Mol. Reprod. Dev.* **50,** 93–102.

Zhou, C., Tsai, S. Y., and Tsai, M. (2000). From apoptosis to angiogenesis: New insights into the roles of nuclear orphan receptors, chicken ovalbumin upstream promoter-transcription factors, during development. *Biochim. Biophys. Acta* **1470,** M63–M68.

Zhu, X. G., Park, K. S., Kaneshige, M., Bhat, M. K., Zhu, Q., Mariash, C. N., McPhie, P., and Cheng, S. Y. (2000). The orphan nuclear receptor EAR-2 is a negative coregulator for thyroid hormone nuclear receptor function. *Mol. Cell. Biol.* **20,** 2604–2618.

2

STRUCTURE AND FUNCTION OF THE GLUCOCORTICOID RECEPTOR LIGAND BINDING DOMAIN

RANDY K. BLEDSOE,* EUGENE L. STEWART,† AND KENNETH H. PEARCE*

*Department of Gene Expression and Protein Biochemistry
†Department of Computational, Analytical, and Structural Sciences
Discovery Research, GlaxoSmithKline
Research Triangle Park, North Carolina 27709

I. Functional Significance of the GR
II. Expression, Purification, and Characterization of the GR LBD
   A. Site-Directed Mutations to Enhance GR LBD Expression
   B. E. coli and Baculovirus Systems
   C. Functional Characterization and Crystallization
III. Fold of the Nuclear Receptor LBD and Specific Features of GR
   A. General Features of the GR LBD Fold
   B. Position of the AF-2 Helix
   C. Location of "Domain-Stabilizing" Mutations
IV. Ligand Recognition by GR
   A. A-Ring Polar Clamp and M604 Interaction Region
   B. C-Ring Polar Interaction Region
   C. 17β D-Ring Interaction Region

D. 17α D-Ring Interaction Region and
      Binding Pocket
   E. Summary of Ligand Recognition
 V. Modes of Ligand Recognition by GR Compared to
    Other Oxosteroid Receptors
   A. Selectivity of Ligands for ERα, AR, and PR
      Over GR
   B. Selectivity of Glucocorticoids and
      Mineralocorticoids for GR and MR
   C. Summary of Ligand Selectivity
 VI. Characteristics of Cofactor Association with the
     GR LBD
 VII. Role of the GR LBD in Chaperone Protein
      Association and Receptor Dimerization
   A. hsp90 and p23
   B. The Putative GR LBD Homodimer Interface
 VIII. Mutations in the GR LBD and Their
       Functional Consequences
   A. Site-Directed Mutagenesis Studies
   B. Role of GR LBD Mutations in Disease
 IX. Progress Toward a Selective GR Modulator
 X. Conclusions and Directions of Future Research
    References

After binding to an activating ligand, such as corticosteroid, the glucocorticoid receptor (GR) performs an impressive array of functions ranging from nuclear translocation, oligomerization, cofactor/kinase/transcription factor association, and DNA binding. One of the central functions of the receptor is to regulate gene expression, an activity triggered by ligand binding. In this role, GR acts as an adapter molecule by encoding the ligand's message within the structural flexibility of the ligand binding domain (LBD). The purpose of this review is to discuss the many structural and functional features of the GR LBD in light of recent successful biochemical and crystallographic studies. Progress in this area of research promises to reveal new strategies and insights allowing for the design of novel drugs to treat inflammatory diseases, diabetic conditions, steroid resistance, and cancers. © 2004 Elsevier Inc.

I. FUNCTIONAL SIGNIFICANCE OF THE GR

It has been known since the early part of the 20th century that corticosteroids, specifically the subclass of steroidal hormones called glucocorticoids, are involved in glucose regulation and also play a significant role in inflammatory processes. Endogenous glucocorticoids, primarily cortisol and corticosterone (Fig. 1A), are synthesized in the adrenal cortex and consequently delivered to target tissues by transport proteins. The realization that these hormones have such a pervasive role in carbohydrate metabolism and rheumatic conditions eventually led to the widespread use of the glucocorticoid-like ligands, such as cortisone and the synthetics dexamethasone (Dex) and prednisolone (Fig. 1B), for treating adrenal insufficiency and inflammatory-related conditions. This class of drugs is still currently prescribed for a wide variety of indications, including inflammation; arthritis; skin, blood, kidney, eye, thyroid, and intestinal disorders (such as colitis); severe allergies; and asthma. Glucocorticoids are also used to help prevent rejection of transplanted organs and to treat certain types of cancer. More recently, the inhaled hydrolyzable steroids, such as fluticasone propionate (Fig. 1B), have proven effective for the management of asthma.

Throughout the years in using glucocorticoids as drugs, it has been revealed that corticosteroid biology is vastly complex, and numerous biochemical processes are influenced in addition to those involving glucose homeostasis and inflammation. For example, glucocorticoids influence protein and fat metabolism, water and electrolyte balance, bone resorption, cardiovascular and kidney function, maintainence of the hypothalmic–pituitary–adrenal axis, and control of the stress response. Several of these effects are actually problematic in glucocortoid therapies. In fact, these undesired side-effects are the driving force behind the quest for a cleaner, more selective glucocorticoid-like drug (Belvisi *et al.*, 2001; Miner, 2002).

Most of the functions of glucocortoid hormones are mediated through interaction with the classical transcription factor known as the glucocorticoid receptor (GR). The gene for this receptor, first cloned in 1985 (Hollenberg *et al.*, 1985), encodes a protein 777 amino acids in length. The expression pattern of GR shows that the receptor is distributed ubiquitously, but is more prevalent in hepatic, nervous system, and muscular tissues. Even though GR is widely distributed, cellular sensitivity to glucocorticoids is complex and dependent on many factors (Bamberger *et al.*, 1996).

GR is a member of a rather large family of proteins known as nuclear receptors (NRs). Thus far, 48 nuclear receptors have been identified in humans. More specifically, based on sequence similarity, GR is a member of the steroid hormone receptor (SHR) subset of these receptors; this subset includes the mineralocorticoid receptor (MR), the progesterone receptor (PR) progestin, and the androgen receptor (AR) as well as the more distantly related estrogen receptor (ER).

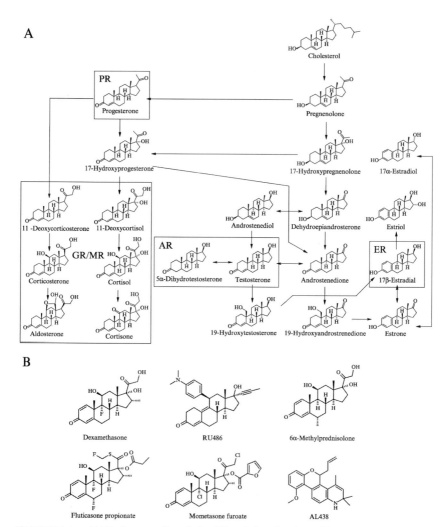

FIGURE 1. (A) Endogenous ligands for GR and the related steroid receptors. Shown is the interconnection between the principal pathways in the biosynthesis of the most important corticosteroids, androgens, and estrogens. Steroids that are the subject of this chapter are highlighted by boxes and labeled by the receptors for which they are most selective. All the steroid ligands shown are derived from the progenitor cholesterol. The corticosteroid ligands, corticosterone and cortisol, are synthesized in the adrenal cortex, whereas most other steroid ligands shown are produced in respective reproductive tissues. (B) Examples of synthetic glucocorticoid ligands. Dexamethasone and prednisolone (6α-methylprednisolone) are two of the most commonly prescribed GR agonist ligands. RU486, primarily a progesterone antagonist, is also a common GR antagonist. Fluticasone and mometasone are GR agonists used in inhalation formulas, and AL438 is a recently described GR agonist with dissociative properties as described in the text.

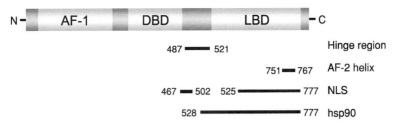

FIGURE 2. Linear schematic of human GR. The three core domains, the activation function-1 (AF-1), DNA binding domain (DBD), and the ligand binding domain (LBD) of GR are shown. The LBD of GR encompasses all the amino acids required for ligand recognition. As discussed and referenced in the text, approximate locations are indicated of some defined regions and functions concerning the LBD, such as the hinge domain, the AF-2 helix, nuclear localization sequences (NLS), and hsp90 binding.

Like most NRs, GR is an assembly of three smaller protein modules (Giguere et al., 1986; Rusconi and Yamamoto, 1987) (Fig. 2). The N-terminal half of GR (roughly amino acids 1–417) contains the activation function-1 (AF-1) activity that governs a constitutive transcriptional role and is the target of several interaction proteins and kinases (Hittelman et al., 1999). Amino acids 418–487 contain the DNA binding domain (DBD) of GR. Crystallographic analysis of this domain shows that this region folds into a zinc finger-type arrangement and makes direct and specific electrostatic interaction with DNA (Luisi et al., 1991). In addition, a well-defined dimer interface is formed as two DBD subunits bind the DNA helix in adjacent major grooves. Following a short stretch of sequence (roughly amino acids 488–520) known as the hinge region, the C-terminal end of GR (amino acids 521–777) contains a folded module that retains the ability to bind to ligands. This region is referred to as the ligand binding domain (LBD).

Each of these domains performs a separate function in tandem to regulate the signal transduction and transcriptional activities of GR. In the absence of a ligand, the majority of the receptor population resides in the cytoplasm, and the receptor is predominantly complexed to heat shock proteins (hsp), such as hsp70, hsp90, and p23 (Pratt, 1998; Pratt and Toft, 1997). One model suggests that ligand binding induces the release of the bound chaperone partners, consequently exposing the nuclear localization sites within the GR LBD. However, the role of hsp partners is probably much more complex, because they have also been implicated in aiding transport and cytoplasmic nuclear shuttling of GR (Defranco et al., 1995; Pratt, 1993) and also participating in an active role within the transcriptional complex (Freeman et al., 2000). Another consequence following ligand association is the formation of either receptor homodimers or heterodimers with related NRs (Savory et al., 2001).

In addition to nuclear translocation and dimerization, the ligand-binding step also triggers the cytoplasmic phosphorylation of serines and threonines in the AF-1 domain (Bodwell et al., 1998). Several kinases, such as mitogen-activated protein kinase (MAPK), cyclin-dependent kinase (Cdk), and c-Jun N-terminal kinase (JNK), have been shown to regulate the activity of ligand-bound GR (Itoh et al., 2002; Krstic et al., 1997; Rogatsky et al., 1998; Wang et al., 2002). Although the regulatory role of GR phosphorylation is well documented, there are few molecular details known about the interplay between ligand binding and phosphorylation.

Once in the nucleus, GR is able to bind to specific DNA promoter segments, commonly referred to as glucocorticoid response elements (GREs), where the associated gene is either up- or down-regulated (Fig. 3). Accessory proteins or coactivators, such as steroid receptor coactivator-1 (SRC-1) and transcriptional intermediary factor 2 (TIF2), are essential for the gene transcription function of GR. Many of these accessory proteins contain chromatin-modifying activities such as histone acetylation (Spencer et al., 1997). These large, multifunctional proteins dock onto the ligand-bound GR primarily through a short, leucine-rich, amphipathic alpha–helix, known as an LXXLL motif or an NR box (Darimont et al., 1998). As described in more detail below, the agonist-bound GR LBD forms a shallow

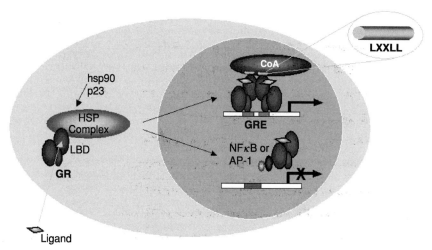

FIGURE 3. Simplified schematic of GR transactivation and transpression activities. GR typically resides in the cytoplasm in complex with heat shock proteins (such as hsp90 and p23) until ligand association stimulates nuclear transport. Two of the most important activities of GR are (1) direct interaction with DNA response elements, such as GREs and negative GREs, to enhance or diminish transcription and (2) interaction with transcription factors (such as NFκB or AP-1) to diminish gene transcription. The LBD plays a direct role in these activities by mediating interactions with coactivators (CoA via the LXXLL NR box), corepressors, and other transcription factors.

hydrophobic groove that can accommodate the leucines of the NR box, thus effectively bridging the two proteins together.

Another ligand-regulated function of the GR LBD is the repression of specific genes. Two mechanisms for this activity have been described. One involves a promoter element, referred to as a negative GRE (nGRE) (Sakai et al., 1988). The mode of interaction on nGREs remains relatively loosely defined, but it is believed that hormone binding by GR induces promoter association and thus competes for additional transcription factors to yield a diminished response (Diamond et al., 1990). Additional mechanisms are also likely including GR heterodimerization with other NRs (Ou et al., 2001) and may involve multiple distal promoter elements (Aslam et al., 1995; Collier et al., 1996; Goodman et al., 1996).

Activated GR can also reduce gene transcription by interfering with parallel signal transduction pathways, such as those involving NFκB and AP-1 (Fig. 3) (Mckay and Cidlowski, 1999). Although several mechanisms have been proposed, most evidence suggests one is more likely: ligand-activated GR directly interacts with the target transcription factor, such as the p65 subunit of NFκB, and thus interferes with the gene transactivation activity of that factor (McKay and Cidlowski, 1998, 1999; Wissink et al., 1997). Therapeutically, this transrepressive activity of GR has been the focus of much investigation since it is believed to drive most of the anti-inflammatory actions associated with glucocorticoids.

Ligand binding by the GR LBD drives the multifaceted functions of the receptor and ultimately regulates signal transduction, gene transcription, and transrepression. The majority of these activities are influenced by the ligand-induced plasticity of the LBD, which ultimately determines various protein–protein and protein–DNA interactions (McEwan et al., 1997). The purpose of this chapter is to outline the structural and functional characteristics of the GR LBD in light of the recent co-crystal structures and new biochemical data. These molecular insights have provided important clues for how ligands help GR transduce signals, and may also provide information for the creation of safer, more effective medications.

II. EXPRESSION, PURIFICATION, AND CHARACTERIZATION OF THE GR LBD

Although GR is an extremely well-validated drug target and has been the focus of intense research, biochemical and structural characterization of GR has been relatively lacking due to problems obtaining purified protein. In the mid-1980s, around the same period in which the gene for GR was cloned (Hollenberg et al., 1985), biochemical studies using crude native protein preparations showed that the receptor was organized in modules and consisted of roughly three domains (Carlstedt-Duke *et al.*, 1987; Giguere *et al.*, 1986). Since then, several examples demonstrating partial purification of GR from recombinant sources have been reported (Alnemri *et al.*, 1991; Caamano *et al.*, 1994; Ohara-Nemoto *et al.*, 1990), and these studies have noted the difficult nature of the protein. Typically for NRs, the LBD is the focus of ligand binding and biochemical studies since it is smaller in size and also retains the entirety of the ligand binding determinates. Recombinant expression and purification of the GR LBD, however, have proven to be problematic. One presumed reason for this difficulty is the well-documented association of GR with heat shock proteins (discussed in more detail in the following paragraphs). Even though obtaining purified GR has proven elusive throughout the years, several other sequence-related NR LBDs, including the progesterone receptor (PR) (Williams and Sigler, 1998) and androgen receptor (AR) (Sack *et al.*, 2001), have been successfully purified, characterized, and crystallized. Information collected from these crystal structures was used in combination with existing GR literature to develop an approach that has recently led to the determination of GR LBD's crystal structure.

A. SITE-DIRECTED MUTATIONS TO ENHANCE GR LBD EXPRESSION

A 1992 study described a yeast-based genetic selection method for identifying "gain of function" mutations in GR (Garabedian and Yamamoto, 1992). One particular mutation, F620S in the rat GR (corresponding to F602S in human GR), was found to dramatically increase reporter activity in yeast. Possible explanations for the increased activity are that the mutation increased the amount of active material produced in yeast or that the mutation increased receptor affinity for the ligand. Interestingly, a structure-based sequence alignment exercise for the NR LBDs, in an attempt to identify possible solvent-exposed hydrophobic residues in GR, also highlighted this particular residue at position 602 (Bledsoe *et al.*, 2002). Only GR has a hydrophobic residue, a phenylalanine, in this position (Fig. 4). In contrast, PR, which is known to express well in bacterial systems, contains a serine at the equivalent position. Inspection of the PR crystal structure shows that the serine residue, located near the helix 4/helix 5(H4/H5) junction in the core of the LBD, is not a direct contributor to the ligand binding pocket. In fact, this residue is directed toward the surface of the protein but is mostly buried by residues on helices 8 and 10 (H8 and H10). This sequence alignment and structural data provided further evidence that the increased efficacy of the GR mutant seen in the earlier yeast experiments most likely resulted from the increased expression and/or stability of the rat GR. In addition, a mutation at position 656 in rat GR, which corresponds to residue 638 in human GR, has been

```
                    Position 602
                         ↓
         GR         WMFLMAFALG
         AR         WMGLMVFAMG
         MR         WMCLSSFALS
         PR         WMSLMVFGLG
         ERα        WLEILMIGLV
         ERβ        WMEVLMMGLM
         RXRα       WNELLIASFS
         PPARγ      VHEIIYTMLA
                      Helix 5
```

FIGURE 4. Alignment of the helix 5 (H5) region of GR with the corresponding region in related nuclear receptors. As shown, GR is the only receptor with a hydrophobic residue at the position 602. As shown by the GR Dex crystal structure, the GR residues W600, M601, M604, A605, and L608 make direct contact with the ligand Dex. The residue at the 602 position is not in the ligand binding pocket (see Fig. 8).

previously shown to increase ligand affinity to create a "super" receptor (Chakraborti *et al.*, 1991). Recently, two separate groups have taken advantage of either one or both of these mutations, F602S and C638D, to improve expression levels of human GR LBD in recombinant systems (Bledsoe *et al.*, 2002, Kauppi *et al.*, 2003).

B. *E. COLI* AND BACULOVIRUS SYSTEMS

Using point mutants, the GR LBD has recently been produced recombinantly in relatively high levels using both *E. coli* and baculovirus expression systems. For bacterial expression, the GR LBD F602S construct aa521–777 was fused to the C-terminal end of a 6xHis-tagged glutathione–S–transferase partner (Bledsoe *et al.*, 2002). The fusion protein was expressed from a typical T7 promoter vector using induction at low temperature. In addition to increased expression of the F602S mutant compared to the wild type, levels were significantly improved by inclusion of Dex during induction. It is likely that the ligand serves to stabilize or nucleate the protein during folding.

Using a slightly longer construct (aa500–777) compared to the bacterial system, baculovirus has also recently been used to produce purified GR LBD (Kauppi *et al.*, 2003). Several combinations of F602S and C638D, as well as the mutation N517D in the hinge region, were used. *Spodoptera frugiperda* (Sf9) cells were infected with a virus containing a recombinant GR LBD construct and were grown for 48 hours also in the presence of ligand. In both systems, slightly altering the solution properties of GR LBD, primarily with point mutations at residues directed toward the surface of the protein, was necessary to allow for appreciable recombinant protein expression and subsequent purification.

C. FUNCTIONAL CHARACTERIZATION AND CRYSTALLIZATION

Because of the need for point mutagenesis to improve recombinant GR LBD expression, several functional studies were completed to determine the possible effect on both ligand and coactivator binding. As mentioned above, F602S (or F620S in rat GR) improved GR activity using a yeast-based reporter assay. However, in a CV-1 cell transient transfection experiment, the F602S mutant had little effect on the ligand-responsive transactivation and transrepression functions of GR (Bledsoe et al., 2002). In addition, the expressed and purified GR F602S LBD protein retained ligand and coactivator peptide binding activities similar to those of the wild type. Using a baculovirus expression system to produce cell lysate containing full length wild type and the F602S mutant GR, a [3H]-Dex filter binding assay showed that both proteins bind ligands with equivalent affinity (Delves and Austin, personal communication, May 2002). As described in more detail in subsequent paragraphs, it is likely that this mutation improves the overall stability and solution properties of the GR LBD fold without having a dramatic effect on ligand or coactivator functions. It is very possible, however, that a mutation at this position may affect other functions of the receptor that have not been fully characterized.

Not only has the purified GR LBD protein been useful for *in vitro*, biochemical characterization, but it has also been used to produce crystals allowing for long sought-after structural determination. In a study by Bledsoe et al. (2002), a GR LBD containing the F602S mutation was used in combination with Dex and a short peptide representing the transcriptional intermediary factor 2 (TIF2) accessory protein. The complex readily crystallized, and the structure was solved at 2.5Å resolution. A report by Kauppi et al. (2003) describes several crystallization successes: protein from the aa500–777 construct with the N517D, F602S, and C638D mutations, along with the antagonist ligand RU486 (Fig. 1B), produced the best diffracting crystals (2.3Å). This latter group also obtained a 2.7Å structure of a mutant GR LBD bound with Dex and a TIF2 peptide.

Studies from these two groups have provided new views concerning the structural features that GR uses to recognize both ligands and coactivators. As a general overview, it is clear that the GR LBD is a remarkably adaptable and flexible structure that helps the receptor bind a variety of ligands and perform a diverse array of signaling activities. The following sections review the molecular aspects of some of these capabilities of the GR LBD.

III. FOLD OF THE NUCLEAR RECEPTOR LBD AND SPECIFIC FEATURES OF GR

The first NR LBD to be solved crystallographically was from the retinoid X receptor alpha (RXRα) (Bourguet *et al.*, 1995). Shortly after, several other structures were solved, including the LBDs for the peroxisome proliferator-activated receptor gamma (PPARγ) (Nolte *et al.*, 1998), ERα (Brzozowski *et al.*, 1997), PR (Williams and Sigler, 1998), and AR (Matias *et al.*, 2000; Sack *et al.*, 2001). Sequence alignments suggested early on that the overall fold of various NR LBDs would be similar. As several structures became available, it was striking to see the degree of similarity in the overall three-dimensional fold of these receptors (Bourguet *et al.*, 2000; Moras and Gronemeyer, 1998).

A. GENERAL FEATURES OF THE GR LBD FOLD

The NR LBDs are comprised primarily of 11 to 12 helices; the helices are arranged to form three separate sheets that are stacked together into a three layer bundle. As shown in Figure 5, the overall fold of the GR LBD follows this same general packing arrangement. Figure 6A shows the front view of this same structure: the GR LBD is oriented to show the ternary complex consisting of Dex and a peptide derived from the TIF2 coactivator (Bledsoe *et al.*, 2002). In total, the GR LBD bound with Dex consists of 11 alpha-helices (9 major) and 4 β-strands that form two short β-sheets. As described in more detail later, it is important to note that certain structural features of the GR LBD can change depending on which ligand is bound to the receptor.

Arranged in antiparallel fashion, the major helices 1 and 3 (H1 and H3) form one side of the bundle while helices 7 and 10 (H7 and H10) form the other. Between these two outer layers are H4, H5, H8 and helix 9 (H9) that for the most part form the top portion of the domain and leave a cavity in the lower half of the protein (Fig. 5), providing the internal space for the lipophilic ligand (Fig. 6B). As seen from the views shown in Figures 5 and 6A, a beta-hairpin motif located between H5 and helix 6 (H6), labeled as β-sheet 1, aids in forming the front portion of the ligand pocket.

B. POSITION OF THE AF-2 HELIX

Structures of NRs bound to agonist ligands show that the most C-terminal alpha–helix, known as the activation function-2 (AF-2) helix, has a universal purpose in forming the lid of the ligand binding pocket. Dex is a well-known potent GR agonist, and AF-2 in the GR/Dex structure is positioned as expected, serving to close the back portion of the ligand pocket by packing against H3, H4, and H10 (Fig. 6A). Stability of the helix

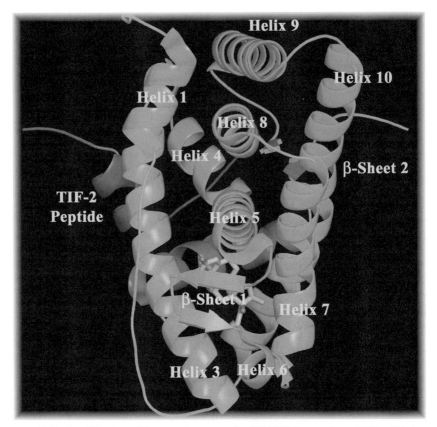

FIGURE 5. Side view of the crystal structure for GR LBD in complex with Dex (Bledsoe et al., 2002). Oriented in this fashion, the three stacked helical sheets can be seen where the first sheet consists of H1 and H3, and the last sheet consists of H7 and H10. The middle layer contains H4, H5, H8, and H9. As shown, the lower half of the domain is occupied by lipophilic ligand. Note that the numbering scheme for helices as shown here is based on historical numbering from previous NR LBD crystal structures and sequence alignments. For example, this GR/Dex structure does not have H2 and H11, but it does contain a short 3' helix between H3 and H4. The AF-2 helix is colored red, and the TIF2 peptide is colored magenta.

in this position is achieved in part by a direct interaction of AF-2 residues, such as L753, with the ligand.

In contrast, the AF-2 helix in the GR/RU486 antagonist structure is displaced such that there is no direct interaction between the residues of AF-2 and the ligand (Fig. 7A) (Kauppi et al., 2003). In this structure, the 11β-dimethylaniline group of RU486 protrudes into the space where AF-2 sits in the agonist structure. This steric interference causes displacement of AF-2, as well as causes a partial unwinding of H10 and formation of an additional

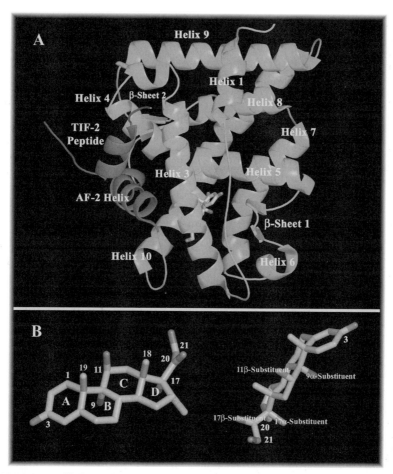

FIGURE 6. (A) Front view of the crystal structure for the GR LBD in complex with Dex (Bledsoe et al., 2002). AF-2 is colored red, and the TIF2 peptide is colored magenta. (B) Classical orientation of Dex with numbering scheme and view of Dex as bound in the GR LBD pocket (left). Key substituents off the steroidal core are labeled (right).

"helix 11." The position of AF-2 and the formation of this additional helix, however, may be influenced by protein packing in the crystalline form.

A second small β-sheet is formed by the two additional β-strands; one of the strands is located between H8 and H9, and the other follows the AF-2 helix. The extended β-strand following AF-2 is most likely key in stabilizing the receptor because removal leads to a compromised receptor that appears unable to bind a ligand (Zhang et al., 1996). The importance of this final β-strand is further seen in the GR/RU486 antagonist structure. Even though AF-2 is extended from the core of the LBD, the second β-sheet remains

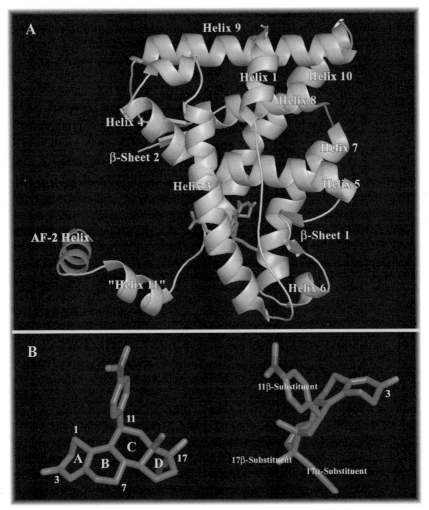

FIGURE 7. (A) Front view of the crystal structure for GR LBD in complex with RU486 (Kauppi *et al.*, 2003). The AF-2 helix is colored red. (B) Classical structure of RU486 with numbering scheme (left) and view of RU486 as bound in the GR LBD pocket (right).

intact and is clearly visible in the crystal structure. However, because electron density between AF-2 and this β-strand is not resolved, it is possible that this strand comes from an adjacent molecule in the crystal.

It is clear from both of these structures that GR ligands have the ability to influence the dynamics of AF-2 and also create other structural changes that depend on the nature of the ligand. These general features are also evident from structures of ERα bound with both an agonist and antagonist ligand (Brzozowski *et al.*, 1997; Shiau *et al.*, 1998). This structural plasticity feature of the GR LBD provides a means to affect the function the receptor via small molecule ligands.

C. LOCATION OF "DOMAIN-STABILIZING" MUTATIONS

For several of these crystal structures, a single phenylalanine to serine mutation at position 602 (F602S) in the human GR LBD played a key role in obtaining protein as described previously. Structurally, position 602 in the human GR LBD lies in the first turn of H5 that is in the middle of the three-layer helix bundle and is oriented away from the ligand binding pocket. As predicted from the PR and AR LBD structures, the position 602 side chain is mostly buried and resides in a predominantly hydrophilic environment. In fact, a hydrophilic cavity is created by the side chains of residues S599, S673, S674, H726, and Y764. In addition, the backbone carbonyls of several amino acids, including L670, Y598, and H726, also contribute to the polar character of the pocket. As shown in Figure 8, several water molecules serve to connect the polar groups within this cavity. It is very likely that this hydrogen bond network helps to stabilize the fold of the GR LBD, and this is perhaps one reason that the F602S mutation improves recombinant expression levels.

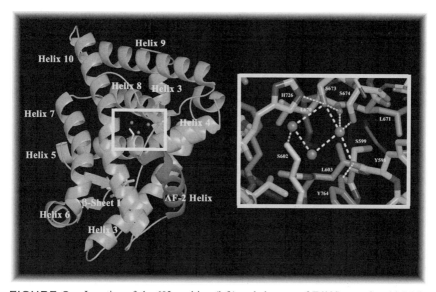

FIGURE 8. Location of the 602 position (left) and close-up of F602S mutation (right) in the GR/Dex structure (Bledsoe *et al.*, 2002). As shown, the serine makes a complex network of hydrogen bonds involving three water molecules. The area surrounding the serine and water molecules is primarily hydrophilic consisting of polar groups from side chains S673, S674, and H726 and Cα-carbonyls from L670 and Y598.

IV. LIGAND RECOGNITION BY GR

The ligand binding pocket in the GR LBD is extremely adaptable and is able to accommodate a diverse set of ligands. The volume of the ligand binding pocket in the GR LBD, using coordinates from the GR/Dex structure, is average (approximately 590Å^3) compared to the pockets in the related NR LBDs. For example, the approximate volume of the pocket in AR, ERα and PR is 420Å^3, 450Å^3, and 560Å^3, respectively (Brzozowski *et al.*, 1997; Sack *et al.*, 2001; Williams and Sigler, 1998), whereas the volume is more expansive in the PPARγ structure, measuring roughly 1440Å^3, (Gampe *et al.*, 2000). In the agonist structure, Dex only occupies about two thirds of the volume in the pocket.

The cavity is mostly lined with hydrophobic residues, but it is clear that both polar and nonpolar residues play specific roles in ligand recognition. The majority of these interaction features can be associated with four distinct regions of the GR ligand binding pocket. These regions refer to specific areas surrounding the bound steroid (Figs. 6B and 7B) the A-ring polar clamp and M604 interactions, C-ring polar interactions, 17β D-ring interactions, and 17α D-ring interactions and pocket. The following sections discuss each of these regions of the binding pocket in the context of the GR crystal structures bound to Dex and RU486.

A. A-RING POLAR CLAMP AND M604 INTERACTION REGION

The polar interactions in the A-ring binding pockets are identical in both the Dex and RU486 structures. This region of the binding pocket consists of a complex network of hydrogen bonds among the 3-oxo group of both steroids, the amide moiety of Q570, the guanidinium group of R611, the backbone carbonyl of M604, and a centrally located water molecule (Fig. 9A). A similar interaction is also observed in other oxosteroid binding receptors, such as AR (Matias *et al.*, 2000; Sack *et al.*, 2001) and PR (Williams and Sigler, 1998), and can also be inferred in mineralocorticoid receptor (MR). This "polar clamp" represents a key interaction for steroid binding among all of these receptors.

Positioning of M604 is dependent on the 19-methyl substituent of the steroid. In the presence of the 19-methyl substituent such as in Dex, M604 adopts a conformation that allows the 19-methyl group to occupy a small pocket above the plane of the steroid. In the absence of this group, such as in RU486, M604 adopts a conformation that occupies this 19-position pocket and positions its ϵ-methyl group proximate to the steroid A-ring. Thus, it appears that M604 is quite mobile, and its conformation can be easily influenced by the presence or absence of substituents above the plane of the steroid A-ring.

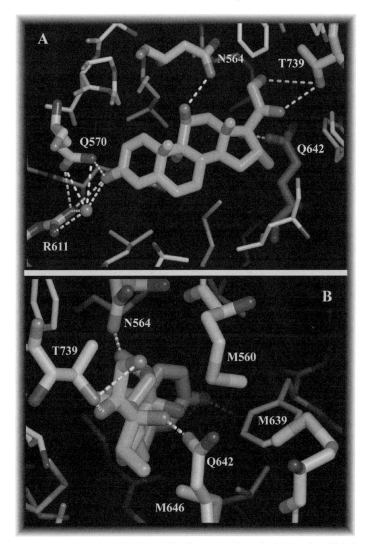

FIGURE 9. (A) View of the GR ligand binding pocket bound with Dex in which the A-ring of the steroid is on the left and the D-ring is on the right. The key hydrogen bonds involving Q570, R611, the 3-oxo steroid group of Dex, and a central water molecule are shown by dashed lines. (B) A view of the GR ligand binding pocket bound with Dex (orthogonal to Figure 9A) illustrating the 17α interactions and the lipophilic 17α pocket. The key interaction between the 17α-hydroxy group of Dex and Q642 is highlighted. Flexibility of the three methionines (M560, M639, and M646) allows for variability in 17α pocket size. Larger 17α substituents can perturb the conformation shown here and further expand this pocket.

B. C-RING POLAR INTERACTION REGION

In the GR/Dex crystal structure, a well-positioned hydrogen bond (1.8Å) between the 11β-hydroxy group and N564 exists (Fig. 9). This bond is a key interaction for ligand binding and may also account for some specificity. As one example, both 11-deoxycortisol and 11-deoxycorticosterone (DOC) are weaker ligands for GR compared to cortisol (Ojasoo *et al.*, 1994; Wolff *et al.*, 1978). The presence of this conserved residue in AR, PR, and MR, however, allows many of these same steroids to bind other oxosteroid NRs. As implied in many of the crystal structures, interaction with this key asparagine seems to be important for the binding and functional activity of all four oxosteroid NRs. It is also possible that stabilization of this asparagine is required for proper scaffolding of the loop preceding AF-2.

When the antagonist RU486 binds to GR, N564 in H3 is displaced from its position observed in the Dex structure by the 11-*N*,*N*-dimethylaniline group of RU486. As discussed earlier, the *N*, *N*-dimethylaniline substituent of RU486 also plays a critical role in the functional activity of GR by sterically displacing AF-2 (Fig. 7A).

C. 17β D-RING INTERACTION REGION

For binding Dex, key polar interactions are made with the substituent in the 17β position (Fig. 9). In particular, the 21-hydroxy group of Dex is in close proximity to both N564 (between 3.5Å and 4.5Å depending on the set of coordinates) and T739 (between 2.5Å and 3.2Å) (Fig. 9B). Based on the structure-activity relationship of other 21-substituted steroids, however, these interactions may not be critical for binding. The 20-carbonyl of the 17β-propanone also interacts with T739 (between 2.8Å and 3.5Å).

Unlike Dex, RU486 contains a much smaller 17β-hydroxy group in this position (Fig. 7B). This substituent, along with AF-2 displacement, has major effects on its 17β D-ring interaction with GR. The displacement of this helix results in regional structural perturbations and loss of interactions with residues T739 and N564. In RU486, the 17β-hydroxy group interacts with Q642 (~2.7Å), which is further stabilized through a secondary interaction with the backbone carbonyl of D638 (~3.1Å). The most significant difference in this region is the presence of three water molecules that act to replace key residues found in the GR/Dex structure. In effect, an indirect interaction between the 17β-hydroxy group of RU486 and N564 is established.

D. 17α D-RING INTERACTION REGION AND BINDING POCKET

In the GR/Dex structure, a ~2.7Å hydrogen bond interaction exists between the 17α-hydroxy group and Q642 (Fig. 9). This key hydrogen bond is likely to contribute to overall domain stability. In the GR/RU486 structure, the 17α interaction with Q642 is not possible due to the presence of a hydrophobic methylacetylene substituent in that position. As previously mentioned, however, a 17β interaction is maintained through the interaction of the 17β-hydroxy group with Q642.

In the Dex crystal structure, the volume of the 17α binding pocket extends well beyond the 17α-hydroxy group. When the appropriate ligand binds, the size of this pocket can be increased through coordinated positional changes of highly flexible methionines (M560, M639, and M646). In fact, expansion of this pocket must occur to accommodate large 17α substituents, such as in fluticasone propionate and mometasone furoate (Fig. 1B).

In the RU486 structure, it appears that the 17α-methylacetylene group induces a conformational change in both M560 and M639. This rearrangement produces a 17α side pocket significantly larger than the one observed in the Dex crystal structure. Because of the flexibility of the residues that compose this region, little protein backbone movement is necessary to accommodate the much larger 17α substituent of RU486. When GR is bound to ligands that have Q642-interacting substituents, such as Dex and RU486, the size of this side pocket may be constricted.

E. SUMMARY OF LIGAND RECOGNITION

As determined experimentally more than 25 years ago (Wolff et al., 1978), and as shown by crystal structures, most of the Dex steroid is enveloped by hydrophobic surfaces in the GR LBD, and key specific hydrophilic interactions appear to play a role in proper ligand docking. In fact, every polar atom on Dex makes an electrostatic bond with either a side chain or backbone atom from a residue in the pocket. The interactions contributing to successful ligand binding exist in four distinct spatial regions in the pocket:

1. The A-ring interactions, involving the guanidinium of R611 and the γ-amide of Q570, dictate the orientation of the steroidal 3-oxosteroid group via a complex and ordered hydrogen-bonding system;
2. Especially in the case of agonist ligands, the 11β-hydroxy group makes a hydrogen bond with the key N564 side chain;
3. The 17α-hydroxy group occupies a rather expansive pocket and, in the case of Dex, makes a hydrogen bond with Q642; and
4. The group off the 17β position has the opportunity to hydrogen bond to residues N564 and T739 (and Q642 in the case of RU486).

Overall, the GR ligand binding pocket is an adaptable and flexible entity that is capable of slight rearrangements to accommodate many related but different ligands. As discussed subsequently, this characteristic, plus the potential to influence the activity profile of the receptor by ligand-induced conformation, is the basis for the search for new therapeutics.

V. MODES OF LIGAND RECOGNITION BY GR COMPARED TO OTHER OXOSTEROID RECEPTORS

This section briefly reviews the major structural activity relationships of endogeneous and synthetic steroids in terms of glucocorticoid receptor (GR) selectivity compared to the other steroid hormone receptors estrogen receptor alpha (ERα), progesterone (PR), androgen receptor (AR) and mineralocorticoid receptor (MR). Based on the available crystal structures of ERα (Brzozowski *et al.*, 1997), PR (Williams and Sigler, 1998), AR (Sack *et al.*, 2001), and GR (Bledsoe *et al.*, 2002; Kauppi *et al.*, 2003), some general rules of molecular recognition for the respective ligands, estradiol, progesterone, dihydrotestosterone (DHT), and Dex, can be developed. In addition, this section presents the principles governing receptor selectivity of corticosterone, cortisol, 11-deoxycorticosterone (DOC), and aldosterone based on the GR structures and a homology model of MR. The purpose of this discussion is to review the basic, principle differences governing ligand selectivity for the human receptors; the discussion is not intended to be a comprehensive analysis. In fact, there is still much to be learned about the energetics governing ligand binding and specificity in this class of receptors.

A. SELECTIVITY OF LIGANDS FOR ERα, AR, AND PR OVER GR

1. Estradiol

In the A-ring "polar clamp" region of the binding pocket for ERα, the 3-hydroxy group of estradiol acts as a hydrogen bond donor to E353. This residue is equivalent in position to Q570 in GR, which acts as a hydrogen bond donor to the 3-ketone group in glucocorticoid-like ligands. This difference in the hydrogen bonding donor–acceptor pattern is a major reason why ERα preferentially binds estradiol over cortisol. In addition, the geometry and aromaticity of the estradiol A-ring also makes this steroid selective for ERα. Planar, aromatic A-rings are mandatory for ERα binding and are only moderately tolerated in GR. It is important to note that estradiol does not have 20 and 21 substituents to interact with the T739 and N564 D-ring "polar clamp" in GR.

2. Dihydrotestosterone/Testosterone (DHT/T)

Although the overall sequence identity between the AR and GR LBDs is high (58%), key differences exist in both proteins that result in good selectivity of DHT/T for AR over GR. One major difference is that the position of the loop between H6 and H7 in AR causes much of the 17α pocket to be filled. The conformation of this loop may influence the selectivity of these compounds. As a result, only smaller steroids with little substitution on the 17α position typically bind to AR with high affinity.

Because many glucocorticoids have only a hydroxyl in the 17α position, however, additional factors must account for selectivity of DHT/T binding in AR. Both AR and GR have mechanisms in place to maintain the orientation of the D-ring of their respective ligands in the receptor binding pocket. Within AR, both T877 and N705 are ideally situated to hydrogen bond with the 17β-hydroxy group of DHT/T. In contrast, N564 in GR, which is equivalent to N705 in AR, forms a hydrogen bond with C-ring 11β substituents of the steroid. The primary stabilization forces for the D-ring in the GR LBD originate from T739, which is in position to interact with the 20-carbonyl and 21-hydroxy of the steroid. In addition, glutamine 642 is in position to hydrogen bond with the 17α-hydroxyl present in Dex (or the 17β-hydroxyl of RU486), thus further stabilizing the D-ring of the steroid. Taken together, these differences result in the ligand selectivity seen between these receptors.

3. Progesterone

As with all receptors in this NR subfamily, the high sequence identity between PR and GR in the LBD (61%) makes achieving selectivity between these two receptors difficult. In fact, both receptors bind progesterone (Alnemri *et al.*, 1991; Wolff *et al.*, 1978). The weaker binding of progesterone to GR, however, can be explained by a combination of differences in the ligand binding determinants. First, a key difference is observed in the 17α region of the ligand binding pocket. In particular, the 17α region in PR does not contain a polar hydrogen bonding residue. As discussed previously, the GR/Dex structures show a key hydrogen bond between the 17α-hydroxy group of Dex (and 17β-hydroxyl of RU486) and Q642. In PR, this interaction is prevented since the corresponding residue is L797. Absence of a 17α-polar group on the progesterone molecule is clearly one reason for the preferred binding to PR versus GR. Because of the lack of a polar hydrogen bonding residue in the 17α position, one might expect that ligands with polar hydrogen bond donors or acceptors in the 17α position of the steroids may not bind PR as effectively. Potent binding of steroids, such as mometasone furoate to PR, suggests, however, that the requirement for hydrogen bonds in the 17α region of the pocket can be overcome with other binding contributions (Austin *et al.*, 2002).

The 17β region, of PR and GR also have some significantly different characteristics created by differences in hydrophobic residues that regulate the pocket size. In GR, these residues are M639 and M560; in PR, the corresponding residues are F794 and L715, respectively. The flexibility of the residues in PR is less compared to GR. Therefore, PR is less accommodating to larger substituents in the 17β position compared to GR. Accordingly, progesterone lacks a 21-hydroxy group, one of the interactions that appears to be important for Dex binding.

Finally, progesterone does not have an 11β-hydroxy group. In the GR/Dex structures, N564 is responsible for the interaction with the polar 11β substituent of the steroid. Interestingly, this asparigine is conserved in PR (N719). One would expect that an interaction with a polar group in the 11β position of a steroid would be possible. PR is apparently more readily able to overcome the absence of this interaction when binding progesterone compared to GR, especially in light of other aforementioned differences.

B. SELECTIVITY OF GLUCOCORTICOIDS AND MINERALOCORTICOIDS FOR GR AND MR

Both GR and MR bind glucocorticoids and mineralocorticoids. GR has a slightly decreased affinity for mineralocorticoids, and MR appears to bind nearly equally well to cortisol and aldosterone (Arriza et al., 1987). This cross-binding phenomenon can be easily explained by the high LBD identity between the two receptors (60%). Based on the GR/Dex structure and alignment with MR, there are only five residues in contact with a ligand that differ between these two related receptors: M604 to S810, G567 to A773, Y735 to F941, M560 to L766, and Q642 to L848 in GR and MR, respectively. However, there are rare examples of GR selectivity; for example, the pyrazolosteroid cortivazol has been shown to selectively bind and activate GR over MR (Schlechte et al., 1985; Yoshikawa et al., 2002).

1. The Influence of Q642 in GR on the Selective Binding of Glucocorticoids and Mineralocorticoids

The slight selectivity of mineralocorticoids for MR appears to be most attributed to the presence of the lipophilic residue L848 in MR rather than the more polar (and potential hydrogen bond acceptor) Q642 in GR. For example, the MR steroids 11-DOC and aldosterone are missing the 17α-hydroxy group that is present in most glucocorticoids. The lack of a polar group at this position in these steroids provides a slightly more favorable packing interaction with the lipophilic L848 in MR. Conversely, it might be expected that the polarity of the 17α-hydroxy group present in many of the glucocorticoids, such as cortisol, would not be detrimental for binding. It is also interesting that MR appears not to bind RU486. Again, this may be influenced in part by the hydrophobic L848 that cannot provide the hydrogen bond seen in the GR structure between Q642 and the 17β-hydroxy group of the molecule.

2. The Influence of N546 in GR and N770 in MR on the Binding of Glucocorticoids and Mineralocorticoids

As with N705 in AR and N719 in PR, N564 in GR and N770 in MR are key residues in the binding of both glucocorticoids and mineralocorticoids to these receptors. The interactions of these steroids with each of the receptor's asparagines, however, are significantly different from those in AR or PR. In the case of GR and MR, a hydrogen bonding interaction may exist between the residue and a 21-hydroxy group present in all of the endogenous glucocorticoids and mineralocorticoids. Although an 11β-hydroxy group is present in many of these steroids, such as cortisol and corticosterone, this substituent does not appear to be essential for binding to MR and, in some cases, to GR.

C. SUMMARY OF LIGAND SELECTIVITY

Although LBD homology in this NR subfamily is high, key and specific differences in pocket residues create a degree of ligand selectivity between these receptors. In the case of each receptor, differences in residue polarity and steric size promote electronic and volumetric differences in binding pockets that increase or decrease the binding affinity of various ligands. Nonoptimized contacts are probably compensated for by certain electronic and steric interactions to maintain maximal surface contact with the receptors. However, in many cases such as for glucocorticoids and mineralocorticoids, the binding selectivity may be only marginal. In fact, because circulating levels of cortisol are several fold higher than those for aldosterone, tissue-specific selectivity may only be obtained via regulatory enzymes (Agarwal and Mirshahi, 1999). For example, in the kidney, 11β-hydroxysteroid dehydrogenase-2 (HSD-2) converts cortisol to the much weaker ligand cortisone and thus effectively clears the way for the high affinity MR ligand aldosterone (Quinkler and Stewart, 2003).

VI. CHARACTERISTICS OF COFACTOR ASSOCIATION WITH THE GR LBD

Ligand-activated transcriptional regulation of nuclear receptors involves the participation of cofactors, which can be either activators or repressors of receptor-mediated transcription (McKenna and O'Malley, 2002; Subramaniam et al., 1999). The primary mechanistic role of these accessory proteins, upon complexing to a ligand-bound NR, is reversible chromatin modification via histone acetylation, in some cases by recruitment of additional coactivators, and general effects on basal transcriptional machinery (Collingwood et al., 1999; Hermanson et al., 2002; Spencer et al., 1997). More than 200 such cofactors for nuclear receptors have been described, and the number continues to grow rapidly. One of the first discovered is the p160 family of coactivators, which includes, in part, SRC-1, SRC-3, CREB-binding protein (CBP), and TIF2 (Leo and Chen, 2000). Still relatively poorly understood is the complex network of interactions involving these sets of coactivators and nuclear receptors (Darimont et al., 1998). There is significant promiscuity in the recognition of coactivators. For example, GR has been reported to interact with a variety of these type proteins, including SRC-1, TIF2, PPARγ coactivator-1 alpha (PGC-1α), CBP, and many others (Jenkins et al., 2002), and these coactivators also interact with other members of the steroid and nuclear receptor families (Darimont et al., 1998; Glass et al., 1997; McKenna et al., 1999). Structurally, this complex cross reaction between NRs and cofactors is simplistically explained by the relative conservation of the coactivator binding cleft in the NR LBD and the use of the LXXLL NR box on coactivators (Feng et al., 1998). Beyond the scope of this review is also the potential role of NR/coactivator interfaces outside the LBD and a discussion about GR's interaction with classical corepressors, such as nuclear receptor corepressor (NCoR).

In the GR LBD/Dex structure, as with other NR LBD crystal structures like ER and PPARg (Brzozowski et al., 1997; Nolte et al., 1998), it is clear the LXXLL region found in coactivators helps mediate the protein–protein interaction. The interaction of receptor with agonist ligand, such as Dex, helps to stabilize the position of the AF-2 helix in a way that a shallow hydrophobic groove is formed between H3 and H4 and AF-2. The LXXLL motif of the coactivator forms an amphipathic helix in which the leucine side chains are presented on one side of the helix. This hydrophobic surface docks into the cleft of the agonist-shaped LBD (Fig. 10). Although a thorough thermodynamic study has not been performed with GR, mutagenesis experiments have shown that the leucines, and not the flanking amino acids, are primarily responsible for the binding (Ding et al., 1998). Some studies, however, have shown that GR preferentially binds to certain LXXLL motifs, such as the third motif of TIF2 (Darimont et al., 1998; Ding et al., 1998). For initial GR LBD crystallography studies, a peptide representing this particular NR box was used.

In the GR LBD/Dex structure, electron density from 11 residues (ALLRYLLDKDD) of the TIF2 peptide is well defined. The peptide orients as expected with the three leucines of the LXXLL motif occupying the hydrophobic cleft. The peptide appears to also make a "charge clamp" interaction with GR LBD side chains in a mode similar to that observed in other NR LBD/peptide complexes. The coactivator peptide forms a $\sim$2.8Å hydrogen bond between K579 (located in H3 of the LBD) and a carbonyl present on the backbone of the coactivator peptide. Glutamate 755, a residue on the AF-2 helix, also interacts via hydrogen bond formation ($\sim$2.9Å) with an amide on the backbone of the coactivator peptide. Both the H3 lysine and

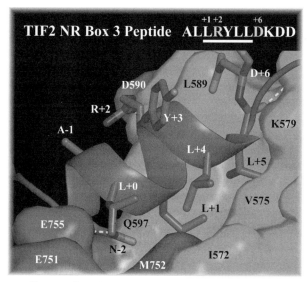

FIGURE 10. Close-up view of the TIF2 coactivator pocket in the GR LBD/Dex structure (Bledsoe et al., 2002). Residues from GR are colored green, the AF2 helix is colored red, and the TIF2 peptide is colored magenta. As shown, the three leucines at the +1, +4, and +5 positions are completely buried by the hydrophobic groove on the GR LBD. In addition, the leucine at position +0 also makes significant VDW contact. The total suface area bound at this interface is 459Å^2. As described in the text, the primary charge clamp involves the side chains K579 and E755. An additional hydrogen bond (∼2.1Å) is made between the coactivator side chain R +2 and the H4 residue D590. Another possible interaction exists between R585 and a backbone carbonyl from the coactivator peptide (∼3.1Å) (not visible in this view). The side chain from the D +6 on the coactivator peptide is also oriented toward R585, but the distance is rather long (>4.4Å) in this structure. These latter two electrostatic interactions, involving GR LBD residues R585 and D590, comprise the "secondary charge clamp."

AF-2 glutamate are conserved across many members of the NR superfamily, suggesting a common mechanism is employed when binding coactivators.

In addition to the hydrophobic contribution of the leucines and the primary "charge clamp" polar interaction, the GR LBD/Dex complex with the TIF2 peptide shows another unique set of interactions that might further stabilize the TIF2 peptide. First, a hydrogen bond interaction that involves an aspartic acid positioned on the first turn of H4 in the LBD is possible. This aspartic acid at position 590 appears to be in position to make a hydrogen bond with the arginine side chain in the TIF2 peptide. Coactivator peptide binding experiments that mutated arginine to a residue from the second LXXLL motif of TIF2, resulted in decreased affinity of the peptide to GR, AR, and PR LBDs, all of which have a conserved aspartic acid at that position (Bledsoe et al., 2002). For ERα, which has a histidine at the position equivalent to D590 in GR, the mutant coactivator peptide binds with the same affinity as the wild-type peptide. This further suggests that hydrogen bonds between GR surface residues and the coactivator play at least a minimal role for proper peptide docking.

The second example of this type of interaction uses R585 on the GR LBD. This arginine lies on a small helical segment designated as H3' that serves to connect H3 and H4. R585 appears to be in a position to make a hydrogen bond with a carbonyl on the coactivator peptide backbone. Taken together, these additional interactions may be the basis for the GR preferentially binding to the third LXXLL motif of TIF2. It is interesting that these two residues are conserved between GR, AR, and PR, whereas MR contains the conservative substitutions of a lysine and a glutamic acid in the same positions.

Although there appear to be additional contacts outside the central leucine amphipathic helix interaction, it is likely that the majority of the binding energy for the GR/cofactor interaction stems from the expansive hydrophobic interaction. The hydrogen bonds in the periphery of this core probably help to govern specificity or selectivity. Further study using larger cofactor fragments should reveal more details about interactions outside this region.

VII. ROLE OF THE GR LBD IN CHAPERONE PROTEIN ASSOCIATION AND RECEPTOR DIMERIZATION

A. HSP90 AND P23

Although a thorough discussion concerning the role of heat shock proteins in GR function is beyond this review, it is worth mentioning a few highlights relating to the ligand binding domain (LBD). It has been realized since the 1980s that ubiquitous chaperone proteins, such as hsp90 and p23, as well as several immunophilins, serve an important role for the ligand binding and signaling functions of GR (Bresnick *et al.*, 1989; Denis *et al.*, 1988a, 1988b; Pratt and Toft, 1997). Examples include maintenance of high affinity ligand interactions and complete abrogation of DNA binding activity when complexed with an apo receptor. Even though the precise molecular function of these heat shock proteins remains relatively poorly understood, the literature well documents their essential role for glucocorticoid receptor function. Both hsp90 and p23 are primarily localized in the cytoplasm along with ligand-free GR. It has been proposed that hsp90 holds GR in an "open" state and is therefore required for ligand association (Bresnick *et al.*, 1989).

Most of the interaction surface for hsp90 has been mapped to the GR LBD and a small region of the hinge that is the N-terminal to the LBD. Several studies have attempted to better define the specific regions governing the GR LBD/hsp90 interaction, but this has proven difficult. One possible reason the studies have encountered obstacles is that the binding surface of hsp90 extends over the entire LBD region and also requires partially or locally unfolded regions. To highlight a few interaction mapping studies, deletion mutagenesis was used to show that residues 616–671 in rat GR (residues 598–653 in human GR) are required for hsp90 binding (Dalman *et al.*, 1991). This same report used peptide competition to show that two sites on mouse GR, residues 587–606 and 624–665 (residues 581–600 and 618–659 in human GR), play a role in hsp90 association. Mutagenesis studies have also identified a short stretch of amino acids (529–536 in human GR) that is required for proper GR LBD/hsp90 complex assembly (Kaul *et al.*, 2002). It is obvious from these results that the interaction between GR LBD and hsp90 is complex and probably contains multiple, discountinuous contact points. The presence of this complex chaperone binding site on the GR LBD is perhaps one reason why the protein has been extremely difficult to express, purify, and characterize.

The F602S mutant of the GR LBD appears to be less dependent on *E. coli* chaperones for expression and stability, and it is possible that F602S may also play a role in mammalian heat shock protein association. In fact, GR containing the F602S mutation is more dispersed into the nucleus compared to the wild type in the absence of a ligand (Bledsoe *et al.*, 2002). One of the remaining mysteries of the GR LBD and other NR LBDs is how a ligand is able to enter and exit the domain. In the GR LBD structures presented to date, there is little to no solvent exposure for the bound ligand, so there must be rather significant structural rearrangements to create an entrance and exit path for the ligand. This is probably one of the essential roles of the heat shock protein in the cytoplasm. As has been previously suggested, it seems plausible that the chaperone holds the GR LBD in a partially folded, more open state that offers more exposure to the ligand binding determinates in the interior pocket. Interestingly, with the purified GR LBD/F602S mutant, the ligand is clearly able to exchange according to ligand binding experiments (Bledsoe *et al.*, 2002).

B. THE PUTATIVE GR LBD HOMODIMER INTERFACE

Prior to solution of the crystal structure for GR, biochemical studies revealed that GR exists as a homodimer even in the absence of DNA (Wrange *et al.*, 1989) and that some regions, including the hinge domain, are responsible for dimer formation (Savory *et al.*, 2001; Wrange *et al.*, 1989). One of the unique features of both GR/Dex crystal structures is the apparent formation of a unique LBD homodimer interface (Bledsoe *et al.*, 2002; Kauppi *et al.*, 2003). Unlike most other NRs, such as RXR, PPARs, HNF4, and the ERs, the GR LBD does not appear to use H10 as a leucine zipper-type helix–helix interface between subunits. In fact, GR lacks the φφKφ (where φ represents any hydrophobic residue) repeat that dominates the coiled–coil interaction of the other receptors. Instead, the two GR LBD subunits make contact through the β turn residues in strands between H5 and H6 and a β-sheet type interaction using the strand between H1 and H3. The apparent primary binding determinates are Q615, P625, and I628.

Whether this observed interaction represents the physiologic dimer interface, especially when in context of the well-defined dimerization of the DNA binding domain (Luisi *et al.*, 1991), remains to be seen. However, some experiments have suggested a critical physiologic role of residues that contribute to this observed interface (Bledsoe *et al.*, 2002). Transient transfection experiments suggest that both I628 and P625 are important for full transactivation activity relative to a wild-type receptor. These results are consistent with a previous study showing that the rat GR with a mutation at P643, a residue analogous to P625 in the human GR, is also defective in transactivation (Caamano *et al.*, 1998). Furthermore, this study also showed that the P643A rat GR retained the normal ability to bind DNA but was destabilized for heterocomplex formation with hsp90. As such, this mutated GR was dramatically deficient in nuclear translocation, and consequently, in transactivation and transrepression.

Because P625 is the central dimerization residue as seen in the GR/Dex structure, this would suggest that the GR dimerization interface might overlap the hsp90 binding site. More detailed structural and functional analyses are required to better define the physiologic receptor dimer interface and its role relative to interaction with chaperones and other accessory proteins. In summary, it is interesting to speculate that the dimer interface observed in the crystal structure also contains the determinates for the hsp90 interaction, particularly if the β-strand and H6 region near the 17α pocket of the ligand is prone to flexibility, plasticity, or even partial unfolding.

VIII. MUTATIONS IN THE GR LBD AND THEIR FUNCTIONAL CONSEQUENCES

A. SITE-DIRECTED MUTAGENESIS STUDIES

Numerous studies have investigated the effect of specific mutations on the function of the GR LBD prior to solution of the crystal structure. This section briefly highlights selected reports and reviews their structural and functional implications regarding the GR LBD crystal structure (summarized in Table I). Of the approximately 24 residues in the GR ligand pocket that make direct contact with Dex, only about 7 residues have been examined to date by site-directed mutagenesis. The fact that a wide variety of steroidal and nonsteroidal ligands can be functionally tested with these mutants complicates these experiments and data interpretation. In other words, a mutation that affects activity with one ligand may show no effect with a related analogue.

One of the most thorough mutagenesis studies, performed by Lind et al. (2000), took a systematic approach to examining the importance of residues surrounding the D-ring of glucocorticoids. Four residues, M560, M639, Q642, and T739, were mutated; the study examined the effects of the mutation on both ligand binding and transactivation for a variety of steroidal ligands. Mutation of Q642 (which makes between a 2.7 Å and 3.3Å hydrogen bond with 17α-hydroxy group of Dex depending on the coordinates used) to alanine resulted in decreased receptor affinity for steroids containing a 17α-hydroxy group, such as cortisol, prednisolone, and Dex. Interestingly, the affinity for steroids without this hydroxy group was not affected by either the Q642A or Q642V mutations. Threonine 739, which makes contact with both the 21-hydroxyl and 20-carbonyl of Dex, was mutated to alanine, and this change reduced binding affinity to both corticosterone and deoxycorticosterone. Changing M560 to leucine had relatively little effect on ligand binding, whereas the M560T mutation decreased binding to corticosterone. The M639V mutation showed reduced affinity for most steroidal ligands tested and also diminished transactivation activity. As noted previously, these two methionines help to form the 17α pocket and are highly mobile side chains.

As already discussed, the ligand binding pocket of the GR LBD contains a complex organization of both hydrophobic and polar interactions. It is clear from the crystal structures and the limited mutagenesis data that the energy for ligand binding and recognition is well dispersed among residues lining the pocket. Further structural and functional work is required to fully appreciate and decifer this complexity.

B. ROLE OF GR LBD MUTATIONS IN DISEASE

The appearance of mutations in the GR gene and the connections of these mutations to a variety of pathological conditions have been documented over roughly the last decade. Primarily missence mutations have been associated with glucocorticoid resistance or insensitivity, a condition that is associated with hypertension, hirutism, steroid-resistant asthma, menstrual irregularities, acne, and acute lymphoblastic leukemia. Although most steroid-insensitive conditions are inherited, there are several reports of acquired resistance attributable to defects in ligand binding by GR (Norbiato et al., 1992; Sher et al., 1994). Mutations have occurred throughout the entire GR gene, but those that are the focus of this discussion are mutations that occur in the LBD. A number of comprehensive reviews exist on the subject of GR mutations and clinical phenotypes (Bray and Cotton, 2003;

TABLE I. Selected GR LBD Point Mutations and Their Reported Functional Effects

GR LBD mutation	Effect on ligand binding	Effect on cellular function	Location in GR LBD structure	Possible reason for effect of mutation based on structure	Reference
M560T	Slight affinity reduction for corticosterone	30-fold reduction in transactivation with triamcinolone acetonide	N-terminal end of H3	VDW interaction with 17α- and 21-OH	Lind et al., 2000
N564A	Large reduction in affinity for steroid (triamcinolone acetonide)	No detectable activity with 11β-OH progesterone or deoxycorticosterone	Middle of H3	Hydrogen bonds with 11-OH and potentially 21-OH	Lind et al., 2000
M565R	1000-fold improved affinity for Dex; no affinity difference with RU486	Enhanced transactivation	Middle portion of H3	Surface residue; may provide improved packing for loop preceding AF-2 helix	Warriar et al., 1994
G567A	Large reduction in affinity for Dex	No transactivation activity	Middle portion of H3	VDW contact with A-ring	Warriar et al., 1994
A573Q	1000-fold improved affinity for Dex; no affinity difference with RU486	No transactivation activity with Dex	Middle portion of H3	Partial surface residue	Warriar et al., 1994
V571M	Increased affinity for aldosterone; no effect on affinity for triamcinolone acetonide	Increased transactivation with aldosterone	Middle portion of H3	Binding determinant for coactivator; hydrophobic environment for C19	Lind et al., 1999
R651A (rat) R633 in human GR	No effect on affinity for Dex	Reduced fold induction from GRE and MMTV reporter with Dex	C-terminal residue of the short H6	No direct contact with ligand; near protein surface; possible effect on putative GR LBD dimer interface	Huang and Simons, 1999
C644G (mouse) C638 in human GR	4-fold increased affinity for triamcinolone acetonide	Reduced transcriptional initiation and chromatin remodeling	N-terminal end of H7	No direct contact with Dex	Sheldon et al., 1999

Mutation	Effect on binding	Effect on transactivation	Location	Structural note	Reference
C656G (rat) C638 in human GR	9-fold increased affinity for Dex	6-fold improved transactivation with Dex	N-terminal end of H7	No direct contact with Dex	Chakraborti et al., 1991
C638S	15-fold and 10-fold higher affinity for Dex and RU486, respectively	4-fold increase in transcriptional activity	N-terminal end of H7	In vicinity (9Å) of D-ring; 17α and 17β substituents	Yu et al., 1995
M639V	Decreased affinity for cortisol and corticosterone	30-fold reduction in transactivation with triamcinolone acetonide	N-terminal end of H7	Long distance VDW interaction with 17α-OH	Lind et al., 2000
Q642A	Decreased affinity for steroids containing 17α-OH (cortisol, Dex, prednisolone)	4-fold enhanced transactivation with triamcinolone acetonide	N-terminal end of H7	Hydrogen bond with 17α-OH	Lind et al., 2000
C643S	10-fold lower affinity for RU486	Little effect on transactivation	N-terminal end of H7	Long distance (4 to 5Å) VDW contact with C-16 methyl of Dex and 17α-methylacetylene in RU486	Yu et al., 1995
R732Q (rat) R714 in human GR	Not reported	Decreased transcriptional activity with Dex and triamcinolone acetonide	N-terminal end of H10	Residue is distal to ligand binding site; surface exposed residue	Sarlis et al., 1999
C665S	No change in Dex affinity; 10-fold lower affinity for RU486; improved binding to aldosterone	Little change in activity with Dex; improved transactivation with aldosterone;	Middle portion of H8	Not in vicinity of ligand; at the top of domain directed toward surface and H9	Yu et al., 1995
T739A	No effect on binding affinity to triamcinolone acetonide; reduced affinity for corticosterone	16-fold reduction in transactivation with triamcinolone acetonide	C-terminal end of H10	Hydrogen bond with 20-carbonyl and VDW with 21-OH	Lind et al., 2000

TABLE II. Mutations Found with the GR LBD That Play a Role in Human Disease

Mutation	Effects on GR function	Disease phenotype	Mutation type	Location in GR LBD structure	Possible reason for effect of mutation based on structure	Reference
I559N	Reduced ligand binding; dominant negative effect	Glucocorticoid resistance; hyperactive HPA	Germ line	N-terminal end of H3	Structural; neighboring effect M560 in the 17α/β region	Karl et al., 1996
V571A	Weak transactivation relative to wild-type GR	Female pseudohermaphroditism; severe hypokalemia	Germ line	Middle portion of H3	Binding determinant for coactivator; forms floor of hydrophobic cleft	Mendonca et al., 2002
V575M	Decreased transactivation and coactivator recruitment; no effect on ligand binding	Steroid-resistant bronchial epithelial cells	Induced in cell culture	C-terminal end of H3	Binding determinant for coactivator; VDW contact with leucines of LXXLL	Kunz et al., 2003
D641V	Approximately 3-fold reduction in Dex binding affinity; decreased transactivation	Glucocorticoid resistance	Germ line	N-terminal end of H7	Surface exposed residue; near 17α pocket	Hurley et al., 1991
G679S	Decreased protein levels; reduced transactivation	Glucocorticoid resistance	Germ line	Loop between H8 and H9	Surface exposed residue; distal to both ligand and coactivator	Ruiz et al., 2001

Mutation	Phenotype	Origin	Location	Structural/Functional Note	Reference
V729I	2-fold decrease in Dex binding affinity; 4-fold decreased transactivation	Glucocorticoid resistance	Germ line	Middle of H10	Structural; residue in core of LBD and against H5	Brufsky et al., 1990; Malchoff et al., 1993
I747M	2-fold decrease in Dex binding affinity; 20- to 30-fold decreased transactivation	Glucocorticoid resistance	Germ line	Loop preceding AF-2 helix	Side chain interacts with 17β group; may affect H12 positioning	Vottero et al., 2002; Kino et al., 2002
L753F	Decreased protein levels; reduction in transactivation	Reduced glucocorticoid sensitivity in leukemic cells	Induced in cell culture	N-terminal end of AF-2 helix	VDW contact with C-11 group; may effect AF-2 helix stabilization	Ashraf and Thompson, 1993; Hillmann et al., 2000
Frameshift in exon 9; 20 additional residues at C-terminal	Decreased receptor number	Lupus nephritis	Somatic in vivo	Additional sequence connected to the AF-2 helix/F-domain	Possible alteration of AF-2 helix positioning and/or coactivator cleft accessibility	Jiang et al., 2001

Bronnegard and Carlstedtduke, 1995; Bronnegard *et al.*, 1996; Lamberts *et al.*, 1996; Leung and Bloom, 2003; Werner and Bronnegard, 1996).

A limited summary of GR LBD mutations that have been implicated in disease, primarily glucocorticoid resistance, are shown in Table II. Interestingly, none of the reported mutations appear to be primary determinants of cortisol binding, such as N564, M604, Q570, Q642, or T739. One exception is the mutation L753F, for which leucine appears to play a critical role in ligand binding via the Van der Waals (VDW) interaction and possible stabilization of AF-2 (Hillmann *et al.*, 2000). As expected, based on the GR LBD structure, a mutation at this position showed dramatic consequences on transactivation function even though the mutation, L753F, is a rather conservative one.

Another group of mutations in steroid-resistant disease involves residues V571 and V575, which are in the middle of H3. A V571A mutation is associated with female pseudohermaphroditism and severe hypokalemia (Mendonca *et al.*, 2002). At a position one helical turn away, the mutation V575M has been identified as the cause for steroid resistance in cultured bronchial epithelial cells (Kunz *et al.*, 2003). Based on the crystal structure, both of these valines are in close vicinity to the coactivator peptide. A mutation most likely will affect coactivator recruitment and, consequently, transactivation function.

Although the effects of previously mentioned mutations are easily rationalized by the crystal structure, a third group of mutations, including I559N, D641V, G679S, and V729I, are more difficult to explain. Most of these residues are at least partially surface exposed and do not reside near either the ligand or coactivator binding sites. Their effects on the function of GR may either be deleterious to the structural integrity of the domain or may be involved in a protein–protein interaction site for a yet-to-be discovered factor. Therapeutically, it is difficult to imagine a way to circumvent these types of mutations with direct receptor targeting. There is a possibility, however, that for some mutations (especially those involved in ligand recognition) further work may identify ligands effective at binding and activating mutant receptors for treatment of glucocorticoid-resistant diseases.

IX. PROGRESS TOWARD A SELECTIVE GR MODULATOR

It is widely believed that a drug with an improved therapeutic ratio (i.e., retaining desired activities while reducing undesired side effects) can possibly be created by differentiating the transactivation and transrepression activities of GR (Adcock, 2003; Belvisi *et al.*, 2001; Coghlan *et al.*, 2003a). Genes encoding proteins involved in inflammatory processes, such as cyclooxygenase-2, interleukin-1, and interleukin-8, are known to be repressed via corticosteroid treatment. The promoters of these genes do not have consensus GRE sequences, but instead are regulated by other transcription factors, namely NFκB and AP-1. A number of mechanisms describing the potential role of the GR LBD in this transrepression activity have been described (De Bosscher et al., 2003; McKay and Cidlowski, 1999). Several studies suggest that GR interacts directly with subunits of NFκB (Caldenhoven et al., 1995; Ray and Prefontaine, 1994; Wissink et al., 1997) and AP-1 (Diamond et al., 1990; Heck et al., 1994; Schule et al., 1990). Another study suggests, however, that transrepression occurs via cross-talk between ligand-regulated GR and protein kinase A (PKA) (Doucas et al., 2000). This group reports that GR interacts with the catalytic subunit of PKA and that this interaction mediates the ability of GR to repress NFκB. It is evident from these studies that the transrepression activity of GR is complex and likely involves multiple mechanisms (Wissink et al., 1998). With further structure and function studies on the GR LBD, the nature of these protein–protein interactions will be better understood, and this information may aid in the discovery of a modulating ligand that can retain full transrepression activity without transactivation.

In addition, as already reviewed, GR interacts with a number of coactivator proteins via the LBD. Understanding the precise role of coactivators in the tissue-specific activities of GR remains an intense area of research. The potential exists for discovery of a ligand that may selectively recruit one or more coactivators. One very recent example of this has been reported by Coghlan et al. (2003b). This group has reported the synthesis and characterization of a novel nonsteroidal ligand, AL438 (Fig. 1B), that demonstrates full anti-inflammatory efficacy with reduced effects on bone metabolism and glucose regulation. This novel ligand also shows an approximately 20-fold selectivity for GR compared to MR. Both transient transfection and glutathione-S-transferase pull-down experiments show that AL438-bound GR is reduced in its ability to interact with the peroxisomal proliferator-activated receptor γ coactivator-1 alpha (PCG-1α). This particular accessory protein has been shown to be important for corticosteroid-driven hepatic gluconeogenesis (Yoon et al., 2001). Interestingly, in a manner similar to prednisolone, AL438 promotes the ability of GR to interact with the GR-interacting protein (GRIP1, the murine counterpart of human TIF2), which is another LXXLL-containing cofactor (Hong et al., 1997). Precisely, how this differential modulation occurs at the molecular level should be an interesting pursuit and should aid future drug discovery efforts.

X. CONCLUSIONS AND DIRECTIONS OF FUTURE RESEARCH

The recently elucidated crystal structures of the GR LBD have allowed us to better understand the complexities of ligand recognition and specificity for this receptor. Not covered in this review are the many interesting and important features of the AF-1 and DBD domains of GR. There is little doubt that the LBD plays a role in function of these domains in ways that we do not yet fully understand. In addition, concerning the LBD, a C-terminal truncated isoform of GR termed GRβ has been described (Bamberger *et al.*, 1995; Oakley *et al.*, 1996). Although still somewhat controversial, this splicing variant, which lacks an AF2 helix, is transcriptionally inactive and has been implicated in some steroid-resistant diseases. Further structural and biochemical studies on GRβ should help us realize the drug target potential of this isoform. Moreover, for GR, structural work on full length or at least longer protein constructs should shed light on the role of LBD function beyond conformation changes in the domain.

Another remaining challenge concerning the GR LBD structure and function regards a better understanding of ligand variety. Features including shape complementarity and polar group position can influence LBD structure and downstream function, such as dimerization and association with factors like the NFkB-subunit p65. There is currently an intense quest for novel, perhaps nonsteroidal, ligands that uniquely occupy the GR LBD pocket, induce slightly different structural conformations, and modulate the receptor in yet unforeseen ways (Coghlan *et al.*, 2002). Obtaining an accurate structural picture of how various ligands are oriented in the pocket, visualizing how each of these complexes influences the overall fold of the domain, and associating this information with the cellular biologic profile of each ligand should aid in the design of more effective and safer medicines.

ACKNOWLEDGMENTS

We would like to thank all our colleagues in our respective departments as well as members of Respiratory Drug Discovery for many very helpful and insightful discussions. In particular, we thank Tom Stanley for his critical review of this manuscript.

REFERENCES

Adcock, I. M. (2003). Glucocorticoids: New mechanisms and future agents. *Current Aller. Asth. Rep.* **3**(3), 249–257.

Agarwal, M. K., and Mirshahi, M. (1999). General overview of mineralocorticoid hormone action. *Pharm. Thera.* **84**(3), 273–326.

Alnemri, E. S., Maksymowych, A. B., Robertson, N. M., and Litwack, G. (1991). Characterization and purification of a functional rat glucocorticoid receptor overexpressed in a baculovirus system. *J. Biol. Chem.* **266**(6), 3925–3936.

Arriza, J. L., Weinberger, C., Cerelli, G., Glaser, T. M., Handelin, B. L., Housman, D. E., and Evans, R. M. (1987). Cloning of human mineralocorticoid receptor complementary DNA: Structural and functional kinship with the glucocorticoid receptor. *Science* **237**(4812), 268–275.

Ashraf, J., and Thompson, E. B. (1993). Identification of the activation-labile gene: A single point mutation in the human glucocorticoid receptor presents as two distinct receptor phenotypes. *Mol. Endocrinol.* **7**(5), 631–642.

Aslam, F., Shalhoub, V., van Wijnen, A. J., Banerjee, C., Bortell, R., Shakoori, A. R., Litwack, G., Stein, J. L., Stein, G. S., and Lian, J. B. (1995). Contributions of distal and proximal promoter elements to glucocorticoid regulation of osteocalcin gene transcription. *Mol. Endocrinol.* **9**(6), 679–690.

Austin, R. J., Maschera, B., Walker, A., Fairbairn, L., Meldrum, E., Farrow, S. N., and Uings, I. J. (2002). Mometasone furoate is a less specific glucocorticoid than fluticasone propionate. *Euro. Resp. J.* **20**(6), 1386–1392.

Bamberger, C. M., Bamberger, A. M., de Castro, M., and Chrousos, G. P. (1995). Glucocorticoid receptor beta, a potential endogenous inhibitor of glucocorticoid action in humans. *J. Clinic. Investig.* **95**(6), 2435–2441.

Bamberger, C. M., Schulte, H. M., and Chrousos, G. P. (1996). Molecular determinants of glucocorticoid receptor function and tissue sensitivity to glucocorticoids. *Endo. Rev.* **17**(3), 245–261.

Belvisi, M. G., Brown, T. J., Wicks, S., and Foster, M. L. (2001). New glucocorticosteroids with an improved therapeutic ratio? *Pulm. Pharma. Thera.* **14**(3), 221–227.

Bledsoe, R. K., Montana, V. G., Stanley, T. B., Delves, C. J., Apolito, C. J., McKee, D. D., Consler, T. G., Parks, D. J., Stewart, E. L., Willson, T. M., Lambert, M. H., Moore, J. T., Pearce, K. H., and Xu, H. E. (2002). Crystal structure of the glucocorticoid receptor ligand binding domain reveals a novel mode of receptor dimerization and coactivator recognition. *Cell* **110**(1), 93–105.

Bodwell, J. E., Webster, J. C., Jewell, C. M., Cidlowski, J. A., Hu, J. M., and Munck, A. (1998). Glucocorticoid receptor phosphorylation: Overview, function, and cell cycle-dependence. *J. of Steroid Biochem. Molec. Biol.* **65**(1–16), 91–99.

Bourguet, W., Germain, P., and Gronemeyer, H. (2000). Nuclear receptor ligand binding domains: Three-dimensional structures, molecular interactions, and pharmacological implications. *Trends in Pharm. Sci.* **21**(10), 381–388.

Bourguet, W., Ruff, M., Chambon, P., Gronemeyer, H., and Moras, D. (1995). Crystal structure of the ligand binding domain of the human nuclear receptor RXR alpha. *Nature.* **375**(6530), 377–382.

Bray, P. J., and Cotton, R. G. H. (2003). Variations of the human glucocorticoid receptor gene (NR3C1): Pathological and *in vitro* mutations and polymorphisms. *Human Mut.* **21**(6), 557–568.

Bresnick, E. H., Dalman, F. C., Sanchez, E. R., and Pratt, W. B. (1989). Evidence that the 90-kDa heat shock protein is necessary for the steroid binding conformation of the L cell glucocorticoid receptor. *J. Biol. Chem.* **264**(9), 4992–4997.

Bronnegard, M., and Carlstedtduke, J. (1995). The genetic basis CF glucocorticoid resistance. *Tren. Endocrin. Metabol.* **6**(5), 160–164.

Bronnegard, M., Stierna, P., and Marcus, C. (1996). Glucocorticoid-resistant syndromes— Molecular basis and clinical presentations. *J. Neuroendocrinol.* **8**(6), 405–415.

Brufsky, A. M., Malchoff, D. M., Javier, E. C., Reardon, G., Rowe, D., and Malchoff, C. D. (1990). A glucocorticoid receptor mutation in a subject with primary cortisol resistance. *Trans. Assoc. Amer. Phys.* **103**, 53–63.

Brzozowski, A. M., Pike, A. C., Dauter, Z., Hubbard, R. E., Bonn, T., Engstrom, O., Ohman, L., Greene, G. L., Gustafsson, J. A., and Carlquist, M. (1997). Molecular basis of agonism and antagonism in the oestrogen receptor. *Nature* **389**(6652), 753–758.

Caamano, C. A., Morano, M. I., Dalman, F. C., Pratt, W. B., and Akil, H. (1998). A conserved proline in the hsp90 binding region of the glucocorticoid receptor is required for hsp90 heterocomplex stabilization and receptor signaling. *J. Biol. Chem.* **273**(32), 20473–20480.

Caamano, C. A., Morano, M. I., Watson, S. J., Jr., Dalman, F. C., Pratt, W. B., and Akil, H. (1994). The functional relevance of the heteromeric structure of corticosteroid receptors. *Ann. NY Acad. Sci.* **746**, 68–77.

Caldenhoven, E., Liden, J., Wissink, S., Van de Stolpe, A., Raaijmakers, J., Koenderman, L., Okret, S., Gustafsson, J. A., and van der Saag, P. T. (1995). Negative cross-talk between RelA and the glucocorticoid receptor: A possible mechanism for the anti-inflammatory action of glucocorticoids. *Molec. Endocrinol.* **9**(4), 401–412.

Carlstedt-Duke, J., Stromstedt, P. E., Wrange, O., Bergman, T., Gustafsson, J. A., and Jornvall, H. (1987). Domain structure of the glucocorticoid receptor protein. *Proc. Natl. Acad. Sci. USA* **84**(13), 4437–4440.

Chakraborti, P. K., Garabedian, M. J., Yamamoto, K. R., and Simons, S. S. (1991). Creation of "super" glucocorticoid receptors by point mutations in the steroid binding domain *J. Biol. Chem.* **266**(33), 22075–22078.

Coghlan, M. J., Elmore, S. W., Kym, P. R., and Kort, M. E. (2002). Selective glucocorticold receptor modulators. *Ann. Rep. Med. Chem.* **37**(37), 167–176.

Coghlan, M. J., Elmore, S. W., Kym, P. R., and Kort, M. E. (2003a). The pursuit of differentiated ligands for the glucocorticoid receptor. *Curr. Topics Med. Chem.* **3**(14), 1617–1635.

Coghlan, M. J., Jacobson, P. B., Lane, B., Nakane, M., Lin, C. W., Elmore, S. W, Kym, P. R., Luly, J. R., Carter, G. W., Turner, R., Tyree, C. M., Hu, J., Elgort, J., and Miner, J. N. (2003b). A novel antiinflammatory maintains glucocorticoid efficacy with reduced side effects. *Molec. Endocrinol.* **17**(5), 860–869.

Collier, C. D., Oshima, H., and Simons, S. S., Jr. (1996). A negative tyrosine aminotransferase gene element that blocks glucocorticoid modulatory element-regulated modulation of glucocorticoid-induced gene expression. *Molec. Endocrinol.* **10**(5), 463–476.

Collingwood, T. N., Urnov, F. D., and Wolffe, A. P. (1999). Nuclear receptors: Coactivators, corepressors, and chromatin remodeling in the control of transcription. *J. Molec. Endocrinol.* **23**(3), 255–275.

Dalman, F. C., Scherrer, L. C., Taylor, L. P., Akil, H., and Pratt, W. B. (1991). Localization of the 90-kDa heat shock protein-binding site within the hormone binding domain of the glucocorticoid receptor by peptide competition. *J. Biol. Chem.* **266**(6), 3482–3490.

Darimont, B. D., Wagner, R. L., Apriletti, J. W., Stallcup, M. R., Kushner, P. J., Baxter, J. D., Fletterick, R. J., and Yamamoto, K. R. (1998). Structure and specificity of nuclear receptor–coactivator interactions. *Genes Develop.* **12**(21), 3343–3356.

De Bosscher, K., Vanden Berghe, W., and Haegeman, G. (2003). The interplay between the glucocorticoid receptor and nuclear factor-kappaB or activator protein-1: Molecular mechanisms for gene repression. *Endocrine Rev.* **24**(4), 488–522.

Defranco, D. B., Madan, A. P., Tang, Y., Chandran, U. R., Xiao, N., and Yang, J. (1995). Nucleocytoplasmic shuttling of steroid receptors. *Vitam. Horm.* **51**, 315–338.

Denis, M., Gustafsson, J. A., and Wikstrom, A. C. (1988a). Interaction of the Mr = 90,000 heat shock protein with the steroid binding domain of the glucocorticoid receptor. *J. Biol. Chem.* **263**(34), 18520–31852.

Denis, M., Poellinger, L., Wikstom, A. C., and Gustafsson, J. A. (1988b). Requirement of hormone for thermal conversion of the glucocorticoid receptor to a DNA-binding state. *Nature* **333**(6174), 686–688.

Diamond, M. I., Miner, J. N., Yoshinaga, S. K., and Yamamoto, K. R. (1990). Transcription factor interactions: Selectors of positive or negative regulation from a single DNA element. *Science* **249**(4974), 1266–1272.

Ding, X. F., Anderson, C. M., Ma, H., Hong, H., Uht, R. M., Kushner, P. J., and Stallcup, M. R. (1998). Nuclear receptor-binding sites of coactivators glucocorticoid receptor interacting protein 1 (GRIP1) and steroid receptor coactivator-1 (SRC-1): Multiple motifs with different binding specificities. *Molec. Endocrinol.* **12**(2), 302–313.

Doucas, V., Shi, Y., Miyamoto, S., West, A., Verma, I., and Evans, R. M. (2000). Cytoplasmic catalytic subunit of protein kinase A mediates cross-repression by NF-kappa B and the glucocorticoid receptor. *Proc. Natl. Acad. of Sci. USA* **97**(22), 11893–11898.
Feng, W., Ribeiro, R. C., Wagner, R. L., Nguyen, H., Apriletti, J. W., Fletterick, R. J., Baxter, J. D., Kushner, P. J., and West, B. L. (1998). Hormone-dependent coactivator binding to a hydrophobic cleft on nuclear receptors. *Science* **280**(5370), 1747–1749.
Freeman, B. C., Felts, S. J., Toft, D. O., and Yamamoto, K. R. (2000). The p23 molecular chaperones act at a late step in intracellular receptor action to differentially affect ligand efficacies. *Genes Develop.* **14**(4), 422–434.
Gampe, R. T.Jr., Montana, V. G., Lambert, M. H., Miller, A. B., Bledsoe, R. K., Milburn, M. V., Kliewer, S. A., Willson, T. M., and Xu, H. E. (2000). Asymmetry in the PPARgamma/RXRalpha crystal structure reveals the molecular basis of heterodimerization among nuclear receptors. *Molec. Cell* **5**(3), 545–555.
Garabedian, M. J., and Yamamoto, K. R. (1992). Genetic dissection of the signaling domain of a mammalian steroid receptor in yeast. *Molec. Biol. Cell* **3**(11), 1245–1257.
Giguere, V., Hollenberg, S. M., Rosenfeld, M. G., and Evans, R. M. (1986). Functional domains of the human glucocorticoid receptor. *Cell* **46**(5), 645–652.
Glass, C. K., Rose, D. W., and Rosenfeld, M. G. (1997). Nuclear receptor coactivators. *Curr. Opin. Cell. Biol.* **9**(2), 222–232.
Goodman, P. A., Medina-Martinez, O., and Fernandez-Mejia, C. (1996). Identification of the human insulin negative regulatory element as a negative glucocorticoid response element. *Molec. Cell. Endocrin.* **120**(2), 139–146.
Heck, S., Kullmann, M., Gast, A., Ponta, H., Rahmsdorf, H. J., Herrlich, P., and Cato, A. C. (1994). A distinct modulating domain in glucocorticoid receptor monomers in the repression of activity of the transcription factor AP-1. *EMBO J.* **13**(17), 4087–4095.
Hermanson, O., Glass, C. K., and Rosenfeld, M. G. (2002). Nuclear receptor coregulators: Multiple modes of modification. *Trends Endocrinol. Metab.* **13**(2), 55–60.
Hillmann, A. G., Ramdas, J., Multanen, K., Norman, M. R., and Harmon, J. M. (2000). Glucocorticoid receptor gene mutations in leukemic cells acquired *in vitro* and *in vivo*. *Cancer Res.* **60**(7), 2056–2062.
Hittelman, A. B., Burakov, D., Iniguez-Lluhi, J. A., Freedman, L. P., and Garabedian, M. J. (1999). Differential regulation of glucocorticoid receptor transcriptional activation via AF-1-associated proteins. *EMBO J.* **18**(19), 5380–5388.
Hollenberg, S. M., Weinberger, C., Ong, E. S., Cerelli, G., Oro, A., Lebo, R., Thompson, E. B., Rosenfeld, M. G., and Evans, R. M. (1985). Primary structure and expression of a functional human glucocorticoid receptor cDNA. *Nature* **318**(6047), 635–641.
Hong, H., Kohli, K., Garabedian, M. J., and Stallcup, M. R. (1997). GRIP1, a transcriptional coactivator for the AF-2 transactivation domain of steroid, thyroid, retinoid, and vitamin D receptors. *Molec. Cell. Biol.* **17**(5), 2735–2744.
Huang, Y., and Simons, S. S., Jr. (1999). Functional analysis of R651 mutations in the putative helix 6 of rat glucocorticoid receptors. *Molec. Cell. Endocrinol.* **158**(1–2), 117–130.
Hurley, D. M., Accili, D., Stratakis, C. A., Karl, M., Vamvakopoulos, N., Rorer, E., Constantine, K., Taylor, S. I., and Chrousos, G. P. (1991). Point mutation causing a single amino acid substitution in the hormone binding domain of the glucocorticoid receptor in familial glucocorticoid resistance. *J. Clinic. Invest.* **87**(2), 680–686.
Itoh, M., Adachi, M., Yasui, H., Takekawa, M., Tanaka, H., and Imai, K. (2002). Nuclear export of glucocorticoid receptor is enhanced by c-Jun N-terminal kinase-mediated phosphorylation. *Mole. Endocrinol.* **16**(10), 2382–2392.
Jenkins, B. D., Pullen, C. B., and Darimont, B. D. (2002). Novel glucocorticoid receptor coactivatior mechanisms. *Trends Endocrinol. Metab.* **12**(3), 122–126.
Jiang, T., Liu, S., Tan, M., Huang, F., Sun, Y., Dong, X., Guan, W., Huang, L., and Zhou, F. (2001). The phase-shift mutation in the glucocorticoid receptor gene: Potential etiologic

significance of neuroendocrine mechanisms in lupus nephritis. *Clinica Chim. Acta* **313**(1–2), 113–117.

Karl, M., Lamberts, S. W., Koper, J. W., Katz, D. A., Huizenga, N. E., Kino, T., Haddad, B. R., Hughes, M. R., and Chrousos, G. P. (1996). Cushing's disease preceded by generalized glucocorticoid resistance: Clinical consequences of a novel, dominant-negative glucocorticoid receptor mutation. *Proc. Assoc. Amer. Phys.* **108**(4), 296–307.

Kaul, S., Murphy, P. J., Chen, J., Brown, L., Pratt, W. B., and Simons, S. S., Jr. (2002). Mutations at positions 547–553 of rat glucocorticoid receptors reveal that hsp90 binding requires the presence, but not defined composition, of a seven amino acid sequence at the amino terminus of the ligand binding domain. *J. Biol. Chem.* **277**(39), 36223–36232.

Kauppi, B., Jakob, C., M, F. A., Yang, J., Ahola, H., Alarcon, M., Calles, K., Engstr, A. M. O., Harlan, J., Muchmore, S., Ramqvist, A. K., Thorell, S., Ohman, L., Greer, J., Gustafsson, J. A., Carlstedt-Duke, J., and Carlquist, M. (2003). The three-dimensional structures of antagonistic and agonistic forms of the glucocorticoid receptor ligand binding domain: RU486 induces a transconformation that leads to active antagonism. *J. Biol. Chem.* **278**(25), 22748–22754.

Kino, T., Vottero, A., Charmandari, E., and Chrousos, G. P. (2002). Familial/sporadic glucocorticoid resistance syndrome and hypertension. *Ann. N.Y. Acad. Sci.* **970**, 101–111.

Krstic, M. D., Rogatsky, I., Yamamoto, K. R., and Garabedian, M. J. (1997). Mitogen-activated and cyclin-dependent protein kinases selectively and differentially modulate transcriptional enhancement by the glucocorticoid receptor. *Molec. Cell. Biol.* **17**(7), 3947–3954.

Kunz, S., Sandoval, R., Carlsson, P., Carlstedt-Duke, J., Bloom, J. W., and Miesfeld, R. L. (2003). Identification of a novel glucocorticoid receptor mutation in budesonide-resistant human bronchial epithelial cells. *Molec. Endocrinol.* **17**(12), 2566–2582.

Lamberts, S. W. J., Huizenga, A. T. M., Delange, P., Dejong, F. H., and Koper, J. W. (1996). Clinical aspects of glucocorticoid sensitivity. *Steroids* **61**(4), 157–160.

Leo, C., and Chen, J. D. (2000). The SRC family of nuclear receptor coactivators. *Gene* **245**(1), 1–11.

Leung, D. Y., and Bloom, J. W. (2003). Update on glucocorticoid action and resistance. *J. Aller. Clin. Immun.* **111**(1), 3–22.

Lind, U., Greenidge, P., Gillner, M., Koehler, K. F., Wright, A., and Carlstedt-Duke, J. (2000). Functional probing of the human glucocorticoid receptor steroid-interacting surface by site-directed mutagenesis. Gln-642 plays an important role in steroid recognition and binding. *J. Biol. Chem.* **275**(25), 19041–19049.

Lind, U., Greenidge, P., Gustafsson, J. A., Wright, A. P., and Carlstedt-Duke, J. (1999). Valine 571 functions as a regional organizer in programming the glucocorticoid receptor for differential binding of glucocorticoids and mineralocorticoids. *J. Biol. Chem.* **274**(26), 18515–18523.

Luisi, B. F., Xu, W. X., Otwinowski, Z., Freedman, L. P., Yamamoto, K. R., and Sigler, P. B. (1991). Crystallographic analysis of the interaction of the glucocorticoid receptor with DNA. *Nature* **352**(6335), 497–505.

Malchoff, D. M., Brufsky, A., Reardon, G., McDermott, P., Javier, E. C., Bergh, C. H., Rowe, D., and Malchoff, C. D. (1993). A mutation of the glucocorticoid receptor in primary cortisol resistance. *J. Clinic. Invest.* **91**(5), 1918–1925.

Matias, P. M., Donner, P., Coelho, R., Thomaz, M., Peixoto, C., Macedo, S., Otto, N., Joschko, S., Scholz, P., Wegg, A., Basler, S., Schafer, M., Egner, U., and Carrondo, M. A. (2000). Structural evidence for ligand specificity in the binding domain of the human androgen receptor. Implications for pathogenic gene mutations. *J. Biol. Chem.* **275**(34), 26164–26171.

McEwan, I. J., Wright, A. P., and Gustafsson, J. A. (1997). Mechanism of gene expression by the glucocorticoid receptor: Role of protein-protein interactions. *Bioessays* **19**(2), 153–160.

McKay, L. I., and Cidlowski, J. A. (1998). Cross-talk between nuclear factor-kappa B and the steroid hormone receptors: Mechanisms of mutual antagonism. *Molec. Endocrinol.* **12**(1), 45–56.

McKay, L. I., and Cidlowski, J. A. (1999). Molecular control of immune/inflammatory responses: Interactions between nuclear factor-kappa B and steroid receptor-signaling pathways. *Endocrine Rev.* **20**(4), 435–459.

McKenna, N. J., Lanz, R. B., and O'Malley, B. W. (1999). Nuclear receptor coregulators: Cellular and molecular biology. *Endocrine Rev.* **20**(3), 321–344.

McKenna, N. J., and O'Malley, B. W. (2002). Minireview: Nuclear receptor coactivators—An update. *Endocrinology* **143**(7), 2461–2465.

Mendonca, B. B., Leite, M. V., de Castro, M., Kino, T., Elias, L. L., Bachega, T. A., Arnhold, I. J., Chrousos, G. P., and Latronico, A. C. (2002). Female pseudohermaphroditism caused by a novel homozygous missense mutation of the GR gene. *J. Clinic. Endocrinol. Metabol.* **87**(4), 1805–1809.

Miner, J. N. (2002). Designer glucocorticoids. *Biochem. Pharma.* **64**(3), 355–361.

Moras, D., and Gronemeyer, H. (1998). The nuclear receptor ligand binding domain: Structure and function. *Curr. Opin. Cell. Biol.* **10**(3), 384–391.

Nolte, R. T., Wisely, G. B., Westin, S., Cobb, J. E., Lambert, M. H., Kurokawa, R., Rosenfeld, M. G., Willson, T. M., Glass, C. K., and Milburn, M. V. (1998). Ligand binding and coactivator assembly of the peroxisome proliferator-activated receptor gamma. *Nature* **395**(6698), 137–143.

Norbiato, G., Bevilacqua, M., Vago, T., Baldi, G., Chebat, E., Bertora, P., Moroni, M., Galli, M., and Oldenburg, N. (1992). Cortisol resistance in acquired immunodeficiency syndrome. *J. Clinic. Endocrinol. Metabol.* **74**(3), 608–613.

Oakley, R. H., Sar, M., and Cidlowski, J. A. (1996). The human glucocorticoid receptor beta isoform. Expression, biochemical properties, and putative function. *J. of Biol. Chem.* **271**(16), 9550–9559.

Ohara-Nemoto, Y., Stromstedt, P. E., Dahlman-Wright, K., Nemoto, T., Gustafsson, J. A., and Carlstedt-Duke, J. (1990). The steroid-binding properties of recombinant glucocorticoid receptor: A putative role for heat shock protein hsp90. *J. Steroid Biochem. Molec. Biol.* **37**(4), 481–490.

Ojasoo, T., Raynaud, J. P., and Doe, J. C. (1994). Affiliations among steroid receptors as revealed by multivariate analysis of steroid binding data. *J. Steroid Biochem. Molec. Biol.* **48**(1), 31–46.

Ou, X. M., Storring, J. M., Kushwaha, N., and Albert, P. R. (2001). Heterodimerization of mineralocorticoid and glucocorticoid receptors at a novel negative response element of the 5-HT1A receptor gene. *J. Biol. Chem.* **276**(17), 14299–14307.

Pratt, W. B. (1993). The role of heat shock proteins in regulating the function, folding, and trafficking of the glucocorticoid receptor. *J. Biol. Chem.* **268**(29), 21455–21458.

Pratt, W. B. (1998). The hsp90-based chaperone system: Involvement in signal transduction from a variety of hormone and growth factor receptors. *Exper. Biol. Med.* **217**(4), 420–434.

Pratt, W. B., and Toft, D. O. (1997). Steroid receptor interactions with heat shock protein and immunophilin chaperones. *Endocrine Rev.* **18**(3), 306–360.

Quinkler, M., and Stewart, P. M. (2003). Hypertension and the cortisol–cortisone shuttle. *J. Clinic. Endocrinol. Metabol.* **88**(6), 2384–2392.

Ray, A., and Prefontaine, K. E. (1994). Physical association and functional antagonism between the p65 subunit of transcription factor NFkappaB and the glucocorticoid receptor. *Proc. Natl. Acad. Sci. USA* **91**(2), 752–756.

Rogatsky, I., Logan, S. K., and Garabedian, M. J. (1998). Antagonism of glucocorticoid receptor transcriptional activation by the c-Jun N-terminal kinase. *Proc. Natl. Acad. of Sci. USA* **95**(5), 2050–2055.

Ruiz, M., Lind, U., Gafvels, M., Eggertsen, G., Carlstedt-Duke, J., Nilsson, L., Holtmann, M., Stierna, P., Wikstrom, A. C., and Werner, S. (2001). Characterization of two novel

mutations in the glucocorticoid receptor gene in patients with primary cortisol resistance. *Clinic. Endocrinol.* **55**(3), 363–371.
Rusconi, S., and Yamamoto, K. R. (1987). Functional dissection of the hormone and DNA binding activities of the glucocorticoid receptor. *EMBO J.* **6**(5), 1309–1315.
Sack, J. S., Kish, K. F., Wang, C., Attar, R. M., Kiefer, S. E., An, Y., Wu, G. Y., Scheffler, J. E., Salvati, M. E., Krystek, S. R., Jr., Weinmann, R., and Einspahr, H. M. (2001). Crystallographic structures of the ligand binding domains of the androgen receptor and its T877A mutant complexed with the natural agonist dihydrotestosterone. *Proc. Natl. Acad. Sci. USA* **98**(9), 4904–4909.
Sakai, D. D., Helms, S., Carlstedt-Duke, J., Gustafsson, J. A., Rottman, F. M., and Yamamoto, K. R. (1988). Hormone-mediated repression: A negative glucocorticoid response element from the bovine prolactin gene. *Genes Develop.* **2**(9), 1144–1154.
Sarlis, N. J., Bayly, S. F., Szapary, D., and Simons, S. S., Jr. (1999). Quantity of partial agonist activity for antiglucocorticoids complexed with mutant glucocorticoid receptors is constant in two different transactivation assays but not predictable from steroid structure. *J. Steroid Biochem. Molec. Biol.* **68**(3–4), 89–102.
Savory, J. G., Prefontaine, G. G., Lamprecht, C., Liao, M., Walther, R. F., Lefebvre, Y. A., and Hache, R. J. (2001). Glucocorticoid receptor homodimers and glucocorticoid–mineralocorticoid receptor heterodimers form in the cytoplasm through alternative dimerization interfaces. *Molec. Cell. Biol.* **21**(3), 781–793.
Schlechte, J. A., Simons, S. S., Jr., Lewis, D. A., and Thompson, E. B. (1985). [3H]cortivazol: A unique high affinity ligand for the glucocorticoid receptor. *Endocrinology* **117**(4), 1355–1362.
Schule, R., Rangarajan, P., Kliewer, S., Ransone, L. J., Bolado, J., Yang, N., Verma, I. M., and Evans, R. M. (1990). Functional antagonism between oncoprotein c-Jun and the glucocorticoid receptor. *Cell* **62**(6), 1217–1226.
Sheldon, L. A., Smith, C. L., Bodwell, J. E., Munck, A. U., and Hager, G. L. (1999). A ligand binding domain mutation in the mouse glucocorticoid receptor functionally links chromatin remodeling and transcription initiation. *Molec. Cell. Biol.* **19**(12), 8146–8157.
Sher, E. R., Leung, D. Y., Surs, W., Kam, J. C., Zieg, G., Kamada, A. K., and Szefler, S. J. (1994). Steroid-resistant asthma. Cellular mechanisms contributing to inadequate response to glucocorticoid therapy. *J. Clinic. Invest.* **93**(1), 33–39.
Shiau, A. K., Barstad, D., Loria, P. M., Cheng, L., Kushner, P. J., Agard, D. A., and Greene, G. L. (1998). The structural basis of estrogen receptor/coactivator recognition and the antagonism of this interaction by tamoxifen. *Cell* **95**(7), 927–937.
Spencer, T. E., Jenster, G., Burcin, M. M., Allis, C. D., Zhou, J., Mizzen, C. A., McKenna, N. J., Onate, S. A., Tsai, S. Y., Tsai, M. J., and O'Malley, B. W. (1997). Steroid receptor coactivator-1 is a histone acetyltransferase. *Nature* **389**(6647), 194–198.
Subramaniam, N., Treuter, E., and Okret, S. (1999). Receptor interacting protein RIP140 inhibits both positive and negative gene regulation by glucocorticoids. *J. Biol. Chem.* **274**(25), 18121–18127.
Vottero, A., Kino, T., Combe, H., Lecomte, P., and Chrousos, G. P. (2002). A novel, C-terminal dominant negative mutation of the GR causes familial glucocorticoid resistance through abnormal interactions with p160 steroid receptor coactivators. *J. Clinic. Endocrin. Metabol.* **87**(6), 2658–2667.
Wang, Z., Frederick, J., and Garabedian, M. J. (2002). Deciphering the phosphorylation "code" of the glucocorticoid receptor *in vivo*. *J. Biol. Chem.* **277**(29), 26573–26580.
Warriar, N., Yu, C., and Govindan, M. V. (1994). Hormone binding domain of human glucocorticoid receptor. Enhancement of transactivation function by substitution mutants M565R and A573Q. *J. Biol. Chem.* **269**(46), 29010–29015.
Werner, S., and Bronnegard, M. (1996). Molecular basis of glucocorticoid-resistant syndromes. *Steroids* **61**(4), 216–221.

Williams, S. P., and Sigler, P. B. (1998). Atomic structure of progesterone complexed with its receptor. *Nature* **393**(6683), 392–396.

Wissink, S., van Heerde, E. C., Schmitz, M. L., Kalkhoven, E., van der Burg, B., Baeuerle, P. A., and van der Saag, P. T. (1997). Distinct domains of the RelA NF-kappaB subunit are required for negative cross-talk and direct interaction with the glucocorticoid receptor. *J. Biol. Chem.* **272**(35), 22278–22284.

Wissink, S., van Heerde, E. C., van der Burg, B., and van der Saag, P. T. (1998). A dual mechanism mediates repression of NFkappaB activity by glucocorticoids. *Molec. Endocrinol.* **12**(3), 355–363.

Wolff, M. E., Baxter, J. D., Kollman, P. A., Lee, D. L., Kuntz, I. D., Bloom, E., Matulich, D. T., and Morris, J. (1978). Nature of steroid-glucocorticoid receptor interactions: Thermodynamic analysis of the binding reaction. *Biochemistry* **17**(16), 3201–3208.

Wrange, O., Eriksson, P., and Perlmann, T. (1989). The purified activated glucocorticoid receptor is a homodimer. *J. Biol. Chem.* **264**(9), 5253–5259.

Yoon, J. C., Puigserver, P., Chen, G., Donovan, J., Wu, Z., Rhee, J., Adelmant, G., Stafford, J., Kahn, C. R., Granner, D. K., Newgard, C. B., and Spiegelman, B. M. (2001). Control of hepatic gluconeogenesis through the transcriptional coactivator PGC-1. *Nature* **413**(6852), 131–138.

Yoshikawa, N., Makino, Y., Okamoto, K., Morimoto, C., Makino, I., and Tanaka, H. (2002). Distinct interaction of cortivazol with the ligand binding domain confers glucocorticoid receptor specificity: Cortivazol is a specific ligand for the glucocorticoid receptor. *J. Biol. Chem.* **277**(7), 5529–5540.

Yu, C., Warriar, N., and Govindan, M. V. (1995). Cysteines 638 and 665 in the hormone binding domain of human glucocorticoid receptor define the specificity to glucocorticoids. *Biochemistry* **34**(43), 14163–14173.

Zhang, S., Liang, X., and Danielsen, M. (1996). Role of the C-terminus of the glucocorticoid receptor in hormone binding and agonist/antagonist discrimination. *Molec. Endocrinol.* **10**(1), 24–34.

3

NUCLEAR RECEPTOR RECRUITMENT OF HISTONE-MODIFYING ENZYMES TO TARGET GENE PROMOTERS

CHIH-CHENG TSAI AND JOSEPH D. FONDELL

Department of Physiology and Biophysics
UMDNJ, Robert Wood Johnson Medical School
Piscataway, New Jersey 08854

I. Overview of Nuclear Receptors
II. Chromatin Structure
III. Histone Modifications in Regulated Transcription
   A. Acetylation
   B. Phosphorylation
   C. Methylation
   D. Patterns and Codes
IV. NR Coactivators with Histone-Modifying Activity
   A. The p160/SRC Family
   B. The p300 and CBP Coactivators
   C. CARM1 and PRMT1
   D. The ASC-2 Complex
V. NR Corepressors with Histone-Modifying Activity
   A. SMRT, NCoR, and SMRTER
   B. RIP140 and LCoR
VI. Conclusion
   References

Nuclear receptors (NRs) compose one of the largest known families of eukaryotic transcription factors and, as such, serve as a paradigm for understanding the fundamental molecular mechanisms of eukaryotic transcriptional regulation. The packaging of eukaryotic genomic DNA into a higher ordered chromatin structure, which generally acts as a barrier to transcription by inhibiting transcription factor accessibility, has a major influence on the mechanisms by which NRs activate or repress gene expression. A major breakthrough in the field's understanding of these mechanisms comes from the recent identification of NR-associated coregulatory factors (i.e., coactivators and corepressors). Although several of these NR cofactors are involved in chromatin remodeling and facilitating the recruitment of the basal transcription machinery, the focus of this chapter is on NR coactivators and corepressors that act to covalently modify the amino-terminal tails of core histones. These modifications (acetylation, methylation, and phosphorylation) are thought to directly affect chromatin structure and/or serve as binding surfaces for other coregulatory proteins. This chapter presents the most current models for NR recruitment of histone-modifying enzymes and then summarizes their functional importance in NR-associated gene expression. © 2004 Elsevier Inc.

I. OVERVIEW OF NUCLEAR RECEPTORS

Nuclear receptors (NR) are structurally related DNA binding transcription factors sharing two conserved modules: a core Cys2–Cys2 zinc Finger DNA binding domain (DBD) and a carboxy-terminal ligand binding domain (LBD) (Fig. 1). Located in the LBD is a highly conserved activation function-2 (AF-2) domain that plays an essential role in facilitating ligand-dependent transcriptional activation (Glass and Rosenfeld, 2000). Genetic studies indicate that NRs are involved in a wide array of biologic processes, including embryonic pattern formation, neuronal development, metamorphosis, sexual differentiation, and metabolic control (Mangelsdorf and Evans, 1995; Thummel, 1995). Several mutant NRs have been implicated in human diseases, including leukemia, breast and prostate cancers, and neurological dysfunctions (Huang and Tindall, 2002; Iwase, 2003; Lin et al., 2001; Wondisford, 2003; Zelent et al., 2001).

Searches for additional NRs in genome databases using the consensus sequences of DBDs and LBDs, resulted in the identification of 48 human NRs (Robinson-Rechavi et al., 2001), 21 Drosophila NRs (Adams et al., 2000), and ~270 C. elegans NRs (Sluder et al., 1999). Because no NRs have been found in nonmetazoans, NRs appear to have evolved only in multicellular organisms. Sequence comparison of the identified NRs reveals that, while most Drosophila NRs share homologues in vertebrates, the majority of C. elegans NRs are unique to that species. Neither the large number of C. elegans NRs nor their divergence from other model systems was

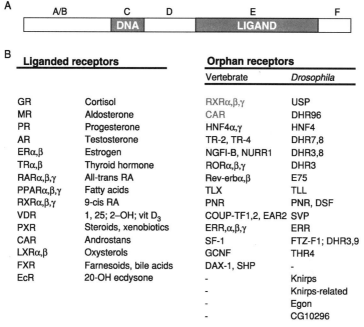

FIGURE 1. Nuclear receptor superfamily. (A) Schematic diagram showing the organization and conserved domains of nuclear hormone receptors. The conserved DNA binding domain (DBD) and the ligand binding domain (LBD) are highlighted in black boxes. Nuclear receptors are also divided into A/B, C, D, E, and F subdomains. (B) A list of known human and *Drosophila* nuclear receptors. The liganded NRs (left) and the orphan NRs (right) are shown is the columns. Most of the orphan NRs are conserved in evolution. The only liganded NR that has been identified in insects is EcR.

expected. The existence of NRs in *C. elegans* has provoked the speculation that they are part of a defensive mechanism that enables *C. elegans* to survive in its environment because the organisms are exposed to many different toxic chemical compounds.

The majority of identified NRs are orphan receptors in that their corresponding cognate ligands have not yet been identified. With the exception of the ecdysone receptor in insects, all NRs currently known to bind ligands are found in vertebrates. This has led to the idea that ancient NRs were all orphan NRs, some of which then derived their ligand-binding abilities during the course of evolution. So far, the known ligands for NRs are steroids, retinoids, thyroid hormones, vitamin D_3 various metabolites, and xenobiotics. Because of the ligand-inducible characteristics of NRs and because NRs are linked to a wide array of biologic pathways and diseases, much effort has been invested in identifying specific NR agonists and antagonists that might be used to develop drugs, or compounds for disease treatment, and for agricultural control.

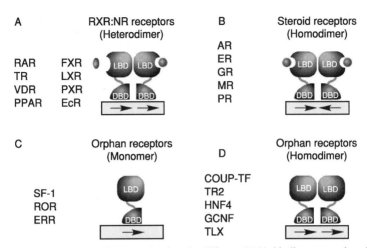

FIGURE 2. Schematic diagrams showing the different DNA binding properties of NRs. (A) Type I liganded NRs, such as RAR and TR, interact with RXR and form a heterodimer on DNA. (B) Type III liganded NRs, such as ER and AR, form a homodimer on DNA. (C) Certain orphan NRs, such as SF-1, bind as a monomer. (D) Type II orphan NRs, such as HNF4, COUP-TF, and TLX, form a homodimer on DNA.

Phylogenetic analysis of the identified NR sequences has defined six subfamilies among the NR superfamily (Laudet, 1997). The NRs in each class share similar properties in binding DNA and in forming distinctive protein–DNA complexes (Fig. 2). For example, class III NRs, such as the estrogen receptor (ER) and the androgen receptor (AR), favor homodimer formation. Class I NRs, such as thyroid hormone receptor (TR), retinoid acid receptor (RAR), and vitamin D receptor (VDR), heterodimerize with the retinoid X receptor (RXR). Class V NRs, such as steroidogenic factor-1 (SF-1), act as monomers. In each case, the resulting NR–DNA complexes act as scaffolds for the assembly of large protein complexes (corepressor or coactivator complexes) that include enzymes shown to covalently modify histones. The aim of this chapter is to give an overview on corepressors, coactivators, and histone-modifying enzymes that are functionally associated with NRs.

II. CHROMATIN STRUCTURE

The genomic DNA in all eukaryotic cells is condensed and packaged by histone and nonhistone proteins into a dynamic ordered structure termed *chromatin*. The basic unit of chromatin is the nucleosome (Fig. 3), which contains approximately 146 base pairs (bp) of DNA wrapped in a left-handed superhelix around an octamer of core histone proteins containing two molecules each of the following histones: H2A, H2B, H3, and H4

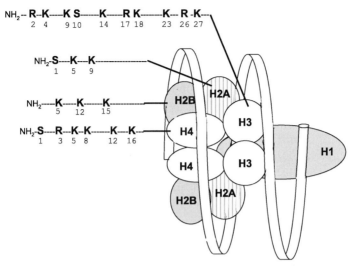

FIGURE 3. Structure and assembly of the nucleosome. The basic unit of chromatin is the nucleosome particle containing 146 bp of DNA wrapped around a core histone octamer containing two molecules each of histones H2A, H2B, H3, and H4. The amino-terminal (NH_2) tails extend radially from the core particle and contain specific sites for post-translational modifications. Specific amino acid residues (lysine, K; serine, S; arginine, R) demonstrated previously to be substrates for post-translational modifications (acetylation, methylation, and phosphorylation) are indicated.

(Luger et al., 1997). H3 and H4 form heterodimers via conserved histone fold motifs and then assemble into (H3–H4)$_2$ tetrameric particles. Histones H2A and H2B likewise heterodimerize and become associated with either side of the tetramer particle, generating the core histone octamer around which the genomic DNA is wound (Horn and Peterson, 2002; Morales et al., 2001; Urnov and Wolffe, 2001) (Fig. 3). Each core histone has a hydrophilic amino-terminal tail containing specific sites for post-translational modifications (see subsequent text). The amino-terminal tails appear to emanate radially out of the nucleosome particle (Hansen, 2002) and, as such, are capable of functional interactions with DNA, adjacent nucleosomes, and specific chromatin-binding cofactors. In addition to core histones, linker histones (e.g., H1) can serve to lock the incoming and outgoing DNA helix to the outside of the core histone octamer, further stabilizing the nucleosome particle.

In the genome, each nucleosome is separated from the next by a stretch of linker DNA varying in length from 10 to 60 bp. This form of DNA packaging (termed *beads-on-a-string*) is considered the primary functional unit of chromatin. *In vivo*, arrays of nucleosomes are packaged into canonical '30-nm' fibers and then further condensed into a higher level of chromatin structure characterized by 80- to 100-nm chromonema fibers

(Horn and Peterson, 2002). Specific nucleosome-nucleosome interactions are essential for the condensation of nucleosome arrays into higher ordered chromatin structures (Hansen, 2002; Luger *et al.*, 1997). In addition, interactions among linker histones are thought to stabilize intramolecular chromatin folding and further promote fiber–fiber interactions (Horn and Peterson, 2002; Ramakrishnan, 1997; Thomas, 1999; Wolffe *et al.*, 1997).

The packaging of genomic DNA into higher ordered chromatin presents an obstacle for regulated gene expression by presumably restricting access of RNA polymerase II and of the basal transcription machinery (Horn and Peterson, 2002; Urnov and Wolffe, 2001). Not surprisingly, activation and repression of transcription is achieved by promoter-specific recruitment of enzymes that disrupt or alter chromatin structure. In general, two types of enzymatic activities are recruited to target promoters. The first type includes adenosine triphosphate (ATP)-dependent chromatin remodeling complexes, such as the SWI/SNF or NURF complexes, which remodel chromatin structure by actively mobilizing nucleosomes (Kingston and Narlikar, 1999; Peterson and Workman, 2000). The second type includes enzymes that covalently modify core histones at specific amino acid residues, typically at the amino-terminal tails (Fig. 3). The modifications, including acetylation, methylation, phosphorylation, and ubiquitination (see section III), are believed to directly affect chromatin structure and/or serve as a binding surface for other coregulatory proteins (Jenuwein and Allis, 2001).

Chromatin structure clearly plays an essential role in attaining appropriate ligand-dependent transcriptional regulation by NRs (Kraus and Wong, 2002; Urnov and Wolffe, 2001). Although the recruitment of ATP-dependent chromatin remodeling complexes to NR-target genes is just beginning to be understood (Belandia and Parker, 2003; Hebbar and Archer, 2003), a great wealth of evidence demonstrates that NRs directly bind to and functionally interact with distinct histone-modifying enzymes and cofactors. An overview of these proteins is presented in Sections IV and V.

III. HISTONE MODIFICATIONS IN REGULATED TRANSCRIPTION

A. ACETYLATION

Acetylation of specific lysine residues occurs most commonly in the amino-terminal tails of core histones H3 and H4 and to a lesser extent in H2A and H2B (Figs. 3 and 4) (Berger, 2002). In the case of H3, the main acetylation sites include lysines 9, 14, 18, and 23 (Fig. 4) (Jenuwein and Allis, 2001; Thorne *et al.*, 1990). The specific residues flanking the preferred lysine acetylation sites appear to be important for targeting distinct histone acetyltransferase (HAT) enzymes (Kimura and Horikoshi, 1998). In general, there is a strong correlation between the acetylation of specific lysine residues

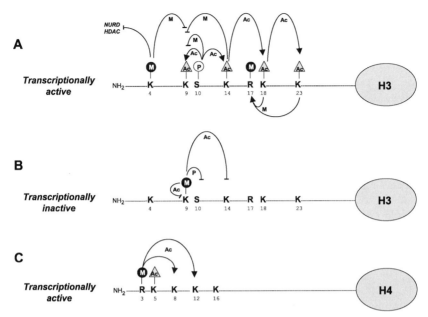

FIGURE 4. Patterns of histone modifications in transcriptionally active or inactive chromatin. Arrows indicate how modifications at specific histone amino acid residues can promote the occurrence of a subsequent modifications at other residues. Blocked arrows indicate the inhibition of a subsequent modification. Modifications include lysine (K) and arginine (R) methylation (M), lysine acetylation (Ac), and serine (S) phosphorylation (P).

in H3 and H4 and activated gene expression (Fig. 4) (Sterner and Berger, 2000; Wade et al., 1997). Accordingly, loss of core histone acetylation has been linked with transcriptional silencing (Kuo and Allis, 1998). Relevant to this review, many of the best-characterized NR transcriptional coactivators and corepressors are HATs and histone deacetylases (HDACs), respectively (see overview in Sections IV and V). Acetylation of specific lysine residues in H3 and H4 also plays an important role in other cellular processes, including DNA replication and cell cycle progression (Howe et al., 2001; Turner and O'Neill, 1995).

B. PHOSPHORYLATION

Histone phosphorylation has been implicated in mitotic chromosome condensation, activated gene expression, response to DNA damage, and induced apoptosis (Cheung et al., 2000a). Phosphorylation of H3 at serine 10 (Figs. 3 and 4) is stimulated by growth factors and mitogens and corresponds with the transcriptional induction of immediate early genes, such as c-fos (Mahadevan et al., 1991; Sassone-Corsi et al., 1999). Two different kinases, ribosomal S6 kinase 2 (RSK-2), and MAP- and stress-activated kinase 1 (MSK-1) are believed to be responsible for phosphorylating H3 *in vivo* in response to activated intracellular signal transduction pathways (Sassone-Corsi et al., 1999; Thomson et al., 1999). It is important to note that phosphorylation of linker H1 has also been shown to influence transcriptional regulation by NRs (Hebbar and Archer, 2003; Kraus and Wong, 2002). Recent studies found that phosphorylated H1 confers a promoter-specific chromatin structure that allows for ligand-dependent activation by the glucocorticoid receptor, whereas loss of H1 phosphorylation confers a refractory state in which the promoter is uninducible (Lee and Archer, 1998). The kinase responsible for H1 phosphorylation in NR-mediated transcriptional activation was shown to be cyclin-dependent kinase 2 (cdk2) (Bhattacharjee et al., 2001).

C. METHYLATION

The amino-terminal tails of core histones can be methylated at specific arginine or lysine residues (Figs. 3 and 4). Histone arginine methylation is commonly associated with activated gene expression (Berger, 2002). Notably, two NR coactivators, CARM1 and PRMT1 (Chen et al., 1999a; Wang et al., 2001b), possess histone methyltransferase (HMT) activity specific for arginine residues; CARM1 specifically methylates arginines 2, 17, and 26 of H3, whereas PRMT1 specifically methylates arginine 3 of H4 (Stallcup, 2001). Methylation is carried out by a family of HMTs containing conserved SET domains (Jenuwein and Allis, 2001). For example, the mammalian SUV39H1 enzyme selectively methylates lysine 9 of H3. This specific modification plays an essential role in heterochromatic gene silencing by facilitating the targeting of the repressor heterochromatin protein 1 (HP1) to chromatin (Kouzarides, 2002). By contrast, the SET9 enzyme specifically methylates lysine 4 of H3 and potentiates transcriptional activation, presumably by antagonizing H3 lysine 9 methylation and, subsequently, promoting H3 lysine acetylation by HATs (Nishioka et al., 2002; Wang et al., 2001a). Similarly ALR-1/2 and HALR, two SET domain proteins containing H3 lysine 4 methylation activity, were recently identified in a distinct coactivator complex that can be targeted to NRs in a ligand-dependent manner (Goo et al., 2003). Consistent with a functional role for histone lysine methylation in NR-regulated gene expression, another recent study shows that transcriptional repression by unliganded TR is associated with a substantial increase in methylation of H3 lysine 9 with a concomitant loss of methylation for H3 lysine 4 (Li et al., 2002).

D. PATTERNS AND CODES

Specific histone modifications typically occur in combination with other distinct types of modifications and are correlated with a specific transcriptional state (Fig. 4) (Berger, 2002; Jenuwein and Allis, 2001). The appearance of different types of modifications at specific promoters tend to occur in ordered patterns (Berger, 2002; Geiman and Robertson, 2002). For example, transcriptionally active chromatin is associated with an initial phosphorylation of H3 serine 10 that precedes and facilitates acetylation of lysines 9 and 14 and methylation of lysine 4 (Fig. 4A) (Cheung *et al.*, 2000b; Lo *et al.*, 2000; Strahl *et al.*, 1999). Interestingly, estrogen-dependent activation of the mammalian pS2 promoter by ER additionally involves acetylation of H3 lysines 18 and 23, followed sequentially by methylation of arginine 17 (Daujat *et al.*, 2002). By contrast, transcriptionally inactive chromatin is associated with deacetylation of H3 lysine 14, followed by methylation of lysine 9 (Fig. 4B). In turn, methylation of lysine 9 antagonizes phosphorylation of serine 10 and acetylation of lysine 14 (Rea *et al.*, 2000). Ordered patterns of amino-terminal modifications on H4 also appear to affect gene expression (Berger, 2002; Jenuwein and Allis, 2001). For instance, ligand-dependent transcriptional activation by the AR is associated with an initial methylation of H4 arginine 3 that precedes and facilitates acetylation of lysines 8 and 12 (Wang *et al.*, 2001b).

The correlation between specific histone modifications and differential transcriptional states is clear, yet the underlying mechanisms by which these modifications influence gene expression remain poorly defined. One model for the function of the modifications is that they directly affect the structure of nucleosomes and, hence, chromatin structure (Berger, 2002). In accordance with the model, specific modifications (e.g., acetylation or phosphorylation) neutralize the overall positive charge in the histone tails, causing changes in the association of the core histone octamer with DNA. This presumably renders the chromatin more "accessible" to various regulatory transcription factors and the RNA polymerase II basal apparatus (Anderson *et al.*, 2001). Furthermore, electrostatic alterations in histone tails may also affect nucleosome-nucleosome interactions, thus generating a more open or closed chromatin structure.

Another model for the functional role of histone modifications is that they serve as epigenetic markers. This concept, termed the *histone code hypothesis*, predicts that distinct histone modifications serve as binding sites for effector proteins (Jenuwein and Allis, 2001). In strong support of this model, several transcriptional regulatory proteins contain conserved *bromo-* and *chromo-*domains that selectively interact with specific modifications in the histone amino-terminal tail. The bromodomain (Dhalluin *et al.*, 1999) selectively interacts with acetylated lysines and is found in SNF2, TAFII250, the mammalian trithorax protein, and several HATs (e.g., GCN5, p300/CBP, PCAF). The chromodomain (Jones *et al.*, 2000) selectively interacts with methylated lysines (and possibly arginines) and is found in HP1, Polycomb (Pc) proteins, the HAT EsaI, and several HMTs (e.g., SUV39H1).

Consistent with the idea that core histone modifications can serve as binding sites for effector molecules, the repressor protein HP1 recognizes the methylated H3 lysine 9 marker through its chromodomain (Fig. 5A) and subsequently facilitates heterochromatin formation (Kouzarides, 2002). Similarly, and consistent with the notion that acetylated core histones facilitate transcriptional initiation (Berger, 2002), a subunit of the general transcription factor complex TFIID (TAFII250) is targeted at acetylated lysine residues via internal bromodomains. It is important to note in this regard that TAFII250, which itself can act as a HAT, contains two tandem bromodomains that likely account for its preferential binding to dual acetylated lysine residues that are appropriately spaced (Fig. 5B) (Jacobson et al., 2000).

Relevant to the observation that sequential chromatin-modifying and chromatin-remodeling events take place at specific target genes (Cosma, 2002), some HATs contain chromodomains (Roth et al., 2001), possibly accounting for why HMT activity precedes HAT activity at specific target genes (Fig. 5C) (Jenuwein and Allis, 2001). Notably, the core ATPases of the yeast and human SWI/SNF complex (SNF2 and hbrm, respectively) contain bromodomains that can stably target the entire complex to acetylated chromatin templates in vitro (Hassan et al., 2001) (Fig. 5D). These findings are consistent with other data showing that at specific yeast and mammalian

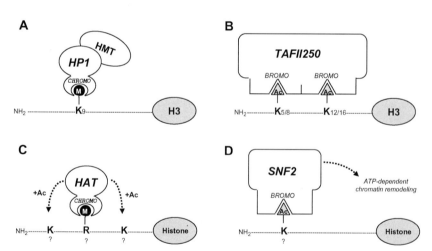

FIGURE 5. Histone modifications as binding sites for effector proteins. Several transcriptional regulatory proteins contain conserved domains (termed bromo- and chromodomains) that selectively interact with specific modifications in the histone amino-terminal tail. (A) The heterochromatin protein 1 (HP1) binds to methylated H3 lysine 9 via its chromodomain and subsequently facilitates heterochromatin formation (Kouzarides 2002). (B) TAFII250 is targeted to acetylated lysine residues via two tandem bromodomains that bind to dual acetylated lysine residues that are appropriately spaced (Jacobson et al., 2000). (C) This diagram shows the proposed interaction between a methylated histone and a chromodomain-containing HAT, followed by acetylation of proximal lysine residues. (D) The core ATPases of the yeast and human SWI/SNF complex (SNF2 and hbrm) contain bromodomains that may stably target the holocomplex to acetylated histones (Hassan et al., 2001).

target genes, the temporal recruitment of HAT activity (and hence histone acetylation) occurs prior to the recruitment of ATP-dependent chromatin remodeling complexes (Agalioti *et al.*, 2000; Dilworth *et al.*, 2000; Reinke *et al.*, 2001). Finally, it is important to note that the yeast Tup1/Ssn6 repressor complex selectively associates with hypoacetylated H3 and H4 (Edmondson *et al.*, 1996), thus indicating that unmodified core histones themselves can act as epigenetic markers.

IV. NR COACTIVATORS WITH HISTONE-MODIFYING ACTIVITY

A. THE P160/SRC FAMILY

Several different types of histone-modifying enzymes can be targeted to NRs in a ligand-dependent manner through conserved adaptor proteins termed the p160/SRC coactivators. Three homologous p160/SRC gene family members have been identified in humans and rodents and are referred to as SRC-1/NCoA-1, TIF2/GRIP1/NCoA-2, and AIB1/RAC3/ACTR/TRAM-1 (Chen, 2000a; Glass and Rosenfeld, 2000; McKenna *et al.*, 1999; Xu and Li, 2003) (Fig. 6A). Remarkably, all three family members can interact with a broad range of NRs in a ligand-dependent fashion and can markedly enhance NR-mediated transcription both *in vitro* and *in vivo*. Each member of this family has a central NR-interaction domain containing three copies of a consensus leucine-rich motif, LXXLL (also termed NR box) (Fig. 6A). Crystallographic and biochemical studies reveal that the surface of a single LXXLL motif directly contacts the ligand-activated AF-2 domain of an NR, thus providing a molecular basis for NR–p160/SRC recruitment (Glass and Rosenfeld, 2000). In addition to NRs, various members of the p160/SRC family can interact with other transcriptional activators including AP-1, SRF and NFκB (reviewed in Xu and Li, 2003).

Members of the p160/SRC family contain a highly conserved basic helix–loop–helix (bHLH) domain and a Per/Arnt/Sim (PAS) homology domain at their amino-terminal (Fig. 6A). The bHLH/PAS domains are believed to mediate homodimeric and heterodimeric interactions and have been implicated as potential target sites for other non-NR activators (McKenna *et al.*, 1999; Xu and Li, 2003). Most important, the p160/SRC family members contain two intrinsic activation domains (AD1 and AD2) that retain their activity when transferred to a heterologous DNA binding domain (Fig. 6A). AD1 serves as a binding site for the HAT enzymes p300 and CBP (Chen *et al.*, 1999b; Korzus *et al.*, 1998; Voegel *et al.*, 1998), and AD2 serves as a binding site for the HMTs CARM1 and PRMT1 (Chen *et al.*, 1999a; Koh *et al.*, 2001). Interestingly, the p160/SRC family members ACTR and SRC-1 have been reported to contain intrinsic HAT activity in their carboxy terminal that is primarily specific for histones H3 and H4 (Chen *et al.*, 1997;

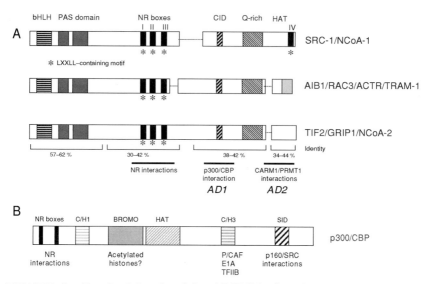

FIGURE 6. Functional domains of the p160/SRC family and p300/CBP. (A) Schematic representation of the three homologous p160/SRC gene family members SRC-1/NCoA-1, TIF2/GRIP1/NCoA-2, and AIB1/RAC3/ACTR/TRAM-1 (see text for references). bHLH:basic helix-loop–helix; PAS:Per/ARNT/Sim homologous domain; CID:CBP/p300-interacting domain; Q-rich:glutamine-rich region; HAT:histone acetyltransferase domain. AD1 and AD2 activation domains 1 and 2. Black boxes indicate consensus LXXLL motifs. (B) Schematic representation of p300/CBP. C/H1 and C/H3:cysteine/histidine-rich regions 1 and 3; BROMO: bromodomain; HAT:histone acetyltransferase; SID:SRC-interacting domain. Proteins and transcription factors proposed to interact with specific domains are indicated.

Spencer et al., 1997). This HAT activity, however, is considerably weaker than the activity of the more potent HAT p300/CBP and p/CAF and does not appear to be essential for the NR-coactivation function (Liu et al., 2001). Taken together, these findings suggest that p160/SRC coactivators have evolved to effectively target histone-modifying enzymes (HATs or HMTs) to NRs in a ligand-regulated manner and, therefore, serve as pivotal adaptor molecules in NR signaling pathways. The p160/SRC–HAT and/or p160/SRC–HMT complexes are in turn recruited to NR-target gene promoters where they then facilitate transcriptional activation.

B. THE P300 AND CBP COACTIVATORS

CBP and p300 serve as essential coactivators for a broad range of transcriptional activators and are among the best-studied HATs in the field of higher eukaryotic transcriptional regulation (Goodman and Smolik, 2000; Roth et al., 2001). The proteins lack direct DNA-binding activity but are recruited to specific target promoters by forming complexes with sequence-specific transcription factors like NRs (Glass and Rosenfeld, 2000; McKenna *et al.*, 1999). Given that p300 and CBP exhibit significant homology and are largely functionally interchangeable (Roth *et al.*, 2001), they are referred to collectively as p300/CBP. p300/CBP proteins contain several distinct and functionally important regions (Fig. 6B), including a conserved HAT domain, a bromodomain, an SRC/p160-interacting domain (SID), and multiple cysteine/histidine (C/H)-rich zinc finger domains.

The intrinsic HAT activity of p300/CBP (Bannister and Kouzarides, 1996; Ogryzko *et al.*, 1996) appears to be essential for p300/CBP's NR coactivator function because mutations in the HAT domain abolish its ability to transactivate *in vitro* and *in vivo* (Korzus *et al.*, 1998; Kraus *et al.*, 1999). All four core histones can serve as specific substrates for p300/CBP–HAT activity (Table I) (Berger, 2002). p300/CBP can also form complexes with other potent HATs, such as the p300/CBP associated factor p/CAF (Yang *et al.*, 1996) via its C/H3 domain (Fig. 6B). These HAT–HAT complexes can additionally promote NR-mediated transcription (Korzus *et al.*, 1998). The C/H3 domain is also a binding site for other potential regulatory targets (e.g., TFIIB, MyoD, Fos, RNA polymerase II), as well as for proteins that appear to regulate p300/CBP function (e.g., the adenovirus E1A protein) (Glass and Rosenfeld, 2000; Roth *et al.*, 2001). The p300/CBP bromodomain (Fig. 6B) has been reported to play an important functional role in facilitating stable p300/CBP interactions with chromatin (Manning *et al.*, 2001). Interestingly, p300/CBP also contains two LXXLL-like NR box motifs at its amino terminal (Fig. 6B) that can facilitate direct NR–p300/CBP interactions (Chakravarti *et al.*, 1996; Kamei *et al.*, 1996). As noted earlier, however, several recent studies indicate that p300/CBP is predominantly targeted to NRs indirectly via its association with SRC/p160 coactivators through its SID domain (Kraus and Wong, 2002).

Although the recruitment of NR–p160/SRC–p300/CBP complexes to specific NR-target genes *in vivo* clearly results in promoter-proximal histone acetylation (Chen *et al.*, 1999b; Shang *et al.*, 2000; Sharma and Fondell, 2002),

TABLE I. NR Coactivators with Histone-Modifying Activity

Cofactor	Modification	Histones modified	Additional interacting cofactors	Effect on transcription
p300/CBP	Lysine acetylation	H2A, H2B, H3, H4	p160/SRC proteins	Activation
SRC-1, ACTR	Lysine acetylation	H3, H4	p300/CBP	Activation
ALR-1/2, HALR	Lysine methylation	H3 lysine 4	ASC-2 Complex	Activation
CARM1	Arginine methylation	H3 arginine 3, 17, 26	p160/SRC proteins	Activation
PRMT1	Arginine methylation	H4 arginine 3	p160/SRC proteins	Activation

the subsequent mechanisms that activate transcription remain unclear. Several lines of evidence suggest that histone acetylation at NR-target promoters may facilitate the ordered recruitment of other types of transcriptional regulatory complexes. First, recent experiments examining RXR/RAR transactivation *in vitro* revealed a sequential requirement for HAT activity followed by the SWI/SNF chromatin remodeling activity (Dilworth et al., 2000). As noted earlier, such recruitment might involve direct stabilizing interactions between the bromodomain of the SWI/SNF subunit SNF2 and acetylated histones (Hassan et al., 2001) or, alternatively, direct interactions between SWI/SNF subunits and NRs (Belandia and Parker, 2003). Second, estrogen-dependent activation of the mammalian pS2 promoter by ER occurs in a stepwise fashion initially involving the recruitment of p300/CBP-HAT activity, followed sequentially by the recruitment of CARM1 HMT activity (Daujat et al., 2002). Third, recruitment of the TRAP/Mediator coactivator complex to specific TR-target promoters occurs after the recruitment of p160/SRC-p300/CBP complexes (Sharma and Fondell, 2002). Given that the TRAP/Mediator complex facilitates direct functional interactions with the RNA polymerase II basal apparatus, this step may represent an endpoint in a multistep NR-coactivator recruitment pathway (Ito and Roeder, 2001).

In addition to facilitating histone acetylation at NR-target gene promoters, p300/CBP can also directly acetylate the p160/SRC coactivator ACTR (Chen et al., 1999b) and probably other members of the p160/SRC family (Sharma and Fondell, 2002). The acetylation of ACTR neutralizes the positive charges of two lysine residues adjacent to the core LXXLL motif and disrupts the association of HAT coactivator complexes with promoter-bound NRs. These findings indicate that p300/CBP may play an important role in attenuating hormone-activated gene expression by exerting so-called factor acetyltransferase (FAT) activity.

C. CARM1 AND PRMT1

The coactivator arginine methyltransferase 1 (CARM1) was first isolated as a cofactor that physically interacts with the AD2 domain of p160/SRC coactivators (Chen et al., 1999a). Subsequent studies found that CARM1 markedly enhanced transcriptional activation by a broad range of NRs but only when p160/SRC proteins are present (Chen et al., 2000b; Koh et al., 2001). CARM1 exhibits extensive homology to a conserved family of protein arginine methyltransferases termed PRMTs (Kouzarides, 2002). Members of the PRMT family form homodimers or homo-oligomers and catalyze the transfer of the methyl moiety from S-adenosyl-L-methionine (AdoMet) onto the guanidino group of arginines in protein substrates. Peptide mapping experiments show that CARM1 can specifically methylate arginines 2, 17, and 26 of histone H3 (Schurter et al., 2001). Similarly, another member of the

PRMT family, PRMT1, was biochemically purified from HeLa cell nuclear extracts and found to specifically methylate arginine 3 of histone H4 (Wang et al., 2001b). PRMT1 is analogous to CARM1 because it can act as a potent NR-coactivator in a manner that requires the presence of a p160/SRC family member. Transient overexpression of both CARM1 and PRMT1 can synergistically activate NR-mediated gene expression (Koh et al., 2001).

Like all members of the PRMT family, CARM1 and PRMT1 share a highly conserved methyltransferase catalytic domain containing the AdoMet and arginine binding sites flanked by a conserved β-barrel-like domain (Zhang et al., 2000) (Fig. 7). Important to note is that point mutagenesis in the AdoMet binding region of CARM1 results in loss of methyltransferase and coactivator activities (Chen et al., 1999a). Deletion mutagenesis of the CARM1 protein further revealed that the p160/SRC binding site and the homo-oligomerization domain significantly overlap with the methyltransferase domain (Teyssier et al., 2002) (Fig. 7). In addition, CARM1 contains a unique carboxy-terminal activation region that appears to function autonomously of HMT activity (Teyssier et al., 2002). Although CARM1 and PRMT1 show functional synergy at NR-target genes, the existence of tertiary p160/SRC–CARM1/PRMT1 complexes has not yet been demonstrated.

Recent studies have clearly correlated arginine-specific H3 methylation with transcriptionally active NR-target genes *in vivo* (Daujat et al., 2002; Ma et al., 2001), yet the functional consequences of these methylation marks

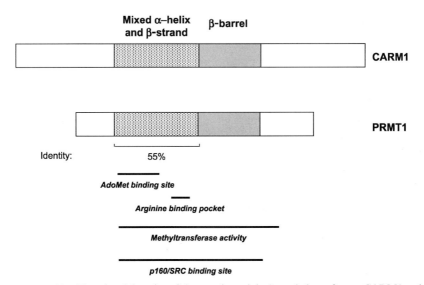

FIGURE 7. Functional domains of the protein arginine's methyltransferases CARM1 and PRMT1. The enzymatic activity is localized in a conserved catalytic domain containing a mixed α–helix and β-strand flanked by a β-barrel domain. AdoMet (S-adenosyl-L-methionine), arginine, and p160/SRC binding sites are shown.

remain unclear. Similar to the situation described earlier for HATs, HMT activity has been proposed to mediate the recruitment and/or functional activity of other types of transcriptional complexes (Kouzarides, 2002). In agreement with this idea, CARM1 cooperates synergistically with p300/CBP to stimulate NR-dependent transcription (Chen et al., 2000b; Lee et al., 2002) and supports the notion of a functional cross-talk between acetylation and methylation. Interestingly, there are differences between the temporal recruitment of CARM1 and PRMT1 relative to the recruitment of HAT activity. As noted earlier, HAT activity has been shown to precede and subsequently facilitate the recruitment of CARM1 activity at an ER-target gene *in vivo* (Daujat et al., 2002), whereas *in vitro* studies suggest that PRMT1 activity precedes and presumably facilitates the recruitment of HAT activity (Wang et al., 2001b). These differences in HMT temporal recruitment may reflect the different substrate specificities of CARM1 and PRMT1. For example, PRMT1 methylation of H4 arginine 3 may be a code for p300/CBP recruitment and activity, and CARM1 methylation of H3 arginine 17 may signal for the recruitment of other, as of yet unknown, coregulatory factors. Another unresolved issue is whether histone methylation can be reversed in a manner similar to the way HDACs counteract HAT activity (see Section V for a discussion). Intriguingly, the existence of these yet-to-be characterized histone demethylases may add another layer of complexity to the regulation of NR-target genes.

Finally, it is important to note that several nonhistone substrates exist for PRMT1 and CARM1. PRMT1 substrates include fibrillarin, nucleolin, several heterogeneous nuclear ribonucleoproteins (hnRNPs), and Sam 68, a substrate for Src kinase (Kouzarides, 2002). PRMT1 has also been reported to methylate the transcriptional activator STAT1 on arginine 31 as a novel requirement for IFN alpha/beta-induced transcription (Mowen et al., 2001). Interestingly, CARM1, but not PRMT1, was recently reported to methylate p300/CBP and modulate its activity (Xu et al., 2001). The methylation occurs on arginine residues proximal to the KIX domain that serves as a binding surface for the transcriptional coactivator CREB. It is proposed that CARM1-mediated methylation of the p300/CBP KIX domain blocks CREB activation by inhibiting the ability of CREB to recruit p300/CBP to target genes.

D. THE ASC-2 COMPLEX

The activating signal cointegrator-2 (ASC-2), which is amplified in some human cancers, was recently identified as a transcriptional coactivator for NRs as well as for other types of transcription factors (Goo et al., 2003). ACS-2, also termed AIB3, NcoA6, RAP250, NRC, PRIP, and TRBP (Hermanson et al., 2002), contains two signature LXXLL motifs (i.e., NR boxes) found in the p160/SRC cofactors and can directly interact with NRs in a ligand-dependent fashion. A recent report found that ASC-2 exists in a multisubunit complex containing the retinoblastoma-binding protein RBQ-3, alpha/beta–tubulins, and trithorax group proteins ALR-1, ALR-2, HALR, and ASH2 (Goo *et al.*, 2003). Interestingly, ALR-1/2 and HALR contain conserved SET domains recently implicated in H3 lysinespecific methylation (Jenuwein and Allis, 2001). Indeed, recombinant ALR-1, HALR, and the purified ASC-2 complex exhibit methylation of H3-lysine 4 *in vitro*, and transactivation by the retinoic acid receptor appears to involve ligand-dependent recruitment of the ASC-2 complex with concomitant methylation of H3-lysine 4 at the promoter region *in vivo*. Thus, the ASC-2 complex may represent a distinct coactivator complex for NRs.

V. NR COREPRESSORS WITH HISTONE-MODIFYING ACTIVITY

A. SMRT, NCoR, AND SMRTER

The silencing mediator of retinoid and thyroid hormone receptors (SMRT) and the nuclear receptor corepressor (NCoR) are two large and related transcriptional corepressors first isolated by virtue of their interaction with RAR and TR (Chen and Evans, 1995; Horlein *et al.*, 1995; Sande and Privalsky, 1996). SMRT and NCoR bind unliganded TR or RAR, and their interactions are disrupted upon TR's or RAR's binding to their respective ligands. Subsequent studies have revealed that SMRT and NCoR also interact with various other nuclear receptors, including VDR, PPARδ, and LXR, and with orphan NRs, such as Rev-ErbA, COUP-TF, RORα, and DAX (McKenna *et al.*, 1999). Recent results have also made it clear that SMRT and/or NCoR also interact with steroid hormone receptors, including ER, AR, and PR (Dotzlaw *et al.*, 2002; Jackson *et al.*, 1997; Liao *et al.*, 2003; Shang *et al.*, 2000; Zhang *et al.*, 1998). In the latter cases, the interactions, however, only take place when the steroid hormone receptors bind their corresponding antagonist. For example, tamoxifen, a known mixed antagonist of ER and compound used for breast cancer treatment, can enhance the interaction between ER and NCoR. Apparently, conformational changes in NR LBD structures that take place in response to agonist or antagonist binding will affect an NR's affinity for SMRT and NCoR (Table II). Nevertheless, these results demonstrate that both SMRT and NCoR are common corepressors of NRs. The interaction between SMRT or NCoR and NRs is dictated by two NR-interacting motifs located at the C-terminal ends of both proteins, with a consensus sequence of L/IXXI/VI, named the CoRNR motif (Hu and Lazar, 1999).

The functions of SMRT and NCoR and the regulatory pathways that these two proteins are involved in are now known to be even broader than previously thought. Various studies have revealed that SMRT and NCoR interact with other types of transcription factors, including CBF1, PLZF,

TABLE II. NR Corepressors

Corepressors	Drosophila homologue	Conserved domains	NR–corepressor interaction
SMRT/NCoR	SMRTER	SANT domain	Disrupted by agonist
			Enhanced by antagonist
RIP140	—		Enhanced by agonist
LCoR	Eip93F	PSQ motif	Enhanced by agonist

Bcl6/LAZ3, ETO, Ski, Stat5, MeCP2, and Kaiso (Hong et al., 1997; Kao et al., 1998; Lin et al., 1998; Melnick et al., 2000; Nakajima et al., 2001; Nomura et al., 1999; Stancheva et al., 2003; Yoon et al., 2003). A recent study also indicated that NCoR is a target of the polyglutamine disease protein Huntingtin (Boutell et al., 1999). The versatility that SMRT and NCoR display in associating with such a wide range of transcription factors indicates that they likely represent basic components in transcriptional repression. Given the large size of these two proteins, it has been proposed that they, like p300/CBP, may serve as platform proteins that integrate signals from multiple transcription factors. Cross-talk among different transcription factors, including NRs, may thus be established through their mutual interactions with SMRT and NCoR.

A similar nuclear protein, called SMRTER, has also been identified in *Drosophila* through a yeast two-hybrid screening using *Drosophila* EcR as bait (Tsai et al., 1999). As is the case for the interactions between SMRT or NCoR and vertebrate NRs, the interaction between SMRTER and EcR is also disrupted by hormone binding. The identification of a SMRT-related factor in *Drosophila* has established that recruiting SMRT-related factors is a conserved property of NRs. Sequence comparison of SMRT, NCoR, and SMRTER (Fig. 8) reveals that they share a conserved domain called the SANT (SWI3, ADA2, NCoR, and TFIIIB-B′) domain (Aasland et al., 1996). Although this domain is repeated in both SMRT and NCoR (SANT1, 2), only one copy of the SANT domain is found in SMRTER. Although the SANT domain was first identified because of its similarity to the DBD of c-Myb, the DNA binding properties of the SANT domain, so far, have not yet been established. Interestingly, the SANT domain has also been found in several other transcriptional cofactors, including CoREST and MTA-1/2 (Andres et al., 1999; Toh et al., 2000) The high sequence conservation of the SANT domain in various transcription factors, ranging from yeast to humans, suggests that this conserved domain likely plays an important role in eukaryotic gene regulation.

Recent studies have revealed two properties of the SANT domain. The first property is that the N-terminal SANT domain of SMRT and NCoR is involved in associating with histone deacetylase 3 (HDAC3) (Guenther et al.,

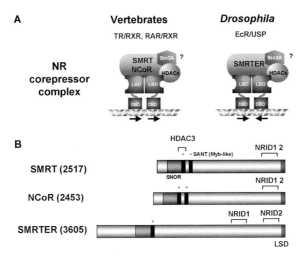

FIGURE 8. Three related corepressors SMRT, NCoR, and SMRTER (A). A schematic diagram showing the conserved NR–corepressor/HDAC complexes in vertebrates and in *Drosophila*. SMRT and NCoR are vertebrate corepressors of TR and RAR; SMRTER is a *Drosophila* corepressor of EcR. The interactions between corepressors and TR or RAR only take place in the absence of a hormone. SMRT, NCoR, and SMRTER also interact with transcriptional corepressor Sin3A in yeast and form complexes with both class I and II HDACs. (B). A schematic diagram showing the functional domains of SMRT, NCoR, and SMRTER. The conserved SANT domain, SNOR motif, and LSD motif positioned in their relative regions in each protein are highlighted with black or gray box. The mapped regions corresponding to HDAC3 and nuclear receptor interacting domain (NRID) are indicated with brackets.

2001; Zhang *et al.*, 2002) and is required for activating its deacetylase activity (Fischle *et al.*, 2002; Guenther *et al.*, 2001). The second finding is that the SANT domain is involved in histone binding. Results were first reported in a study that showed the SANT domain of Ada2 can augment the HAT activity of the associated GCN5 and the binding of GCN5 to histone tail. This observation was later confirmed by a study on the SANT2 domain of SMRT (Yu *et al.*, 2003), showing that the SANT2 domain can bind histone directly. Interestingly, SANT2 of SMRT prefers to bind unacetylated histone over the acetylated form. More surprisingly, this property was not observed for the SANT1 domain of SMRT, indicating that each of these two SANT domains encodes distinctive properties. The preferential binding of the SMRT SANT2 domain to unacetylated histone tails suggest that it can block the binding of HATs to histones. Together with the finding that the SMRT SANT 1 domain binds HDAC3, it appears that the two SANT domains of SMRT and NCoR can synergize with each other to promote and maintain histone deacetylation.

Beyond the SANT domain, SMRT, NCoR, and SMRTER also share additional conserved domains or motifs, including a highly charged SNOR (*SMRT* and *NCoR*) domain; an ITS motif, a D/ER stretch, and a octapeptides GSI motif; and a LSD motif at the C-terminal end (Tsai et al., 1999). Currently, other than the LSD motif, little is known about the exact functions of these conserved motifs among SMRT and its related factors. Interestingly, a *C. elegans* protein, called GEX-interacting protein-8 (Gei-8), also encodes the SNOR, SANT, GSI, and LSD-like motifs. The presence of these hallmark motifs of SMRT-related factors in Gei-8, whose function has not been defined, raises the possibility that Gei-8 may represent a corepressor for *C. elegans's* NRs.

One aspect of the LSD motif's function was recently uncovered through the identification of an LSD-binding factor, SMRT/HDAC1 associated repressor protein (SHARP) (Shi et al., 2001). SHARP is related to Split ends (Spen)—a *Drosophila* protein involved in Wingless signaling (Lin et al., 2003). Both proteins share repeated NRA recognition motifs (RPMs) and a highly conserved C-terminal Spen paralog and ortholog C-terminal (SPOC) domain (Ariyoshi and Schwabe, 2003). The interaction between SMRT and SHARP is mediated through the conserved SPOC domain (Ariyoshi and Schwabe, 2003; Shi et al., 2001). Another recent study indicated that SHARP associates with SMRT interacting factors, including CBF1 and HDACs (Oswald et al., 2002). It is thus possible that one property of SHARP is to stabilize the interactions between SMRT and its associated proteins, therefore ensuring that SMRT can efficiently carry out its potent transcriptional repression effects. Unexpectedly, SHARP also binds steroid receptor RNA coactivator via its RPM motifs and suppresses SRA-potentiated transcriptional activity (Shi et al., 2001). This latter finding suggests the following scenario for SHARP: not only does it enhance the repression caused by SMRT or NCoR, but it also actively attenuates ligand-induced transcription by disrupting the activities of coactivator(s).

Both the SMRT and NCoR complexes are estimated to be 1.5–3MDa in size (Guenther et al., 2000; Li et al., 2000), suggesting that SMRT and NCoR associate with multiple protein components. Biochemical purification and characterization of these SMRT and NCoR associating proteins have identified HDAC3 (Guenther et al., 2000; Li et al., 2000), transducing beta-like protein 1 (TBL1), and TBL1 related protein (TBLR1) as common components (Guenther et al., 2000; Li et al., 2000; Yu et al., 2003). Further characterization of TBL1 and TBLR1, which are related WD40 repeat proteins, has revealed their selective affinity for histones H2B and H4 (Yu et al., 2003). It was found that the histone-binding ability of TBL1 and TBLR1 is important for their transcriptional repression effect. The involvement of TBL1 and TBR1 in the TR repression pathway was further revealed by the siRNA approach. Repression by TR was efficiently relieved when cells were treated with siRNA against both proteins (a less significant effect was observed when only single siRNA was used). This *in vivo* result suggests a redundant yet essential role for TBL1 and TBLR1 in mediating the repression brought about by TR. A similar effect on TR repression was also observed when siRNA against HDAC3 was used. Together, these results reveal that TBL, TBLR1, and HDAC3 are integral components in the SMRT or NCoR complex and are critical for the transcriptional repression caused by NRs.

Intriguingly, many known interacting factors of SMRT and/or NCoR are not present in the purified SMRT and NCoR protein complex; these include many of the transcription factors mentioned above, such as Sin3A (Alland et al., 1997; Nagy et al., 1997) and class II HDACs (Huang et al., 2000; Kao et al., 2000). The lack of detectable Sin3A in purified SMRT and NCoR complexes was particularly surprising given the fact that Sin3A interacts strongly with SMRT and NCoR in yeast. Moreover, a similar interaction profile also exists for SMRTER and *Drosophila* Sin3A (Pile and Wassarman, 2000; Tsai et al., 1999). It is possible that the discrepancy reflects an unfavorable binding condition for Sin3A and SMRT or NCoR in protein purification procedures or, alternatively, that the interaction between Sin3A and SMRT or NCoR only exists in certain cell lines. Future studies will need to resolve these puzzling yet suggestive discordant results.

B. RIP140 AND LCoR

Although agonist binding to NRs is mostly associated with recruitment of coactivators, recent studies show that agonist binding also results in corepressor binding. This section describes two of these agonist-binding dependent corepressors: RIP140 and LCoR (Cavailles et al., 1995; Fernandes et al., 2003) (Table II). Although RIP140 and LCoR are structurally unrelated, they share many similar properties. Most notably, they both encode transcriptional corepressors, they associate with HDACs and C-terminal binding protein (CtBP), and their association with NRs is enhanced by hormone binding. CtBP is known to be a corepressor for a number of transcription factors such as E1A, *Knirp* (Chinnadurai, 2002).

RIP140 was first identified as a coactivator of a chimeric ER in a ligand-dependent manner (Cavailles et al., 1995). Later studies indicated that RIP140 also interacts with TR2 constitutively (Lee et al., 1998). In addition, RIP140 interacts with TR, RAR, RXR, and PPAR in a ligand-dependent manner (Treuter et al., 1998). Unexpectedly, in a GAL4 reporter system, RIP140 was found to encode a transcriptional repressor instead of a transcriptional activator (Lee et al., 1998). This finding has reversed the previous view that RIP140 is a coactivator of NRs. This landmark finding was of particular significance, since it provides a new model for how coregulators interact with NRs.

The interaction between RIP140 with NRs, such as RAR, is mediated through a unique motif at the C-terminal region of the protein (Wei et al., 2001), although the constitutive binding appears to be mediated through the repeated LXXLL motifs (Farooqui et al., 2003). Mutations in the NR-interacting motif in RIP140 compromise its ability to suppress an

RA-responsive reporter gene, suggesting that RIP140 indeed functions as an NR corepressor. Moreover, RIP140 suppresses RA receptor-mediated RA induction in a dosage-dependent manner (Wei *et al.*, 2001). Transcriptional repression by RIP140 has been attributed to its interaction with HDAC1, HDAC3 (Wei *et al.*, 2000), and CtBP (Vo *et al.*, 2001).

LCoR is another NR cofactor that was first isolated as a ligand-dependent interacting factor of ERα LBD in a yeast two-hybrid screening (Fernandes *et al.*, 2003). Its interaction with ER in yeast and in mammalian cell cultures takes place in an estradiol binding-dependent manner. Subsequent studies have further indicated that LCoR interacts with various other nuclear receptors, including GR, PR, and VDR, in a ligand binding-dependent fashion. LCoR bears a repeated PSQ motif, similar to those found in other proteins such as *Drosophila* Eip93F and *Drosophila* pipsqueak. Since the PSQ helix-loop–helix domain is implicated in DNA binding for pipsqueak, this leads to the speculation that LCoR may bind DNA as well. Coupling LCoR to the GAL4 DBD reveals that it functions not as a transcriptional activator but as a transcriptional repressor. The LCoR possesses repressive properties is also supported by the finding that it interacts with CtBP through tandem consensus CtBP-interacting motifs. Mutation of CtBP binding sites in LCoR compromises the capacity of LCoR to repress ligand-dependent transcription by ERα, indicating that CtBP is an integral component of an LCoR pathway. Furthermore, LCoR also interacts selectively with HDAC3 or HDAC6 but not with HDAC1 or HDAC4.

The interactions between RIP140 or LCoR and NRs are mediated through motifs similar to the LXXLL motif. Because this motif resembles the NR motif found in many coactivators, it has been postulated that in addition to RIP140's and LCoR's active roles in repressing gene transcription, they compete with coactivators in binding the hydrophobic pocket in the LBD. This competitive property of RIP140 and LCoR is ligand-dependent. The combination of both properties could be the key reason for the rapid attenuation of transcription immediately after agonist-induced transactivation.

VI. CONCLUSION

The genomic DNA of eukaryotic cells is packaged by histones into a condensed ordered structure termed chromatin. One of the predominant mechanisms used by NRs to activate or repress target-gene transcription is via the recruitment of coregulatory factors capable of covalently modifying the amino terminal ends of histones. These modifications, including acetylation and deacetylation, methylation, and in some cases phosphorylation, are thought to alter chromatin structure and to additionally serve as epigenetic markers (i.e., a histone code) for the subsequent recruitment of other effector proteins. Members of the p160/SRC family of coactivators serve as pivotal adaptor molecules by targeting HATs and HMTs to NRs in the presence of cognate ligand where they facilitate target gene activation. By contrast, corepressor complexes containing SMRT/NCoR (or related) proteins can associate with NRs in the absence of ligand, or in the presence of specific antagonists, and facilitate transcriptional repression at target genes. Importantly, NR corepressor complexes contain HDAC activity that apparently reverses the effects of HAT action mediated by NR coactivators. In the future, the identity and functional relevance of the effector proteins that read the histone code will likely be revealed. Moreover, future studies should lead to a better understanding of how histone modifying enzymes functionally interact with other types of NR coregulatory complexes (e.g., TRAP/Mediator, SWI/SNF, 19S proteosome complex) to regulate transcription in a promoter- and tissue-specific manner.

REFERENCES

Aasland, R., Stewart, A. F., and Gibson, T. (1996). The SANT domain: A putative DNA-binding domain in the SWI/SNF and ADA complexes, the transcriptional corepressor NCoR, and TFIIIB. *Trends Biochem. Sci.* **21**, 87–88.

Adams, M. D., Celniker, S. E., Holt, R. A., Evans, C. A., Gocayne, J. D., Amanatides, P. G., Scherer, S. E., Li, P. W., Hoskins, R. A., Galle, R. F. *et al.* (2000). The genome sequence of *Drosophila melanogaster*. *Science* **287**, 2185–2195.

Agalioti, T., Lomvardas, S., Parekh, B., Yie, J., Maniatis, T., and Thanos, D. (2000). Ordered recruitment of chromatin-modifying and general transcription factors to the IFN-beta promoter. *Cell* **103**, 667–678.

Alland, L., Muhle, R., Hou, H., Jr., Potes, J., Chin, L., Schreiber-Agus, N., and DePinho, R. A. (1997). Role for NCoR and histone deacetylase in Sin3-mediated transcriptional repression. *Nature* **387**, 49–55.

Anderson, J. D., Lowary, P. T., and Widom, J. (2001). Effects of histone acetylation on the equilibrium accessibility of nucleosomal DNA target sites. *J. Mol. Biol.* **307**, 977–985.

Andres, M. E., Burger, C., Peral-Rubio, M. J., Battaglioli, E., Anderson, M. E., Grimes, J., Dallman, J., Ballas, N., and Mandel, G. (1999). CoREST: A functional corepressor required for regulation of neural-specific gene expression. *Proc. Natl. Acad. Sci. USA* **96**, 9873–9878.

Ariyoshi, M., and Schwabe, J. W. (2003). A conserved structural motif reveals the essential transcriptional repression function of Spen proteins and their role in developmental signaling. *Genes Dev.* **17**, 1909–1920.

Bannister, A. J., and Kouzarides, T. (1996). The CBP coactivator is a histone acetyltransferase. *Nature* **384**, 641–643.

Belandia, B., and Parker, M. G. (2003). Nuclear receptors: A rendezvous for chromatin remodeling factors. *Cell* **114**, 277–280.

Berger, S. L. (2002). Histone modifications in transcriptional regulation. *Curr. Opin. Genet. Dev.* **12**, 142–148.

Bhattacharjee, R. N., Banks, G. C., Trotter, K. W., Lee, H. L., and Archer, T. K. (2001). Histone H1 phosphorylation by Cdk2 selectively modulates mouse mammary tumor virus transcription through chromatin remodeling. *Mol. Cell. Biol.* **21**, 5417–5425.
Boutell, J. M., Thomas, P., Neal, J. W., Weston, V. J., Duce, J., Harper, P. S., and Jones, A. L. (1999). Aberrant interactions of transcriptional repressor proteins with the Huntington's disease gene product, Huntingtin. *Hum. Mol. Genet.* **8**, 1647–1655.
Cavailles, V., Dauvois, S., L'Horset, F., Lopez, G., Hoare, S., Kushner, P. J., and Parker, M. G. (1995). Nuclear factor RIP140 modulates transcriptional activation by the estrogen receptor. *EMBO J.* **14**, 3741–3751.
Chakravarti, D., LaMorte, V. J., Nelson, M. C., Nakajima, T., Schulman, I. G., Juguilon, H., Montminy, M., and Evans, R. M. (1996). Role of CBP/P300 in nuclear receptor signalling. *Nature* **383**, 99–103.
Chen, D., Huang, S. M., and Stallcup, M. R. (2000b). Synergistic, p160 coactivator-dependent enhancement of estrogen receptor function by CARM1 and p300. *J. Biol. Chem.* **275**, 40810–40816.
Chen, D., Ma, H., Hong, H., Koh, S. S., Huang, S. M., Schurter, B. T., Aswad, D. W., and Stallcup, M. R. (1999a). Regulation of transcription by a protein methyltransferase. *Science* **284**, 2174–2177.
Chen, H., Lin, R. J., Schiltz, R. L., Chakravarti, D., Nash, A., Nagy, L., Privalsky, M. L., Nakatani, Y., and Evans, R. M. (1997). Nuclear receptor coactivator ACTR is a novel histone acetyltransferase and forms a multimeric activation complex with P/CAF and CBP/p300. *Cell* **90**, 569–580.
Chen, H., Lin, R. J., Xie, W., Wilpitz, D., and Evans, R. M. (1999b). Regulation of hormone-induced histone hyperacetylation and gene activation via acetylation of an acetylase. *Cell* **98**, 675–686.
Chen, J. D. (2000a). Steroid/nuclear receptor coactivators. *Vitam. Horm.* **58**, 391–448.
Chen, J. D., and Evans, R. M. (1995). A transcriptional corepressor that interacts with nuclear hormone receptors. *Nature* **377**, 454–457.
Cheung, P., Allis, C. D., and Sassone-Corsi, P. (2000a). Signaling to chromatin through histone modifications. *Cell* **103**, 263–271.
Cheung, P., Tanner, K. G., Cheung, W. L., Sassone-Corsi, P., Denu, J. M., and Allis, C. D. (2000b). Synergistic coupling of histone H3 phosphorylation and acetylation in response to epidermal growth factor stimulation. *Mol. Cell* **5**, 905–915.
Chinnadurai, G. (2002). CtBP, an unconventional transcriptional corepressor in development and oncogenesis. *Mol. Cell* **9**, 213–224.
Cosma, M. P. (2002). Ordered recruitment: Gene-specific mechanism of transcription activation. *Mol. Cell* **10**, 227–236.
Daujat, S., Bauer, U. M., Shah, V., Turner, B., Berger, S., and Kouzarides, T. (2002). Cross-talk between CARM1 methylation and CBP acetylation on histone H3. *Curr. Biol.* **12**, 2090–2097.
Dhalluin, C., Carlson, J. E., Zeng, L., He, C., Aggarwal, A. K., and Zhou, M. M. (1999). Structure and ligand of a histone acetyltransferase bromodomain. *Nature* **399**, 491–496.
Dilworth, F. J., Fromental-Ramain, C., Yamamoto, K., and Chambon, P. (2000). ATP-driven chromatin remodeling activity and histone acetyltransferases act sequentially during transactivation by RAR/RXR *in vitro*. *Mol. Cell* **6**, 1049–1058.
Dotzlaw, H., Moehren, U., Mink, S., Cato, A. C., Iniguez Lluhi, J. A., and Baniahmad, A. (2002). The amino terminus of the human AR is target for corepressor action and antihormone agonism. *Mol. Endocrinol.* **16**, 661–673.
Edmondson, D. G., Smith, M. M., and Roth, S. Y. (1996). Repression domain of the yeast global repressor Tup1 interacts directly with histones H3 and H4. *Genes Dev.* **10**, 1247–1259.

Farooqui, M., Franco, P. J., Thompson, J., Kagechika, H., Chandraratna, R. A., Banaszak, L., and Wei, L. N. (2003). Effects of retinoid ligands on RIP140: Molecular interaction with retinoid receptors and biological activity. *Biochemistry* **42**, 971–979.

Fernandes, I., Bastien, Y., Wai, T., Nygard, K., Lin, R., Cormier, O., Lee, H. S., Eng, F., Bertos, N. R., Pelletier, N. *et al.* (2003). Ligand-dependent nuclear receptor corepressor LCoR functions by histone deacetylase-dependent and -independent mechanisms. *Mol. Cell* **11**, 139–150.

Fischle, W., Dequiedt, F., Hendzel, M. J., Guenther, M. G., Lazar, M. A., Voelter, W., and Verdin, E. (2002). Enzymatic activity associated with class II HDACs is dependent on a multiprotein complex containing HDAC3 and SMRT/NCoR. *Mol. Cell* **9**, 45–57.

Geiman, T. M., and Robertson, K. D. (2002). Chromatin remodeling, histone modifications, and DNA methylation—How does it all fit together? *J. Cell. Biochem.* **87**, 117–125.

Goo, Y. H., Sohn, Y. C., Kim, D. H., Kim, S. W., Kang, M. J., Jung, D. J., Kwak, E., Barlev, N. A., Berger, S. L., Chow, V. T. *et al.* (2003). Activating signal cointegrator 2 belongs to a novel steady-state complex that contains a subset of trithorax group proteins. *Mol. Cell. Biol.* **23**, 140–149.

Glass, C. K., and Rosenfeld, M. G. (2000). The coregulator exchange in transcriptional functions of nuclear receptors. *Genes Dev.* **14**, 121–141.

Goodman, R. H., and Smolik, S. (2000). CBP/p300 in cell growth, transformation, and development. *Genes Dev.* **14**, 1553–1577.

Guenther, M. G., Barak, O., and Lazar, M. A. (2001). The SMRT and NCoR corepressors are activating cofactors for histone deacetylase 3. *Mol. Cell. Biol.* **21**, 6091–6101.

Guenther, M. G., Lane, W. S., Fischle, W., Verdin, E., Lazar, M. A., and Shiekhattar, R. (2000). A core SMRT corepressor complex containing HDAC3 and TBL1, a WD40-repeat protein linked to deafness. *Genes Dev.* **14**, 1048–1057.

Hansen, J. C. (2002). Conformational dynamics of the chromatin fiber in solution: Determinants, mechanisms, and functions. *Annu. Rev. Biophys. Biomol. Struct.* **31**, 361–392.

Hassan, A. H., Neely, K. E., and Workman, J. L. (2001). Histone acetyltransferase complexes stabilize SWI/SNF binding to promoter nucleosomes. *Cell* **104**, 817–827.

Hebbar, P. B., and Archer, T. K. (2003). Chromatin remodeling by nuclear receptors. *Chromosoma* **111**, 495–504.

Hermanson, O., Glass, C. K., and Rosenfeld, M. G. (2002). Nuclear receptor coregulators: Multiple modes of modification. *Trends. Endocrinol. Metab.* **13**, 55–60.

Hong, S. H., David, G., Wong, C. W., Dejean, A., and Privalsky, M. L. (1997). SMRT corepressor interacts with PLZF and with the PML-retinoic acid receptor alpha (RARalpha) and PLZF–RARalpha oncoproteins associated with acute promyelocytic leukemia. *Proc. Natl. Acad. Sci. USA* **94**, 9028–9033.

Horlein, A. J., Naar, A. M., Heinzel, T., Torchia, J., Gloss, B., Kurokawa, R., Ryan, A., Kamei, Y., Soderstrom, M., Glass, C. K. *et al.* (1995). Ligand-independent repression by the thyroid hormone receptor mediated by a nuclear receptor corepressor. *Nature* **377**, 397–404.

Horn, P. J., and Peterson, C. L. (2002). Molecular biology. Chromatin higher order folding—Wrapping up transcription. *Science* **297**, 1824–1827.

Howe, L., Auston, D., Grant, P., John, S., Cook, R. G., Workman, J. L., and Pillus, L. (2001). Histone H3-specific acetyltransferases are essential for cell cycle progression. *Genes Dev.* **15**, 3144–3154.

Hu, X., and Lazar, M. A. (1999). The CoRNR motif controls the recruitment of corepressors by nuclear hormone receptors. *Nature* **402**, 93–96.

Huang, E. Y., Zhang, J., Miska, E. A., Guenther, M. G., Kouzarides, T., and Lazar, M. A. (2000). Nuclear receptor corepressors partner with class II histone deacetylases in a Sin3-independent repression pathway. *Genes Dev* **14,** 45–54.
Huang, H., and Tindall, D. J. (2002). The role of the androgen receptor in prostate cancer. *Crit. Rev. Eukaryot. Gene Expr.* **12,** 193–207.
Ito, M., and Roeder, R. G. (2001). The TRAP/SMCC/Mediator complex and thyroid hormone receptor function. *Trends Endocrinol. Metab.* **12,** 127–134.
Iwase, H. (2003). Molecular action of the estrogen receptor and hormone dependency in breast cancer. *Breast Cancer* **10,** 89–96.
Jackson, T. A., Richer, J. K., Bain, D. L., Takimoto, G. S., Tung, L., and Horwitz, K. B. (1997). The partial agonist activity of antagonist-occupied steroid receptors is controlled by a novel hinge domain-binding coactivator L7/SPA and the corepressors NCoR or SMRT. *Mol. Endocrinol.* **11,** 693–705.
Jacobson, R. H., Ladurner, A. G., King, D. S., and Tjian, R. (2000). Structure and function of a human TAFII250 double bromodomain module. *Science* **288,** 1422–1425.
Jenuwein, T., and Allis, C. D. (2001). Translating the histone code. *Science* **293,** 1074–1080.
Jones, D. O., Cowell, I. G., and Singh, P. B. (2000). Mammalian chromodomain proteins: Their role in genome organisation and expression. *Bioessays* **22,** 124–137.
Kamei, Y., Xu, L., Heinzel, T., Torchia, J., Kurokawa, R., Gloss, B., Lin, S. C., Heyman, R. A., Rose, D. W., Glass, C. K., and Rosenfeld, M. G. (1996). A CBP integrator complex mediates transcriptional activation and AP-1 inhibition by nuclear receptors. *Cell* **85,** 403–414.
Kao, H. Y., Downes, M., Ordentlich, P., and Evans, R. M. (2000). Isolation of a novel histone deacetylase reveals that class I and class II deacetylases promote SMRT-mediated repression. *Genes Dev.* **14,** 55–66.
Kao, H. Y., Ordentlich, P., Koyano-Nakagawa, N., Tang, Z., Downes, M., Kintner, C. R., Evans, R. M., and Kadesch, T. (1998). A histone deacetylase corepressor complex regulates the Notch signal transduction pathway. *Genes Dev.* **12,** 2269–2277.
Kimura, A., and Horikoshi, M. (1998). How do histone acetyltransferases select lysine residues in core histones? *FEBS Lett.* **431,** 131–133.
Kingston, R. E., and Narlikar, G. J. (1999). ATP-dependent remodeling and acetylation as regulators of chromatin fluidity. *Genes Dev.* **13,** 2339–2352.
Koh, S. S., Chen, D., Lee, Y. H., and Stallcup, M. R. (2001). Synergistic enhancement of nuclear receptor function by p160 coactivators and two coactivators with protein methyltransferase activities. *J. Biol. Chem.* **276,** 1089–1098.
Korzus, E., Torchia, J., Rose, D. W., Xu, L., Kurokawa, R., McInerney, E. M., Mullen, T. M., Glass, C. K., and Rosenfeld, M. G. (1998). Transcription factor-specific requirements for coactivators and their acetyltransferase functions. *Science* **279,** 703–707.
Kouzarides, T. (2002). Histone methylation in transcriptional control. *Curr. Opin. Genet. Dev.* **12,** 198–209.
Kraus, W. L., Manning, E. T., and Kadonaga, J. T. (1999). Biochemical analysis of distinct activation functions in p300 that enhance transcription initiation with chromatin templates. *Mol. Cell. Biol.* **19,** 8123–8135.
Kraus, W. L., and Wong, J. (2002). Nuclear receptor-dependent transcription with chromatin. Is it all about enzymes? *Eur. J. Biochem.* **269,** 2275–2283.
Kuo, M. H., and Allis, C. D. (1998). Roles of histone acetyltransferases and deacetylases in gene regulation. *Bioessays* **20,** 615–626.
Laudet, V. (1997). Evolution of the nuclear receptor superfamily: Early diversification from an ancestral orphan receptor. *J. Mol. Endocrinol.* **19,** 207–226.
Lee, C. H., Chinpaisal, C., and Wei, L. N. (1998). Cloning and characterization of mouse RIP140, a corepressor for nuclear orphan receptor TR2. *Mol. Cell. Biol.* **18,** 6745–6755.

Lee, H. L., and Archer, T. K. (1998). Prolonged glucocorticoid exposure dephosphorylates histone H1 and inactivates the MMTV promoter. *EMBO J.* **17**, 1454–1466.
Lee, Y. H., Koh, S. S., Zhang, X., Cheng, X., and Stallcup, M. R. (2002). Synergy among nuclear receptor coactivators: Selective requirement for protein methyltransferase and acetyltransferase activities. *Mol. Cell. Biol.* **22**, 3621–3632.
Li, J., Lin, Q., Yoon, H. G., Huang, Z. Q., Strahl, B. D., Allis, C. D., and Wong, J. (2002). Involvement of histone methylation and phosphorylation in regulation of transcription by thyroid hormone receptor. *Mol. Cell. Biol.* **22**, 5688–5697.
Li, J., Wang, J., Nawaz, Z., Liu, J. M., Qin, J., and Wong, J. (2000). Both corepressor proteins SMRT and NCoR exist in large protein complexes containing HDAC3. *EMBO J.* **19**, 4342–4350.
Liao, G., Chen, L. Y., Zhang, A., Godavarthy, A., Xia, F., Ghosh, J. C., Li, H., and Chen, J. D. (2003). Regulation of androgen receptor activity by the nuclear receptor corepressor SMRT. *J. Biol. Chem.* **278**, 5052–5061.
Lin, H. V., Doroquez, D. B., Cho, S., Chen, F., Rebay, I., and Cadigan, K. M. (2003). Split ends is a tissue/promoter specific regulator of Wingless signaling. *Development* **130**, 3125–3135.
Lin, R. J., Nagy, L., Inoue, S., Shao, W., Miller, W. H., Jr., and Evans, R. M. (1998). Role of the histone deacetylase complex in acute promyelocytic leukemia. *Nature* **391**, 811–814.
Lin, R. J., Sternsdorf, T., Tini, M., and Evans, R. M. (2001). Transcriptional regulation in acute promyelocytic leukemia. *Oncogene* **20**, 7204–7215.
Liu, Z., Wong, J., Tsai, S. Y., Tsai, M. J., and O'Malley, B. W. (2001). Sequential recruitment of steroid receptor coactivator-1 (SRC-1) and p300 enhances progesterone receptor-dependent initiation and reinitiation of transcription from chromatin. *Proc. Natl. Acad. Sci. USA* **98**, 12426–12431.
Lo, W. S., Trievel, R. C., Rojas, J. R., Duggan, L., Hsu, J. Y., Allis, C. D., Marmorstein, R., and Berger, S. L. (2000). Phosphorylation of serine 10 in histone H3 is functionally linked *in vitro* and *in vivo* to Gcn5-mediated acetylation at lysine 14. *Mol. Cell* **5**, 917–926.
Luger, K., Mader, A. W., Richmond, R. K., Sargent, D. F., and Richmond, T. J. (1997). Crystal structure of the nucleosome core particle at 2.8Å resolution. *Nature* **389**, 251–260.
Ma, H., Baumann, C. T., Li, H., Strahl, B. D., Rice, R., Jelinek, M. A., Aswad, D. W., Allis, C. D., Hager, G. L., and Stallcup, M. R. (2001). Hormone-dependent, CARM1-directed, arginine-specific methylation of histone H3 on a steroid-regulated promoter. *Curr. Biol.* **11**, 1981–1985.
Mahadevan, L. C., Willis, A. C., and Barratt, M. J. (1991). Rapid histone H3 phosphorylation in response to growth factors, phorbol esters, okadaic acid, and protein synthesis inhibitors. *Cell* **65**, 775–783.
Mangelsdorf, D. J., and Evans, R. M. (1995). The RXR heterodimers and orphan receptors. *Cell* **83**, 841–850.
Manning, E. T., Ikehara, T., Ito, T., Kadonaga, J. T., and Kraus, W. L. (2001). p300 forms a stable, template-committed complex with chromatin: Role for the bromodomain. *Mol. Cell. Biol.* **21**, 3876–3887.
McKenna, N. J., Lanz, R. B., and O'Malley, B. W. (1999). Nuclear receptor coregulators: Cellular and molecular biology. *Endocr. Rev.* **20**, 321–344.
Melnick, A. M., Westendorf, J. J., Polinger, A., Carlile, G. W., Arai, S., Ball, H. J., Lutterbach, B., Hiebert, S. W., and Licht, J. D. (2000). The ETO protein disrupted in t(8;21)-associated acute myeloid leukemia is a corepressor for the promyelocytic leukemia zinc finger protein. *Mol. Cell. Biol.* **20**, 2075–2086.
Morales, V., Giamarchi, C., Chailleux, C., Moro, F., Marsaud, V., Le Ricousse, S., and Richard-Foy, H. (2001). Chromatin structure and dynamics: Functional implications. *Biochimie* **83**, 1029–1039.

Mowen, K. A., Tang, J., Zhu, W., Schurter, B. T., Shuai, K., Herschman, H. R., and David, M. (2001). Arginine methylation of STAT1 modulates IFNalpha/beta-induced transcription. *Cell* **104**, 731–741.

Nagy, L., Kao, H. Y., Chakravarti, D., Lin, R. J., Hassig, C. A., Ayer, D. E., Schreiber, S. L., and Evans, R. M. (1997). Nuclear receptor repression mediated by a complex containing SMRT, mSin3A, and histone deacetylase. *Cell* **89**, 373–380.

Nakajima, H., Brindle, P. K., Handa, M., and Ihle, J. N. (2001). Functional interaction of STAT5 and nuclear receptor corepressor SMRT: Implications in negative regulation of STAT5-dependent transcription. *EMBO J.* **20**, 6836–6844.

Nishioka, K., Chuikov, S., Sarma, K., Erdjument-Bromage, H., Allis, C. D., Tempst, P., and Reinberg, D. (2002). SET9, a novel histone H3 methyltransferase that facilitates transcription by precluding histone tail modifications required for heterochromatin formation. *Genes Dev.* **16**, 479–489.

Nomura, T., Khan, M. M., Kaul, S. C., Dong, H. D., Wadhwa, R., Colmenares, C., Kohno, I., and Ishii, S. (1999). Ski is a component of the histone deacetylase complex required for transcriptional repression by Mad and thyroid hormone receptor. *Genes Dev.* **13**, 412–423.

Ogryzko, V. V., Schiltz, R. L., Russanova, V., Howard, B. H., and Nakatani, Y. (1996). The transcriptional coactivators p300 and CBP are histone acetyltransferases. *Cell* **87**, 953–959.

Oswald, F., Kostezka, U., Astrahantseff, K., Bourteele, S., Dillinger, K., Zechner, U., Ludwig, L., Wilda, M., Hameister, H., Knochel, W., Liptay, S., and Schmid, R. M. (2002). SHARP is a novel component of the Notch/RBP–Jkappa signalling pathway. *EMBO J.* **21**, 5417–5426.

Peterson, C. L., and Workman, J. L. (2000). Promoter targeting and chromatin remodeling by the SWI/SNF complex. *Curr. Opin. Genet. Dev.* **10**, 187–192.

Pile, L. A., and Wassarman, D. A. (2000). Chromosomal localization links the SIN3–RPD3 complex to the regulation of chromatin condensation, histone acetylation, and gene expression. *EMBO J.* **19**, 6131–6140.

Ramakrishnan, V. (1997). Histone H1 and chromatin higher-order structure. *Crit. Rev. Eukaryot. Gene Expr.* **7**, 215–230.

Rea, S., Eisenhaber, F., O'Carroll, D., Strahl, B. D., Sun, Z. W., Schmid, M., Opravil, S., Mechtler, K., Ponting, C. P., Allis, C. D., and Jenuwein, T. (2000). Regulation of chromatin structure by site-specific histone H3 methyltransferases. *Nature* **406**, 593–599.

Reinke, H., Gregory, P. D., and Horz, W. (2001). A transient histone hyperacetylation signal marks nucleosomes for remodeling at the PHO8 promoter *in vivo*. *Mol. Cell* **7**, 529–538.

Robinson-Rechavi, M., Carpentier, A. S., Duffraisse, M., and Laudet, V. (2001). How many nuclear hormone receptors are there in the human genome? *Trends Genet.* **17**, 554–556.

Roth, S. Y., Denu, J. M., and Allis, C. D. (2001). Histone acetyltransferases. *Annu. Rev. Biochem.* **70**, 81–120.

Sande, S., and Privalsky, M. L. (1996). Identification of TRACs (T3 receptor-associating cofactors), a family of cofactors that associate with, and modulate the activity of, nuclear hormone receptors. *Mol. Endocrinol.* **10**, 813–825.

Sassone-Corsi, P., Mizzen, C. A., Cheung, P., Crosio, C., Monaco, L., Jacquot, S., Hanauer, A., and Allis, C. D. (1999). Requirement of Rsk-2 for epidermal growth factor-activated phosphorylation of histone H3. *Science* **285**, 886–891.

Schurter, B. T., Koh, S. S., Chen, D., Bunick, G. J., Harp, J. M., Hanson, B. L., Henschen-Edman, A., Mackay, D. R., Stallcup, M. R., and Aswad, D. W. (2001). Methylation of histone H3 by coactivator-associated arginine methyltransferase 1. *Biochemistry* **40**, 5747–5756.

Shang, Y., Hu, X., DiRenzo, J., Lazar, M. A., and Brown, M. (2000). Cofactor dynamics and sufficiency in estrogen receptor-regulated transcription. *Cell* **103**, 843–852.

Sharma, D., and Fondell, J. D. (2002). Ordered recruitment of histone acetyltransferases and the TRAP/Mediator complex to thyroid hormone-responsive promoters in vivo. *Proc. Natl. Acad. Sci. USA* **99,** 7934–7939.

Shi, Y., Downes, M., Xie, W., Kao, H. Y., Ordentlich, P., Tsai, C. C., Hon, M., and Evans, R. M. (2001). Sharp, an inducible cofactor that integrates nuclear receptor repression and activation. *Genes Dev.* **15,** 1140–1151.

Sluder, A. E., Mathews, S. W., Hough, D., Yin, V. P., and Maina, C. V. (1999). The nuclear receptor superfamily has undergone extensive proliferation and diversification in nematodes. *Genome Res.* **9,** 103–120.

Spencer, T. E., Jenster, G., Burcin, M. M., Allis, C. D., Zhou, J., Mizzen, C. A., McKenna, N. J., Onate, S. A., Tsai, S. Y., Tsai, M. J., and O'Malley, B. W. (1997). Steroid receptor coactivator-1 is a histone acetyltransferase. *Nature* **389,** 194–198.

Stallcup, M. R. (2001). Role of protein methylation in chromatin remodeling and transcriptional regulation. *Oncogene* **20,** 3014–3020.

Stancheva, I., Collins, A. L., van den Veyver, I. B., Zoghbi, H., and Meehan, R. R. (2003). A mutant form of MeCP2 protein associated with human Rett syndrome cannot be displaced from methylated DNA by notch in *Xenopus* embryos. *Mol. Cell* **12,** 425–435.

Sterner, D. E., and Berger, S. L. (2000). Acetylation of histones and transcription-related factors. *Microbiol. Mol. Biol. Rev.* **64,** 435–459.

Strahl, B. D., Ohba, R., Cook, R. G., and Allis, C. D. (1999). Methylation of histone H3 at lysine 4 is highly conserved and correlates with transcriptionally active nuclei in *Tetrahymena*. *Proc. Natl. Acad. Sci. USA* **96,** 14967–14972.

Teyssier, C., Chen, D., and Stallcup, M. R. (2002). Requirement for multiple domains of the protein arginine methyltransferase CARM1 in its transcriptional coactivator function. *J. Biol. Chem.* **277,** 46066–46072.

Thomas, J. O. (1999). Histone H1: Location and role. *Curr. Opin. Cell. Biol.* **11,** 312–317.

Thomson, S., Clayton, A. L., Hazzalin, C. A., Rose, S., Barratt, M. J., and Mahadevan, L. C. (1999). The nucleosomal response associated with immediate-early gene induction is mediated via alternative MAP kinase cascades: MSK1 as a potential histone H3/HMG-14 kinase. *EMBO J.* **18,** 4779–4793.

Thorne, A. W., Kmiciek, D., Mitchelson, K., Sautiere, P., and Crane-Robinson, C. (1990). Patterns of histone acetylation. *Eur. J. Biochem.* **193,** 701–713.

Thummel, C. S. (1995). From embryogenesis to metamorphosis: The regulation and function of *Drosophila* nuclear receptor superfamily members. *Cell* **83,** 871–877.

Toh, Y., Kuninaka, S., Endo, K., Oshiro, T., Ikeda, Y., Nakashima, H., Baba, H., Kohnoe, S., Okamura, T., Nicolson, G. L., and Sugimachi, K. (2000). Molecular analysis of a candidate metastasis-associated gene, MTA1: Possible interaction with histone deacetylase 1. *J. Exp. Clin. Cancer Res.* **19,** 105–111.

Treuter, E., Albrektsen, T., Johansson, L., Leers, J., and Gustafsson, J. A. (1998). A regulatory role for RIP140 in nuclear receptor activation. *Mol. Endocrinol.* **12,** 864–881.

Tsai, C. C., Kao, H. Y., Yao, T. P., McKeown, M., and Evans, R. M. (1999). SMRTER, a *Drosophila* nuclear receptor coregulator, reveals that EcR-mediated repression is critical for development. *Mol. Cell* **4,** 175–186.

Turner, B. M., and O'Neill, L. P. (1995). Histone acetylation in chromatin and chromosomes. *Semin. Cell. Biol.* **6,** 229–236.

Urnov, F. D., and Wolffe, A. P. (2001). A necessary good: Nuclear hormone receptors and their chromatin templates. *Mol. Endocrinol.* **15,** 1–16.

Vo, N., Fjeld, C., and Goodman, R. H. (2001). Acetylation of nuclear hormone receptor-interacting protein RIP140 regulates binding of the transcriptional corepressor CtBP. *Mol. Cell. Biol.* **21,** 6181–6188.

Voegel, J. J., Heine, M. J., Tini, M., Vivat, V., Chambon, P., and Gronemeyer, H. (1998). The coactivator TIF2 contains three nuclear receptor-binding motifs and mediates transactivation through CBP binding-dependent and -independent pathways. *EMBO J.* **17**, 507–519.

Wade, P. A., Pruss, D., and Wolffe, A. P. (1997). Histone acetylation: Chromatin in action. *Trends Biochem. Sci.* **22**, 128–132.

Wang, H., Cao, R., Xia, L., Erdjument-Bromage, H., Borchers, C., Tempst, P., and Zhang, Y. (2001a). Purification and functional characterization of a histone H3 lysine 4-specific methyltransferase. *Mol. Cell.* **8**, 1207–1217.

Wang, H., Huang, Z. O., Xia, L., Feng, Q., Erdjument-Bromage, H., Strahl, B. D., Briggs, S. D., Allis, C. D., Wong, J., Tempst, P., and Zhang, Y. (2001b). Methylation of histone H4 at arginine 3 facilitating transcriptional activation by nuclear hormone receptor. *Science* **293**, 853–857.

Wei, L. N., Farooqui, M., and Hu, X. (2001). Ligand-dependent formation of retinoid receptors, receptor-interacting protein 140 (RIP140), and histone deacetylase complex is mediated by a novel receptor-interacting motif of RIP140. *J. Biol. Chem.* **276**, 16107–16112.

Wei, L. N., Hu, X., Chandra, D., Seto, E., and Farooqui, M. (2000). Receptor-interacting protein 140 directly recruits histone deacetylases for gene silencing. *J. Biol. Chem.* **275**, 40782–40787.

Wolffe, A. P., Khochbin, S., and Dimitrov, S. (1997). What do linker histones do in chromatin? *Bioessays* **19**, 249–255.

Wondisford, F. E. (2003). Thyroid hormone action: Insight from transgenic mouse models. *J. Investig. Med.* **51**, 215–220.

Xu, J., and Li, Q. (2003). Review of the *in vivo* functions of the p160 steroid receptor coactivator family. *Mol. Endocrinol.* **17**, 1681–1692.

Xu, W., Chen, H., Du, K., Asahara, H., Tini, M., Emerson, B. M., Montminy, M., and Evans, R. M. (2001). A transcriptional switch mediated by cofactor methylation. *Science* **294**, 2507–2511.

Yang, X. J., Ogryzko, V. V., Nishikawa, J., Howard, B. H., and Nakatani, Y. (1996). A p300/CBP-associated factor that competes with the adenoviral oncoprotein E1A. *Nature* **382**, 319–324.

Yoon, H. G., Chan, D. W., Reynolds, A. B., Qin, J., and Wong, J. (2003). NCoR mediates DNA methylation-dependent repression through a methyl CpG binding protein Kaiso. *Mol. Cell.* **12**, 723–734.

Yu, J., Li, Y., Ishizuka, T., Guenther, M. G., and Lazar, M. A. (2003). A SANT motif in the SMRT corepressor interprets the histone code and promotes histone deacetylation. *EMBO J.* **22**, 3403–3410.

Zelent, A., Guidez, F., Melnick, A., Waxman, S., and Licht, J. D. (2001). Translocations of the RARalpha gene in acute promyelocytic leukemia. *Oncogene* **20**, 7186–7203.

Zhang, J., Kalkum, M., Chait, B. T., and Roeder, R. G. (2002). The NCoR–HDAC3 nuclear receptor corepressor complex inhibits the JNK pathway through the integral subunit GPS2. *Mol. Cell* **9**, 611–623.

Zhang, X., Jeyakumar, M., Petukhov, S., and Bagchi, M. K. (1998). A nuclear receptor corepressor modulates transcriptional activity of antagonist-occupied steroid hormone receptor. *Mol. Endocrinol.* **12**, 513–524.

Zhang, X., Zhou, L., and Cheng, X. (2000). Crystal structure of the conserved core of protein arginine methyltransferase PRMT3. *EMBO J.* **19**, 3509–3519.

4

Corepressor Recruitment by Agonist-Bound Nuclear Receptors

John H. White,[*,†] Isabelle Fernandes,[*] Sylvie Mader,[†,‡] and Xiang-Jiao Yang[†]

*Departments of *Physiology and †Medicine, McGill University, McIntyre Medical Sciences Bldg, Montreal, Quebec H3G 1Y6, Canada, and ‡Department of Biochemistry, Université de Montréal Montreal, Quebec H3C 3J7, Canada*

I. The Nuclear Receptor Superfamily
  A. *Identification and Cloning of Receptor cDNAs*
  B. *Nuclear Receptor Ligands*
  C. *Domain Organization and Function of Nuclear Receptors*
II. Coregulatory Proteins in Hormone-Dependent Regulation of Transcription
  A. *Coactivators*
  B. *Corepressors*
III. Histone Deacetylases in Regulation of Gene Expression
  A. *Human Classes I and II HDACs in Transcriptional Repression by Nuclear Receptors*
  B. *Class IIb HDACs 6 and 10*
  C. *Recruitment of HDACs by LCoR and RIP140*
IV. LCoR and RIP140 Recruit the Corepressor C-Terminal Binding Protein (CtBP)

V. Potential Roles of LCoR and RIP140 in
   Hormone-Dependent Receptor Function
VI. Concluding Remarks
   References

Members of the nuclear receptor superfamily are ligand-regulated transcription factors that are composed of a series of conserved domains. These receptors are targets of a wide range of lipophilic signaling molecules that modulate many aspects of physiology and metabolism. Binding of cognate ligands to receptors induces a conformational change in the ligand binding domain (LBD) that creates a pocket for recruitment of coregulatory proteins, which are essential for ligand-dependent regulation of transcription. Several coregulatory proteins that interact with hormone-bound receptors contain characteristic helical LXXLL motifs, known as nuclear receptor (NR) boxes. Generally, ligand binding to receptors is associated with activation of transcription, and most of the NR box-containing proteins characterized to date are coactivators. However, a full understanding of the function of hormone-bound receptors must also incorporate their recruitment of corepressors. The recent identification of ligand-dependent corepressor (LCoR) is a case in point. LCoR contains a single NR box that mediates its hormone-dependent interaction with several nuclear receptors. It functions as a molecular scaffold that recruits several proteins that function in transcriptional repression. Remarkably, although the two proteins share only very limited homology, LCoR and another NR box-containing corepressor RIP140 recruit similar cofactors implicated in transcriptional repression, suggesting many parallels in their mechanisms of action. Corepressors such as LCoR and RIP140 may function in negative feedback loops to attenuate hormone-induced transactivation, act more transiently as part of a cycle of cofactors recruited to target promoters by ligand-bound receptors, or function in hormone-induced target gene repression. © 2004 Elsevier Inc.

I. THE NUCLEAR RECEPTOR SUPERFAMILY

A. IDENTIFICATION AND CLONING OF RECEPTOR cDNAs

Nuclear receptors (NRs) are ligand-regulated transcription factors whose activities are controlled by a range of lipophilic extracellular signals. These receptors have been characterized by a variety of experimental approaches over the last few decades. The first nuclear receptors were identified by classical endocrinology techniques as primary intracellular targets of steroid and thyroid hormones (Green and Chambon, 1988; Hollenberg and Evans, 1985; McKenna and O'Malley, 2002). However, the techniques used to identify novel nuclear receptors changed radically in the mid-1980s with the cloning of the first receptor cDNAs, whose sequences revealed a series of conserved domains (Green and Chambon, 1988; Krust *et al.*, 1986). Subsequently, intensive cDNA screening efforts in both academia and the pharmaceutical industry based on domain homology yielded several additional nuclear receptors. Indeed, cDNAs of all 48 members of the human superfamily (Robinson-Rechavi *et al.*, 2003) were cloned prior to sequencing of the human genome. These new members were called orphan receptors because of their unknown ligand binding specificity.

B. NUCLEAR RECEPTOR LIGANDS

The identification and characterization of steroid and thyroid hormones arose from decades, and in some cases centuries, of efforts to understand endocrine physiology, and these studies preceded the identification of the hormones' receptors. In stark contrast, the rapid and intensive efforts to determine the ligand binding specificity of orphan receptor ligands essentially represented examples of classical endocrinology in reverse (Kliewer *et al.*, 1999). Many synthetic and physiologic ligands were identified by high-throughput screening of compound libraries through the use of receptor expression vectors and reporter gene assays, where gene expression was dependent on ligand-activated receptor function. These efforts revealed that signaling through nuclear receptors impinges on numerous developmental processes, as well as on a wide range of critical physiologic and metabolic functions. Ligands identified using these approaches include vitamin A metabolites (retinoids), cholesterol metabolites, bile acids, specific prostaglandins, and numerous xenobiotics (Chawla *et al.*, 2001; McKenna and O'Malley 2002). Not surprisingly, nuclear receptors are targets of numerous drugs, including synthetic steroid hormone agonists and antagonists, thiazolidenedione antidiabetics, and modifiers of cholesterol and bile acid metabolism. Synthetic analogs of vitamins A and D and activators of peroxisomal proliferator-activated receptors (PPARs) have anticancer properties (Altucci and Gronemeyer, 2001; Leibowitz and Kantoff, 2003; Lin and White, 2004), as do antagonists of the estrogen and androgen receptors (Hirawat *et al.*, 2003; McDonnell *et al.*, 2002).

C. DOMAIN ORGANIZATION AND FUNCTION OF NUCLEAR RECEPTORS

Structure and function studies have revealed that nuclear receptors share a series of conserved domains of A through F (Fig. 1). N-terminal A/B regions of receptors are highly variable. Although receptors such as the vitamin

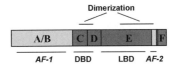

FIGURE 1. Schematic representation of the domain structures of nuclear receptors. See text for details.

D receptor (VDR) have little or no A/B region sequence, A/B domains of many receptors contain transactivation functions (activating function-1; AF-1) that cooperate with another activating function, AF-2, located in the C-terminal ligand binding domain (LBD) (Tora *et al.*, 1989). The most highly conserved region among receptors is region C. Domain swap experiments (Green and Chambon, 1987) showed that region C contains the DNA binding domain (DBD). The DBD contains two zinc finger motifs that fold into a single structural domain (Schwabe and Rhodes, 1991) with an α-helical "reading head" containing three amino acids (the P box) responsible for DNA sequence recognition (Mader *et al.*, 1989; Umesono and Evans, 1989). Nuclear receptors recognize a series of specific, related DNA sequences known collectively as hormone response elements, usually composed of so-called half-sites arranged as either direct repeats (DRs) or palindromes. Receptor DNA binding specificities are based on their capacity to recognize response elements with distinct half-site sequences, orientations, and spacings (Khorasanizadeh and Rastinejad, 2001; Sanchez *et al.*, 2002).

Receptor LBDs, located in region E are critical not only for their ligand binding properties but also because of their control over receptor dimerization and transcriptional regulation. Crystal structures of LBDs have revealed highly conserved α-helical structures in spite of only modest sequence conservation (Bourget *et al.*, 1995; Brzozowski *et al.*, 1997; Renaud *et al.*, 1995; Wagner *et al.*, 1995). Comparisons of hormone-free and hormone-bound structures have shown that ligand binding induces a striking conformational change that reorients the C-terminal AF-2 helix (helix 12, H12) with respect to the main body of the LBD (Renaud and Moras, 2000 (Fig. 2)). The AF-2 helix is critical for LBD function because its deletion or mutation abolishes ligand-dependent transcriptional regulation (Danielian *et al.*, 1992). It is now clear that the orientation of the AF-2 helix controls whether receptors recruit coactivators or corepressors (discussed later in this chapter).

Different receptors have distinct ligand binding, DNA binding, and transcriptional regulation properties (Chawla *et al.*, 2001). Steroid receptors homodimerize through strong dimerization domains in LBDs and weaker domains in DBDs (Kumar and Chambon, 1988). Closely related steroid (or steroid-like) receptors can also heterodimerize (Cowley *et al.*, 1997; Liu *et al.*, 1995; Pettersson *et al.*, 1997; Savory *et al.*, 2001; Seel *et al.*, 1998).

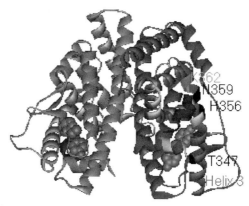

FIGURE 2. Ribbon diagram of the crystal structure of the hormone-bound ERα LBD homodimer. Helix 3 (H3) is in pink, and the cocrystallized GRIP1/TIF2 NR box peptide (purple) is shown. Key contact residues between the NR box peptide and the ERα LBD, including K362 at the head of H3, are in yellow. The diagram also indicates residues of H3 (in black): mutation disrupts binding of LCoR but not TIF-2 (see Fernandes *et al.* (2003) for details).

Steroid receptors recognize palindromic response elements with highest affinity due in large part to "toe-to-toe" dimerization by DBDs. Several nonsteroid receptors, including those for retinoic acid (RAR), vitamin D (VDR), thyroid hormone (TR), cholesterol metabolites (LXR and FXR), and prostaglandins, (PPARs), function as heterodimers with retinoid X receptors (RXRs), which in most instances act as nonsignaling partners (Germain *et al.*, 2002). Most nonsteroid receptor heterodimers recognize DNA response elements in the form of direct repeats of PuG(G/T)TCA motifs. Different heterodimeric combinations recognize direct repeats differing in half-site spacing through distinct "heel-to-toe" interactions between DBDs (Zechel *et al.*, 1994).

II. COREGULATORY PROTEINS IN HORMONE-DEPENDENT REGULATION OF TRANSCRIPTION

By the mid-1990s, it became clear that regulation of transcription by nuclear receptors required specific factors, so-called coregulatory proteins, that functioned between DNA-bound receptors and recruitment of RNA polymerase II and its ancillary transcription factors. The search for coregulators was driven largely by two sets of findings. First, gene transfer experiments showed that different hormone-bound receptors competed for limiting quantities of specific factors that were not required for transcription from heterologous promoters (Meyer *et al.*, 1989), a phenomenon called *squelching* (Gill and Ptashne, 1988). In addition, there was the increasing awareness among researchers studying transcriptional regulation in general and among those in the nuclear receptor field in particular that chromatin structure represented a barrier to transcriptional initiation. This barrier indicated that certain factors recruited by receptors were essential for the post-translational modification and remodeling of chromatin required for initiation of transcription. Collectively, these findings led to several screens for novel factors that interacted with receptors in a hormone-dependent manner. In the years since 1995, a variety of coregulatory proteins have been identified that control transcriptional regulation by nuclear receptors (Glass and Rosenfeld, 2000; McKenna and O'Malley, 2002). They are classified as coactivators or corepressors depending on whether they promote or inhibit initiation of target gene transcription in gene transfer experiments.

A. COACTIVATORS

Coactivators identified to date are remarkable in both their number and diversity, suggesting that transcriptional activation occurs through recruitment of multiple factors acting sequentially or in combination (Rosenfeld and Glass, 2001). Coactivators are essential for histone modifications, chromatin remodeling, and recruitment of RNA polymerase and 'ancillary factors necessary for initiation of transcription. Much work remains to understand the functional integration of such a wide array of factors in the control of hormone-dependent transcription. Coactivators include widely expressed factors, such as the p160 family of proteins, CREB binding protein (CBP) and related factor p300, DRIP/TRAP complex components, tissue-specific coactivators such as PGC-1, and factors with RNA binding motifs (Glass and Rosenfeld, 2000; McKenna and O'Malley, 2002). Structure and function studies have revealed that many coactivators interact directly with ligand-bound receptors through LXXLL motifs, known as NR boxes (Heery *et al.*, 1997; Voegel *et al.*, 1996). Crystallographic studies have shown that alpha-helical NR boxes are oriented in a hydrophobic pocket of hormone-bound LBDs. These LBDs contain the repositioned AF-2 helix by a charge clamp that was formed by conserved residues in helices 3 and 12 (H3 and H12) (Renaud and Moras, 2000; Shiau *et al.*, 1998).

1. Recruitment of Histone Acetyltransferase Activity by Coactivators

The most extensively characterized coactivators are the p160 proteins, SRC1/NCoA1, TIF2/GRIP1 and pCIP/AIB1/RAC3/ACTR/TRAM-1 (Anzick *et al.*, 1997; Chakravarti *et al.*, 1996; Chen *et al.*, 1997; Hong *et al.*, 1996; Onate *et al.*, 1995; Voegel *et al.*, 1996). The p160 coactivator binding recruits other factors essential for transactivation, including CBP and its homologue p300 (Glass and Rosenfeld, 2000; McKenna and O'Malley, 2002). Several coactivators, including CBP/p300 and associated factor p/CAF, possess histone acetyltransferase (HAT) activity. HAT activity essentially caps positively charged lysine residues in histone N-terminal tails and loosens their association with DNA, facilitating subsequent chromatin remodeling and access of the transcriptional machinery to promoters. The finding that coactivation is associated with histone acetylation is entirely consistent with a wealth of literature over the years demonstrating that acetylated histones are associated with actively transcribed genes (Kornberg and Lorch, 1992; Turner, 1991).

2. Dynamic Cycling of Receptors and Coactivators on Hormone-Regulated Promoters

Chromatin immunoprecipitation (ChIP) studies have revealed that p160 proteins are recruited to estrogen-responsive promoters *in vivo* within minutes of adding a hormone and suggest that the proteins represent the first wave of factors recruited by DNA-bound receptors (Shang *et al.*, 2000). Moreover, binding of p160s to target promoters precedes histone acetylation, consistent with the recruitment of HAT complexes. Remarkably, however, these and other studies (e.g., Becker *et al.*, 2002) have revealed that receptor and coregulator binding to promoters is much more dynamic than anticipated. Even in the continuous presence of a hormone, receptors and recruited factors cycle on and off target promoters (Burakov *et al.*, 2002; Reid *et al.*, 2003; Shang *et al.*, 2000). Intriguingly, this cycling is accompanied by waves of acetylation and, significantly, deacetylation of promoter histones. Moreover, other work has suggested that p160-containing HAT complexes and chromatin remodeling components of the DRIP/TRAP complex are recruited reciprocally during the cycling of factors associated with DNA-bound receptors (Burakov *et al.*, 2002).

B. COREPRESSORS

1. NCoR and SMRT

Similar to coactivators, screening efforts during the last few years have produced a rich harvest of corepressors (Fernandes and White, 2003; Glass and Rosenfeld, 2000). Not surprisingly, given the association of coactivation and HAT activity, many corepressors identified to date function at least in part by recruiting histone deacetylase (HDAC) activity (see following section). The most extensively characterized are nuclear receptor corepressor (NCoR) and silencing mediator for retinoid acid and thyroid hormone receptors (SMRT) which were isolated as factors mediating transcriptional repression by thyroid hormone receptors (TRs) and retinoid acid receptors (RARs) (Chen and Evans, 1995; Horlein *et al.*, 1995). Unlike many receptors that bind to DNA in the presence of ligand, TRs and RARs interact with DNA in a ligand-independent manner in the presence of RXR heterodimeric partners, where they repress target gene expression. Upon binding of the hormone, receptors are switched from transcriptional repressors to activators. Consistent with this ligand-dependent switch, NCoR and SMRT interact with RARs and TRs in the absence, but not the presence of a hormone, in contrast to coactivators. Accordingly, NCoR and SMRT do not interact with LBDs when the AF-2 helix is in the agonist-bound conformation. Rather, they recognize LBDs in a hormone-free or, in some cases, antagonist-bound conformation through LXXI/HIXXXI/L motifs that resemble extended NR boxes (Perissi *et al.*, 1999). Ligand binding induces movement of the AF-2 helix, displacement of NCoR and SMRT, and subsequent recruitment of NR box-containing cofactors.

2. NR Box-Containing Corepressors

Hormone binding, particularly by steroid hormone receptors, is associated with activation of target gene transcription and, not surprisingly, recruitment of coactivators. A model of receptor action where only coactivators are recruited to agonist-bound receptors, however, cannot account for all of the cofactors identified because several corepressors contain one or more NR boxes. Among the first coregulators isolated was transcriptional intermediary protein 1α (TIF1α) (Le Douarin *et al.*, 1995), an NR box-containing factor that functioned as a transcriptional repressor when fused to the DBD of the GAL4 transcription factor. Although the repressor activity of TIF1α can be blocked by the HDAC inhibitor trichostatin A (TSA) (Nielsen *et al.*, 1999), its role in transcriptional regulation remains unclear. Unlike its related factor TIF1β, TIF1α does not associate with factors, such as heterochromatin protein 1 (HP1), that are associated with heterochromatin formation.

Nuclear receptor SET-domain containing protein 1 (NSD1) is unusual in that it contains both NR boxes and LXXI/HIXXXI/L motifs. The large 2588 amino acid NSD protein interacts with several receptors in both ligand-dependent and ligand-independent manners (Huang *et al.*, 1998). Regulation of the NSD function is likely to be very complex because it contains multiple independent coactivation and corepression domains, suggesting that their functions may be selectively modified by several signal transduction pathways.

RIP140 was initially characterized as a coactivator (Cavailles *et al.*, 1995) that interacted with agonist-bound receptors through LXXLL motifs (Heery *et al.*, 1997). However, subsequent work showed that RIP140 functioned as a corepressor that competes with p160 proteins for binding to agonist-bound LBDs, blocking coactivation *in vivo* (Eng *et al.*, 1998; Lee *et al.*, 1998; Miyata *et al.*, 1998; Treuter *et al.*, 1998). Although RIP140 is widely expressed, gene ablation studies have shown that it is particularly important for normal development of female reproductive tissue. Although RIP140 null mice are viable, females are infertile as a result of failure of mature follicles to release oocytes at ovulation (White *et al.*, 2000). Luteinization occurs normally, and the phenotype resembles that of luteinized unruptured follicle syndrome that is associated with infertility in women. These findings indicate that suppression of nuclear receptor signaling is essential for coordinated control of ovarian function and specifically that RIP140 action is essential for oocyte release.

Recently, a ligand-dependent corepressor, LCoR, was identified in a screen for proteins that interacted with the estrogen receptor α (ERα) LBD in the presence of a hormone (Fernandes et al., 2003). Gene transfer experiments revealed that LCoR repressed hormone-dependent transactivation of several receptors in a concentration-dependent manner. LCoR transcripts are widely expressed in both fetal and adult human tissues. Moreover, EST data revealed that the highly similar mouse homologue of LCoR is expressed as early as the two-cell stage of embryonic development (Fernandes et al., 2003). Highest levels of LCoR expression were detected in placental tissue, and in situ hybridization studies revealed specific expression of LCoR in the syncytiotrophoblast layer of near-term placenta, a critical site of steroid hormone signaling. LCoR contains a single LXXLL motif that is critical for LCoR's agonist-dependent interaction with receptors. Moreover, as detailed below, structure and function studies have revealed remarkable functional parallels between LCoR and RIP140 despite very limited sequence homology.

3. Differential Binding of LCoR, RIP140, and p160 Proteins to Coactivator Binding Pockets of Nuclear Receptor LBDs

Several structure and function studies have shown that p160 proteins, RIP140, and LCoR recognize the same LBD coactivator binding pocket formed in the presence of hormone, but they do not make identical amino acid contacts. Crystallographic studies of the ERα LBD have shown that NR-box peptides make critical interactions with residues of H3 and H12 (the AF-2 helix). The integrity of H12 is crucial for interaction of all NR box-containing proteins. Site-directed mutagenesis of the mouse ERα LBD showed that a highly conserved lysine residue in H3 of the LBD (K366; K362 in human ERα) was essential for both recruitment of p160 proteins and transactivation in the presence of a hormone (Henttu et al., 1997). Similarly, the integrity of K362 of human ERα was essential for recruitment of LCoR (Fernandes et al., 2003). Hormone-dependent binding of RIP140 to mouse ERα, however, was not disrupted by mutagenesis of K366 (Henttu et al., 1997).

H3 residues in addition to K362 can also play critical roles in cofactor binding. The structure of the ERα LBD co-crystallized with an NR box peptide of GRIP1/TIF2 revealed contacts between the α-helical peptide and residues of the so-called static portion of the coactivator binding pocket in H3 (Shiau et al., 1998). A comparison of binding in vitro of TIF2 and LCoR to a series of ERα LBD H3 mutants derived from random mutagenesis revealed that LCoR and TIF2 recognize distinct residues on H3 and suggested that LCoR interacts with an extended portion of the helix

(Fernandes et al., 2003). It is therefore possible that LCoR amino acids outside the core NR box make contacts with nuclear receptor LBDs that would contribute to the affinity of LCoR binding (Fig. 2).

III. HISTONE DEACETYLASES IN REGULATION OF GENE EXPRESSION

Histone deacetylases (HDACs) are widely associated with transcriptional repression. HDACs and their roles in transcriptional regulation have been subjects of intensive study over the last few years. To date, dozens of proteins with HDAC activity have been identified in a wide range of organisms (Grozinger and Schreiber, 2002; Khochbin, 2001; Yang and Seto, 2003). Their roles in transcriptional repression arise from the fact that histone deacetylation frees the epsilon amino groups of modified lysines, thereby strengthening the interaction of histones with negatively charged DNA and stabilizing nucleosomal and higher order chromatin structures (Hansen, 2002; Horn and Peterson, 2002). Moreover, association with chromatin of other protein classes responsible for transcriptional silencing and heterochromatin formation (e.g., DNA methylases and HP1) is sensitive to the acetylation state (Grewall and Elgin, 2002; Taddei et al., 2001).

A. HUMAN CLASSES I AND II HDACs IN TRANSCRIPTIONAL REPRESSION BY NUCLEAR RECEPTORS

A significant portion of the understanding of human HDAC function has come from studies in yeast. HDACs have been grouped into different classes based on their homology to the yeast protein's reduced potassium dependency 3 (Rpd3), yeast histone deacetylases 1 (Hda1), and silent information regulator 2 (Sir2) (Grozinger and Schreiber, 2002; Kurdistani and Grunstein, 2003). In humans, Rpd3 homologues (class I) include HDACs 1, 2, 3, 8, and 11 (Fig. 3A). HDAC3 is a component of NCoR- and SMRT-containing complexes; HDACs 1 and 2, which heterodimerize, are components of multiple complexes, including a distinct NCoR- and SMRT-containing complex (Rosenfeld and Glass, 2001).

Human HDACs homologous to yeast Hda1 (class II enzymes) function catalytically very similarly to class I enzymes. They differ considerably, however, both structurally and functionally. They have been divided into two subclasses (Fig. 3B): HDACs 4, 5, 7, and 9 (class IIa) and HDACs 6 and 10 (class IIb) (Verdin et al., 2003; Yang and Seto, 2003). Class IIa members have been clearly implicated in transcriptional corepression, and all are known to interact with one or more DNA binding transcription factors. For example, HDACs 4 and 7 interact directly with the class I enzyme HDAC3

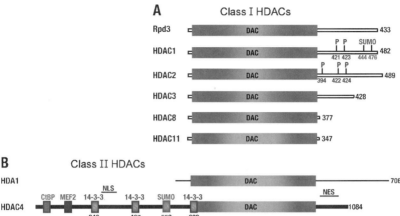
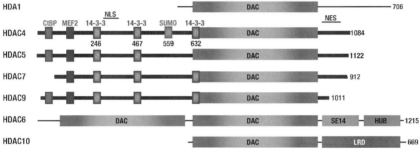

FIGURE 3. Domain organization of class I/II HDACs. (A) Comparison of Rpd3 with related human HDACs. Total residues of each deacetylase are shown at the right. Catalytic domains are boxed and labeled with DAC. P represents the CK2 phosphorylation site, and SUMO is the sumoylation site. Among the human HDACs, HDAC11 is most divergent. (B) Comparison of Hda1 and related human HDACs. HDACs 4, 5, 7, and 9 also display similarity in noncatalytic regions (bold lines). Among them, HDAC7 is most divergent. CtBP-, MEF2- and 14-3-3-binding motifs are depicted with small boxes. Other motifs or domains are labeled as follows: NLS: nuclear localization signal; NES: nuclear export signal; SE14: SE-containing tetradecapeptide repeats; HUB: an HDAC6/USP3/BRAP2-like ubiquitin-binding zinc finger; and LRD: leucine-rich domain.

and are also components of the complexes containing HDAC3 and the NCoR/SMRT complex (Fischle et al., 2001, 2002). Finally, it is important to note that seven class III HDACs homologous to yeast Sir 2 (sirtuins) have been identified in the human genome. They are distinct from class I and II enzymes in that they require NAD^+ for catalytic activity (Grozinger and Schreiber, 2002). Class III HDACs however, have yet to be associated with nuclear receptor-mediated transcriptional repression.

B. CLASS IIB HDACS 6 AND 10

Less is known about the potential roles of HDACs 6 and 10 in transcriptional corepression. HDAC6 is unusual because it contains tandem deacetylase domains (Fig. 3). It is largely cytoplasmic in most cells (Khochbin *et al.*, 2001), and its activity has been shown to regulate cytoplasmic microtubule function (Hubbert *et al.*, 2002; Matsuyama *et al.*, 2002; Zhang *et al.*, 2003). Its cytoplasmic location is determined by the presence of a potent nuclear export signal at the N-terminal of the protein (Verdel *et al.*, 2000). However, HDAC6 can become partially nuclear in B16 melanoma cells induced to differentiate (Verdel *et al.*, 2000), suggesting that it may regulate gene expression under some conditions. HDAC10 was recently identified independently by four laboratories (Fischer *et al.*, 2002; Guardiola and Yao, 2002; Kao *et al.*, 2002; Tong *et al.*, 2002). It is most similar to HDAC6, although one of its two deacetylase domains is inactive (Guardiola and Yao, 2002). HDAC10 is considered to be both cytoplasmic and nuclear, although reports vary on the relative distribution between the two compartments. It is also noteworthy that that HDAC10 may heterodimerize with HDAC3, raising the possibility that it may be a component of HDAC3-containing corepressor complexes (Tong *et al.*, 2002).

C. RECRUITMENT OF HDACs BY LCoR AND RIP140

LCoR and RIP140 recruit similar cofactors, revealing remarkable parallels in their mechanisms of action (Fig. 4). Corepression by LCoR and RIP140 can be blocked by the HDAC inhibitor TSA, and both LCoR and RIP140 interact directly with HDACs. RIP140 has been shown to interact directly with class I HDACs 1 and 3 (Wei *et al.*, 2000). LCoR interacts directly with HDACs 3 and 6 but not HDACs 1 and 4 *in vitro* and interacts with endogenous LCoR coimmunoprecipitated with HDACs 3 and 6 (Fernandes *et al.*, 2003). The interaction of LCoR with HDAC6 is significant because it suggests a role for HDAC6 in transcriptional

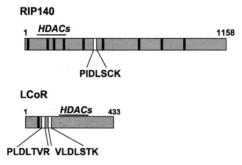

FIGURE 4. Schematic representations of the primary structures of RIP140 and LCoR. The 11 NR boxes of RIP140 and the LCoR NR box are represented by black bars. Positions of CtBP binding motifs are indicated by white boxes, and their sequences are indicated below each structure. HDAC binding domains are underlined and listed at their proper locations.

repression. More recent work has shown that LCoR also interacts with HDAC10 *in vitro* and that HDACs 3, 6, and 10 interact with distinct domains of the protein (Fig. 3 and unpublished results).

IV. LCoR AND RIP140 RECRUIT THE COREPRESSOR C-TERMINAL BINDING PROTEIN (CtBP)

The corepressor CtBP interacts with a number of transcription factors. It was originally identified as a factor that interacted with the C-terminal of the adenoviral oncoprotein E1A, and it was shown that mutations in the CtBP binding motif of E1A increase its oncogenicity (Chinnadurai, 2002). Subsequently, highly homologous CtBP2 was identified from EST databases (Katsanis and Fisher, 1998). Recently, biochemical studies have shown that CtBPs are components of a large complex of proteins containing several histone-modifying enzymes consistent with the function of CtBP as corepressors (Shi *et al.*, 2003). These include HDACs 1 and 2, histone methyltransferases, and two chromodomain-containing proteins. In other studies, several proteins that interact with CtBPs through a common motif with the consensus P/VLDLS/TXK/R were identified using a yeast two-hybrid screening (Vo *et al.*, 2001). Significantly, one of these proteins was RIP140, and mutational analysis showed that function of its CtBP binding motif was essential for corepressor activity.

Remarkably, sequence analysis of LCoR revealed tandem N-terminal motifs PLDLTVR and VLDLSTK (Fig. 3), both of which are homologous to the P/VLDLS/TXK/R consensus for CtBP binding proteins (Fernandes *et al.*, 2003). *In vitro* binding studies and coimmunoprecipitation experiments revealed a strong interaction of CtBP with LCoR that is disrupted only when both motifs are mutated (Fernandes *et al.*, 2003 and unpublished results). Immunocytochemical studies revealed a substantial overlap of CtBP and LCoR in discrete nuclear bodies.

The sensitivity to TSA of CtBP corepression is promoter-specific and transcription factor-specific (Chinnadurai, 2002). The binding of CtBPs may thus help explain the observation that the sensitivity of corepression by LCoR to the HDAC inhibitor trichostatin A (TSA) is receptor-dependent. Although inhibition of estrogen- and glucocorticoid-dependent transcription was TSA-sensitive, corepression of transactivation by a progesterone receptor (PR) or a vitamin D receptor and repression by GAL–LCoR fusions were largely TSA-resistant (Fernandes *et al.*, 2003). Moreover, while corepression of ERα transactivation was only partially disrupted by mutation of the CtBP motifs of LCoR, corepression of PR function was completely blocked by the same mutations, indicating that the contribution of CtBP and its associated proteins' corepression by LCoR is receptor-dependent.

V. POTENTIAL ROLES OF LCoR AND RIP140 IN HORMONE-DEPENDENT RECEPTOR FUNCTION

The existence of corepressors such as LCoR and RIP140 that recognize agonist-bound receptors through LXXLL motifs is perhaps counterintuitive. What then are their potential biochemical and physiologic roles? Several experiments have shown that nuclear receptor function is modulated by signals other than ligand binding (McKenna and O'Malley, 2002). There are numerous examples of the effects of phosphorylation on nuclear receptor function (Shao and Lazar, 1999). MAP kinase signaling stimulates ERα function by phosphorylation of Ser118 (Kato *et al.*, 1995). Coactivation of ERα by p160/AIB1 is enhanced by MAP kinase phosphorylation of an AIB1 activation domain, which enhances recruitment of p300 and associated HAT activity (Font de Mora and Brown, 2000). There may, therefore, exist signals that counteract the effects of MAP kinase signaling on ERα function and enhance the recruitment and/or corepressor activity of LCoR or RIP140, thus attenuating receptor activity. These may simply alter the balance between the activities of MAP kinases and MAP kinase phosphatases, or there may be independent signaling events that enhance corepressor function to the detriment of coactivation. These pathways would thus modulate competition between ligand-dependent coactivators and corepressors for hormone-bound receptors over relatively long-term periods (several cycles of transcriptional initiation).

RIP140 expression is inducible by several nuclear receptor ligands including estrogens, vitamin D_3, and retinoids (Freemantle *et al.*, 2002; Inoue *et al.*, 2002; Lin *et al.*, 2002; Soulez and Parker, 2001), suggesting that it may function as part of a negative feedback loop. This notion was supported by recent findings that RIP140 attenuated ligand-dependent transactivation by retinoid receptors. Knockdown of RIP140 expression with siRNA substantially enhanced retinoid-induced transactivation of both reporter genes and endogenous retinoid target genes (White *et al.*, 2003). Moreover, retinoic acid-induced neuronal differentiation and growth arrest were enhanced in knockdown experiments in p19 embryonal carcinoma cells in these studies. Whether LCoR functions similarly remains to be seen. However, initial studies have not found any evidence for direct regulation of LCoR expression by retinoids or estrogens (unpublished results).

It is also possible that ligand-dependent corepressors may intervene more transiently to control transcription. Current models suggest that multiple factors required for initiation of transcription are recruited by transcription factors sequentially. ERα and p160s rapidly associate with promoters (Shang *et al.*, 2000); as detailed previously, even in the presence of hormone, receptors and recruited factors cycle on and off target promoters (Becker *et al.*, 2002; McNally *et al.*, 2000; Shang *et al.*, 2000). This cycling is accompanied by waves of histone acetylation and deacetylation. Recruitment of corepressors such as LCoR or RIP140 may contribute to promoter deacetylation. It will therefore be important to analyze corepressor recruitment to hormone-responsive promoters by chromatin immunoprecipitation (ChIP) assays to determine whether the recruitment contributes to the cycles of histone acetylation and deacetylation observed during hormone-dependent transcription.

The recent wave of papers describing microarray analyses of hormone-regulated gene expression have emphasized that hormone-bound receptors also repress expression of several target genes (Inoue *et al.*, 2002; Lin *et al.*, 2002; Soulez and Parker, 2001; Wan and Nordeen, 2002). Target gene repression likely entails sets of protein–protein or protein–DNA interactions that are distinct from those associated with transactivation. NR box-containing corepressors may therefore be selectively recruited to repressed genes. Recent studies with the DEAD box RNA helicase DP97 have supported this notion. DP97, which has corepressor activity, is recruited by hormone-bound ERα to repressed genes (Rajendran *et al.*, 2003). Knockdown of DP97 expression attenuated repression of genes inhibited by hormone-bound ERα.

VI. CONCLUDING REMARKS

The field has learned much about coregulatory proteins recruited by nuclear receptors since the identification of the first factors in 1995. Much work remains to be done, however, particularly regarding the roles of NR box-containing corepressors in controlling nuclear receptor-regulated transcription. Development of the molecular tools necessary for chromatin immunoprecipitation (ChIP) assays will be very valuable in determining when NR box-containing corepressors, such as RIP140 and LCoR, intervene in complex molecular events underlying hormone-regulated transcription. In addition, gene knockdown techniques coupled with high-throughput screening approaches, including microarray analyses, will be critical for determining the genes targeted by these factors and for providing further insights into their physiologic functions.

ACKNOWLEDGMENTS

This work was supported by grants from the Canadian Institutes of Health Research (CIHR) and the National Cancer Institute of Canada to John H. White, Sylvie Mader, and Xiang-Jiao Yang. Isabelle Fernandes held a postdoctoral fellowship from the CIHR.

REFERENCES

Altucci, L., and Gronemeyer, H. (2001). The promise of retinoids to fight against cancer. *Nature Rev. Cancer* **1**, 181–193.
Anzick, S. L., Kononen, J., Walker, R. L., Azorsa, D. O., Tanner, M. M., Guan, X. Y., Sauter, G., Kallioniemi, O. P., Trent, J. M., and Meltzer, P. S. (1997). AIB1, a steroid receptor coactivator amplified in breast and ovarian cancer. *Science* **277**, 965–968.
Becker, M., Baumann, C., John, S., Walker, D. A., Vigneron, M., McNally, J. G., and Hager, G. L. (2002). Dynamic behavior of transcription factors on a natural promoter in living cells. *EMBO Rep.* **3**, 1188–1194.
Bourget, W., Ruff, M., Chambon, P., Gronemeyer, H., and Moras, D. (1995). Crystal structure of the ligand binding domain of the human nuclear receptor RXRa. *Nature* **375**, 377–382.
Burakov, D., Crofts, L. A., Chang, C. P. B., and Freedman, L. P. (2002). Reciprocal recruitment of DRIP/Mediator and p160 coactivator complexes *in vivo* by estrogen receptor. *J. Biol. Chem.* **277**, 14359–14362.
Brzozowski, A. M., Pike, A. C., Dauter, Z., Hubbard, R. E., Bonn, T., Engstrom, O., Ohman, L., Greene, G. L., Gustafsson, J. A., and Carlquist, M. (1997). Molecular basis of agonism and antagonism in the oestrogen receptor. *Nature* **389**, 753–758.
Cavaillès, V., Dauvois, S., L'Horset, F., Lopez, G., Hoare, S., Kushner, P. J., and Parker, M. G. (1995). *EMBO J.* **14**, 3741–3751.
Chakravarti, D., LaMorte, V. J., Nelson, M. C., Nakajima, T., Schulman, I. G., Jugulon, H., Montminy, M., and Evans, R. M. (1996). Role of CBP/P300 in nuclear receptor signalling. *Nature* **383**, 99–103.
Chawla, A., Repa, J., Evans, R. M., and Mangelsdorf, D. J. (2001). Nuclear receptors and lipid physiology: Opening the X-files. *Science* **294**, 1866–1870.
Chen, J. D., and Evans, R. M. (1995). A transcriptional corepressor that interacts with nuclear hormone receptors. *Nature* **377**, 454–457.
Chen, H., Jin, R. J., Schitz, R. S., Chakravarti, D., Nash, A., Nagy, L., Privalsky, M. L., Nakatani, Y., and Evans, R. M. (1997). Nuclear receptor coactivator ACTR is a novel histone acetyltransferase and forms a multimeric activation complex with P/CAF and CBP/p300. *Cell* **90**, 569–580.
Chinnadurai, G. (2002). CtBP, an unconventional transcriptional corepressor in development and oncogenesis. *Mol. Cell* **9**, 213–224.
Cowley, S. M., Hoare, S., Mosselman, S., and Parker, M. G. (1997). Estrogen receptors alpha and beta form heterodimers on DNA. *J. Biol. Chem.* **272**, 19858–19862.
Danielian, P. S., White, R., Lees, J. A., and Parker, M. G. (1992). Identification of a conserved region required for hormone-dependent transcriptional activation by steroid hormone receptors. *EMBO J.* **11**, 1025–1033.
Eng, F. C. S., Barsalou, A., Akutsu, N., Mercier, I., Zechel, C., Mader, S., and White, J. H. (1998). Different classes of coactivators recognize distinct but overlapping sites on the estrogen receptor ligand binding domain. *J. Biol. Chem.* **273**, 28371–28377.
Fernandes, I., Bastien, Y., Wai, T., Nygard, K., Lin, R., Cormier, O., Lee, H. S., Eng, F., Bertos, N. R., Pelletier, N., Mader, S., Han, V. K. M., Yang, X. J., and White, J. H. (2003). Ligand-dependent corepressor LCoR functions by histone deacetylase-dependent and -independent mechanisms. *Mol. Cell* **11**, 139–150.
Fernandes, I., and White, J. H. (2003). Agonist-bound nuclear receptors: Not just targets of coactivators. *J. Mol. Endocrinol.* **31**, 1–7.
Fischer, D. D., Cai, R., Bhatia, U., Asselbergs, F. A. M., Song, C. Z., Terry, R., Trogani, N., Widmer, R., Atadja, P., and Cohen, D. (2002). Isolation and characterization of a novel class II histone deacetylase, HDAC10. *J. Biol. Chem.* **277**, 6656–6666.

Fischle, W., Dequiedt, F., Fillion, M., Hendzel, M. J., Voelter, W., and Verdin, E. (2001). Human HDAC7 histone deacetylase activity is associated with HDAC3 in vivo. *J. Biol. Chem.* **276,** 35826–35835.

Fischle, W., Dequiedt, F., Fillion, M., Hendzel, M. J., Guenther, M. G., Lazar, M. A., Voelter, W., and Verdin, E. (2002). Enzymatic activity associated with class II HDACs is dependent on a multiprotein complex containing HDAC3 and SMRT/NCoR. *Mol. Cell* **9,** 45–57.

Font de Mora, J., and Brown, M. (2000). AIB1 is a signaling conduit for kinase-mediated growth factor signaling to the estrogen receptor. *Mol. Cell. Biol.* **20,** 5041–5047.

Freemantle, S. J., Kerley, J. S., Olsen, S. L., Gross, R. H., and Spinella, M. J. (2002). Developmentally related candidate retinoic acid target genes regulated early during neuronal differentiation of human embryonal carcinoma. *Oncogene* **21,** 2880–2889.

Germain, P., Iyer, J., Zechel, C., and Gronemeyer, H. (2002). Coregulator recruitment and the mechanism of retinoic acid receptor synergy. *Nature* **415,** 187–192.

Gill, G., and Ptashne, M. (1988). Negative effect of the transcriptional activator GAL4. *Nature* **334,** 721–724.

Glass, C. K., and Rosenfeld, M. G. (2000). The coregulator exchange in transcriptional functions of nuclear receptors. *Genes Dev.* **14,** 121–141.

Green, S., and Chambon, P. (1987). Estradiol induction of a glucocorticoid-responsive gene by a chimeric receptor. *Nature* **325,** 75–78.

Green, S., and Chambon, P. (1988). Nuclear receptors enhance our understanding of transcriptional regulation. *Trends Genet.* **4,** 309–314.

Grewall, S. L., and Elgin, S. C. (2002). Heterochromatin: New possibilities for the inheritance of structure. *Curr. Opin. Genet. Dev.* **12,** 178–187.

Grozinger, C. M., and Schreiber, S. L. (2002). Deacetylase enzymes: Biological functions and the use of small molecule inhibitors. *Chem. Biol.* **9,** 3–16.

Guardiola, A. R., and Yao, T. P. (2002). Molecular cloning and characterization of a novel histone deacetylase HDAC10. *J. Biol. Chem.* **277,** 3350–3356.

Hansen, J. C. (2002). Conformational dynamics of the chromatin fiber in solution: Determinants, mechanisms, and functions. *Annu. Rev. Biophys. Biomol. Struct.* **31,** 361–369.

Heery, D. M., Kalkhoven, E., Hoare, S., and Parker, M. G. (1997). A signature motif in transcriptional coactivators mediates binding to nuclear receptors. *Nature* **387,** 733–736.

Henttu, P. M. A., Kalkhoven, E., and Parker, M. G. (1997). AF-2 activity and recruitment of steroid receptor coactivator-1 to the estrogen receptor depend on a lysine residue conserved in nuclear receptors. *Mol. Cell. Biol.* **17,** 1832–1839.

Hirawat, S., Budman, D. R., and Kreis, W. (2003). The androgen receptor: Structure, mutations, and antiandrogens. *Cancer Investig.* **21,** 400–417.

Hollenberg, S. M., Weinberger, C., Ong, E. S., Cerelli, G., Oro, A., Lebo, R., Thompson, E. B., Rosenfeld, M. G., and Evans, R. M. (1985). Primary structure and expression of a functional human glucocorticoid receptor cDNA. *Nature* **318,** 635–641.

Hong, H., Kohli, K., Trivedi, A., Johnson, D. L., and Stallcup, M. R. (1996). GRIP1, a novel mouse protein that serves as a transcriptional coactivator in yeast for the hormone binding domains of steroid receptors. *Proc. Nat. Acad. Sci. USA* **93,** 4948–4952.

Horlein, A. J., Naar, A. M., Heinzel, T., Torchia, J., Gloss, B., Kurokawa, R., Ryan, A., Kamei, Y., Soderstrom, M., Glass, C. K., and Rosenfeld, M. G. (1995). Ligand-independent repression by the thyroid hormone receptor mediated by a nuclear receptor corepressor. *Nature* **377,** 397–404.

Horn, P. J., and Peterson, C. L. (2002). Chromatin higher order folding-wrapping up transcription. *Science* **297,** 1824–1827.

Huang, N. W., vom Baur, E., Garnier, J. M., Lerouge, T., Vonesch, J. L., Lutz, Y., Chambon, P., and Losson, R. (1998). Two distinct nuclear receptor interaction domains in NSD1, a novel SET protein that exhibits characteristics of both corepressors and coactivators. *EMBO J.* **17,** 3398–3412.

Hubbert, C., Guardiola, A., Shao, R., Kawaguchi, Y., Ito, A., Nixon, A., Yoshida, M., Wang, X. F., and Yao, T. P. (2002). HDAC6 is a microtubule-associated deacetylase. *Nature* **417,** 455–458.
Inoue, A., Yoshida, N., Omoto, Y., Oguchi, S., Yamori, T., Kiyama, R., and Hayashi, S. (2002). Development of cDNA microarray for expression profiling of estrogen-responsive genes. *J. Mol. Endocrinol.* **29,** 175–192.
Kao, H. Y., Lee, C. H., Komarov, A., Han, C. C., and Evans, R. M. (2002). Isolation and characterization of mammalian HDAC10, a novel histone deacetylases. *J. Biol. Chem.* **277,** 187–193.
Kato, S., Endoh, H., Masuhiro, Y., Kitamoto, T., Uchyama, S., Sasaki, H., Masushige, S., Gotoh, Y., Hishida, E., Kawashima, H., Metzger, D., and Chambon, P. (1995). Activation of the estrogen receptor through phosphorylation by mitogen-activated protein kinase. *Science* **270,** 1491–1494.
Katsanis, N., and Fisher, E. M. C. (1998). A novel C-terminal binding protein (*CTBP2*) is closely related to *CTBP1*, an adenovirus E1A-binding protein, and maps to human chromosome 21q21.3. *Genomics* **47,** 294–299.
Khochbin, S., Verdel, A., Lemercier, C., and Seigneurin-Berny, D. (2001). Functional significance of histone deacetylase diversity. *Curr. Opinion Genet. Dev.* **11,** 162–166.
Khorasanizadeh, S., and Rastinejad, F. (2001). Nuclear-receptor interactions on DNA-response elements. *Trends Biochem. Sci.* **26,** 384–390.
Kliewer, S. A., Lehmann, J. M., and Willson, T. M. (1999). Orphan nuclear receptors: Shifting endocrinology into reverse. *Science* **284,** 757–760.
Kornberg, R. D., and Lorch, Y. (1992). Chromatin structure and transcription. *Ann. Rev. Cell. Biol.* **8,** 563–697.
Krust, A., Green, S., Argos, P., Kumar, V., Walter, P., Bornert, J. M., and Chambon, P. (1986). Chicken estrogen receptor sequence—Homology with v-erbA and the human estrogen and glucocorticoid receptors. *EMBO J.* **5,** 891–897.
Kumar, V., and Chambon, P. (1988). The estrogen receptor binds tightly to its responsive element as a ligand-induced homodimer. *Cell* **55,** 145–156.
Kurdistani, S. K., and Grunstein, M. (2003). Histone acetylation and deacetylation in yeast. *Nat. Rev. Mol. Cell. Biol.* **4,** 276–284.
Le Douarin, B., Zechel, C., Garnier, J. M., Lutz, Y., Tora, L., Pierrat, P., Heery, D., Gronemeyer, H., Chambon, P., and Losson, R. (1995). The N-terminal part of TIF1, a putative mediator of the ligand-dependent activation function (AF-2) of nuclear receptors, is fused to B-ref in the oncogenic protein T18. *EMBO J.* **14,** 2020–2033.
Lee, C. H., Chinpaisal, C., and Wei, L. N. (1998). Cloning and characterization of mouse RIP140, a corepressor for nuclear orphan receptor TR2. *Mol. Cell. Biol.* **18,** 6745–6755.
Leibowitz, S. B., and Kantoff, P. W. (2003). Differentiating agents and the treatment of prostate cancer: Vitamin D_3 and peroxisome proliferator-activated receptor gamma ligands. *Semin. Oncol.* **30,** 698–708.
Lin, R., Nagai, Y., Sladek, R., Bastien, Y., Ho, J., Petrecca, K., Sotiropoulou, G., Diamandis, E. P., Hudson, T., and White, J. H. (2002). Expression profiling in squamous carcinoma cells reveals pleiotropic effects of vitamin D_3 signaling on cell proliferation, differentiation, and immune system regulation. *Mol. Endocrinol.* **16,** 1243–1256.
Lin, R., and White, J. H. (2004). The pleiotropic actions of vitamin D. *BioEssays* **26,** 21–28.
Liu, W. H., Wang, J., Sauter, N. K., and Pearce, D. (1995). Steroid receptor heterodimerization demonstrated *in vitro* and *in vivo*. *Proc. Nat. Acad. Sci. USA* **92,** 12480–12484.
Mader, S., Kumar, V., Duverneuil, H., and Chambon, P. (1989). Three amino acids of the estrogen are essential to its ability to distinguish an estrogen from a glucocortiocoid responsive element. *Nature* **338,** 271–274.
Matsuyama, A., Shimazu, T., Sumida, Y., Saito, A., Yoshimatsu, Y., Seigneurin-Berny, D., Osada, H., Komatsu, Y., Nishino, N., Khochbin, S., Horinouchi, S., and Yoshida, M.

(2002). In vivo destabilization of dynamic microtubules by HDAC6-mediated deacetylation. *EMBO J.* **21,** 6820–6831.

McDonnell, D. P., Connor, C. E., Wijayaratne, A., Chang, C. Y., and Norris, J. D. (2002). Definition of the molecular and cellular mechanisms underlying the tissue-selective agonist/antagonist activities of selective estrogen receptor modulators. *Recent Prog. Horm. Res.* **57,** 295–316.

McKenna, N. J., and O'Malley, B. W. (2002). Combinatorial control of gene expression by nuclear receptors and coregulators. *Cell* **108,** 465–474.

McNally, J. G., Muller, W. G., Walker, D., Wolford, R., and Hager, G. L. (2000). The glucocorticoid receptor: Rapid exchange with regulatory sites in living cells. *Science* **287,** 1262–1265.

Meyer, M. E., Gronemeyer, H., Turcotte, B., Bocquel, M. T., Tasset, D., and Chambon, P. (1989). Steroid hormone receptors compete for factors that mediate their enhancer function. *Cell* **57,** 433–442.

Miyata, K. S., McCaw, S. E., Meertens, L. M., Patel, H. V., Rachubinski, R. A., and Capone, J. P. (1998). Receptor-interacting protein 140 interacts with and inhibits transactivation by peroxisome proliferator-activated receptor alpha and liver-X-receptor alpha. *Mol. Cell. Endocrinol.* **146,** 69–76.

Nielsen, A. L., Ortiz, J. A., You, J., Oulad-Abdelghani, M., Khechumian, R., Gansmuller, A., Chambon, P., and Losson, R. (1999). Interaction with members of the heterochromatin protein 1 (HP1) family and histone deacetylation are differentially involved in transcriptional silencing by members of the TIF1 family. *EMBO J.* **18,** 6385–6395.

Onate, S. A., Tsai, S. Y., Tsai, M. J., and O'Malley, B. (1995). Sequence and characterization of a coactivator for the steroid hormone receptor superfamily. *Science* **270,** 1354–1357.

Perissi, V., Staszewski, L. M., McInerney, E. M., Kurokawa, R., Krones, A., Rose, D. W., Lambert, M. H., Milburn, M. V., Glass, C. K., and Rosenfeld, M. G. (1999). Molecular determinants of nuclear receptor-corepressor interaction. *Genes Dev.* **13,** 3198–3208.

Pettersson, K., Grandien, K., Kuiper, G. G. J., and Gustafsson, J. A. (1997). Mouse estrogen receptor beta forms estrogen response element-binding heterodimers with estrogen receptor alpha. *Mol. Endocrinol.* **11,** 1486–1496.

Rajendran, R. R., Nye, A. C., Frasor, J., Balsara, R. D., Martini, P. G. V., and Katzenellenbogen, B. S. (2003). Regulation of nuclear receptor transcriptional activity by a novel DEAD box RNA helicase (DP97). *J. Biol. Chem.* **278,** 4628–4638.

Reid, G., Hubner, M. R., Metivier, R., Brand, H., Denger, S., Manu, D., Beaudouin, J., Ellenberg, J., and Gannon, F. (2003). Cyclic, proteasome-mediated turnover of unliganded and liganded ER alpha on responsive promoters is an integral feature of estrogen signaling. *Mol. Cell* **11,** 695–707.

Renaud, J. P., Rochel, N., Chambon, P., Gronemeyer, H., and Moras, D. (1995). Crystal structure of the RAR-γ ligand binding domain bound to all-trans retinoic acid. *Nature* **378,** 681–689.

Renaud, J. P., and Moras, D. (2000). Structural studies on nuclear receptors. *Cell. Mol. Life Sci.* **57,** 1748–1769.

Robinson-Rechavi, M., Garcia, H. E., and Laudet, V. (2003). The nuclear receptor superfamily. *J. Cell. Sci.* **116,** 585–586.

Rosenfeld, M. G., and Glass, C. K. (2001). Coregulator codes of transcriptional regulation by nuclear receptors. *J. Biol. Chem.* **276,** 36865–36868.

Sanchez, R., Nguyen, D., Rocha, W., White, J. H., and Mader, S. (2002). Diversity in the mechanisms of gene regulation by estrogen receptors. *BioEssays* **24,** 244–254.

Savory, J. G. A., Prefontaine, G. G., Lamprecht, C., Liao, M. M., Walther, R. F., Lefebvre, Y. A., and Hache, R. J. G. (2001). Glucocorticoid receptor homodimers and glucocorticoid-mineralocorticoid receptor heterodimers form in the cytoplasm through alternative dimerization interfaces. *Mol. Cell. Biol.* **21,** 781–793.

Schwabe, J. W. R., and Rhodes, D. (1991). Beyond zinc fingers—Steroid hormone receptors have a novel structural motif for DNA recognition. *Trends Biochem. Sci.* **16**, 291–297.

Seel, W. G., Hanstein, B., Brown, M., and Moore, D. D. (1998). Inhibition of estrogen receptor action by the orphan receptor SHP (short heterodimer partner). *Mol. Endocrinol.* **12**, 1551–1557.

Shang, Y., Xiao, H., DiRenzo, J., Lazar, M. A., and Brown, M. (2000). Cofactor dynamics and sufficiency in estrogen receptor-regulated transcription. *Cell* **103**, 843–852.

Shao, D., and Lazar, M. (1999). Modulating nuclear receptor function: May the phos be with you. *J. Clin. Invest.* **103**, 1617–1618.

Shi, Y. J., Sawada, J., Sui, G. C., Affar, E. B., Whetstine, J. R., Lan, F., Ogawa, H., Luke, M. P. S., Nakatani, Y., and Shi, Y. (2003). Coordinated histone modifications mediated by a CtBP corepressor complex. *Nature* **422**, 735–738.

Shiau, A. K., Barstad, D., Loria, P. M., Cheng, L., Kushner, P. J., Agard, D. A., and Greene, G. L. (1998). The structural basis of estrogen receptor/coactivator recognition and the antagonism of this interaction by tamoxifen. *Cell* **95**, 927–937.

Soulez, M., and Parker, M. G. (2001). Identification of novel oestrogen receptor target genes in human ZR75-1 breast cancer cells by expression profiling. *J. Mol. Endocrinol.* **27**, 259–274.

Taddei, A., Maison, C., Roche, D., and Almouzni, G. (2001). Reversible disruption of pericentric heterochromatin and centromere function by inhibiting deacetylases. *Nat. Cell. Biol.* **3**, 114–120.

Tong, J. J., Liu, J. H., Bertos, N. R., and Yang, X. J. (2002). Identification of HDAC10, a novel class II human histone deacetylase containing a leucine-rich domain. *Nucl. Acid. Res.* **30**, 1114–1123.

Tora, L., White, J. H., Brou, C., Tasset, D., Webster, N., Scheer, E., and Chambon, P. (1989). The human estrogen receptor has two independent nonacidic transcriptional activation functions. *Cell* **59**, 477–487.

Treuter, E., Albrektsen, T., Johansson, L., Leers, J., and Gustafsson, J. A. (1998). A regulatory role for RIP140 in nuclear receptor activation. *Mol. Endocrinol.* **12**, 864–881.

Turner, B. M. (1991). Histone acetylation and control of gene expression. *J. Cell. Sci.* **99**, 13–20.

Umesono, K., and Evans, R. M. (1989). Determinants of target gene specificity for steroid-thyroid hormone receptors. *Cell* **57**, 1139–1146.

Verdel, A., Curtet, S., Brocard, M.-P., Rousseaux, S., Lemercier, C., Yoshida, M., and Khochbin, S. (2000). Active maintenance of mHDA2/mHDAC6 histone-deacetylase in the cytoplasm. *Curr. Biol.* **10**, 747–749.

Verdin, E., Dequiedt, F., and Kasler, H. G. (2003). Class II histone deacetylases: Versatile regulators. *Trends Genet.* **19**, 286–293.

Vo, N., Fjeld, C., and Goodman, R. H. (2001). Acetylation of nuclear hormone receptor-interacting protein RIP140 regulates its interaction with CtBP. *Mol. Cell. Biol.* **21**, 6181–6188.

Voegel, J. J., Heine, M. J. S., Zechel, C., and Chambon, P. (1996). The coactivator TIF2 contains three nuclear receptor-binding motifs and mediates transactivation through CBP binding-dependent and -independent pathways. *EMBO J.* **13**, 3667–3675.

Wagner, R. L., Apriletti, J. W., McGrath, M. E., West, B. L., Baxter, J. D., and Fletterick, R. J. (1995). A structural role for hormone in the thyroid hormone receptor. *Nature* **378**, 690–697.

Wan, Y. H., and Nordeen, S. K. (2002). Overlapping but distinct gene regulation profiles by glucocorticoids and progestins in human breast cancer cells. *Mol. Endocrinol.* **16**, 1204–1214.

Wei, L. N., Hu, X. L., Chandra, D., Seto, E., and Farooqui, M. (2000). Receptor-interacting protein 140 directly recruits histone deacetylases for gene silencing. *J. Biol. Chem.* **275**, 40782–40787.

White, K. A., Yore, M. M., Warburton, S. L., Vaseva, A. V., Rieder, E., Freemantle, S. J., and Spinella, M. J. (2003). Negative feedback at the level of nuclear receptor coregulation: Self-limitation of retinoid signaling by RIP140. *J. Biol. Chem.* **278**, 43889–43892.

White, R., Leonardsson, G., Rosewell, I., Jacobs, M. A., Milligan, S., and Parker, M. (2000). The nuclear receptor corepressor Nrip1 (RIP140) is essential for female fertility. *Nature Med.* **6**, 1368–1374.

Yang, X. J., and Seto, E. (2003). Collaborative spirit of histone deacetylases in regulating chromatin structure and gene transcription. *Curr. Opin. Genet. Dev.* **13**, 143–153.

Zechel, C., Shen, X. Q., Chen, J. Y., Chen, Z. P., Chambon, P., and Gronemeyer, H. (1994). The dimerization interfaces formed between the DNA binding domains of RXR, RAR, and TR determine the binding specificity and polarity of the full-length receptors to direct repeats. *EMBO J.* **13**, 1425–1433.

Zhang, Y., Li, N., Caron, C., Matthias, G., Hess, D., Khochbin, S., and Matthias, P. (2003). HDAC6 interacts with and deacetylates tubulin and microtubules *in vivo*. *EMBO J.* **22**, 1168–1179.

5

PHARMACOLOGY OF NUCLEAR RECEPTOR–COREGULATOR RECOGNITION

RAJESH S. SAVKUR, KELLI S. BRAMLETT,
DAVID CLAWSON, AND THOMAS P. BURRIS

*Lilly Research Laboratories, Eli Lilly and Company
Indianapolis, Indiana 46285*

I. Introduction
   A. Historical Perspective
   B. Coregulators Mechanisms of Action
II. Coregulators Involved in Nuclear Receptor Action
   A. Coactivators
   B. Corepressors
III. Nuclear Receptor–Coregulator Recognition
   A. Coactivators
   B. Corepressors
IV. Pharmacology of Nuclear Receptor–Coregulator Recognition
V. Conclusion
   References

The nuclear receptor (NR) superfamily comprises approximately 50 members that are responsible for regulating a number of physiologic processes in humans, including metabolism, homeostasis, and reproduction. Included in the superfamily are the receptors for steroids, lipophilic vitamins, bile acids, retinoids, and various fatty acids. NRs exert their action as transcription factors that directly bind to the promoters of target genes and regulate their rate of transcription. To modulate transcription, however, NRs must recruit a number of accessory coregulators known as corepressors and coactivators. These coregulators harbor a variety of activities, such as the ability to modify chromatin structure, interact with basal transcriptional machinery, and modify RNA splicing. Recent studies have revealed that the pharmacological characteristics of various NR ligands are regulated by their ability to modulate the coregulator interaction profile of an NR. © 2004 Elsevier Inc.

I. INTRODUCTION

A. HISTORICAL PERSPECTIVE

The nuclear receptor (NR) superfamily's nearly 50 transcription factors include receptors for a variety of hydrophobic bioactive compounds, such as steroids, thyroid hormones, vitamin A and D derivatives, oxysterols, various lipids, and certain xenobiotics. Also included in the superfamily are a number of orphan receptors and may not necessarily require a ligand for activity. NRs have a conserved domain structure that includes an amino-terminal variable A/B region, a highly conserved DNA binding domain (DBD, C region), a hinge region (D region), and a ligand binding domain (LBD, E region), which is the second most conserved region of this superfamily. Several receptors also contain an F region whose function is not well characterized. NRs primarily regulate transcription of their target genes by recognizing specific DNA sequences (response elements) in the promoters of target genes. The steroid receptors, classified as type I receptors, recognize DNA as a homodimer, whereas type II receptors recognize DNA as a heterodimer with the common NR partner retinoid X receptor (RXR). Other receptors, particularly in the orphan class, may recognize DNA as a monomer.

Once bound to the DNA response element, NRs increase the rate of preinitiation complex formation, thus stimulating the rate of transcription. Specific regions in the NRs that are responsible for activation of transcription are known as activation functions (AF). The LBD contains a ligand-regulated AF, which is known as AF-2, and the A/B region of at least some receptors contains a constitutively active AF (AF-1). A number of NRs, such as the thyroid hormone receptor (TR), are able to actively silence gene transcription of their respective target genes in the absence of a ligand. In addition to the classic mechanisms of transcriptional activation or silencing, some NRs can modulate transcription independent of their DNA binding activity by interacting with other classes of transcription factors that affect their transactivation activity.

Transcription of genes by RNA polymerase II (Pol II) begins with the formation of a preinitiation complex. This preinitiation complex is formed by a defined order of assembly of general transcription factors and Pol II at the promoter. Although NRs have increased the formation rate for this complex of basal transcription factors and Pol II at the promoter, how they accomplish this is not well understood. NRs have been demonstrated to interact with components of the basal transcription factor complex, including TFIIB, TFIID, and TATA binding protein associated factors (TAFs); however, it is evident that the process requires a multitude of other factors. The initial concept of transcriptional coactivators or adaptors is based on the observation that overexpression of the GAL4 transcription factor in yeast squelches (inhibits) the transcription of genes lacking GAL4 response elements (Gill and Ptashne, 1988). This squelching also occurs when a mutant of GAL4 lacking the DBD is overexpressed. In mammalian cells, expression of a GAL4–VP16 chimera or wild-type VP16 also squelches the expression of other nontarget genes (Sadowski *et al.*, 1988; Triezenberg *et al.*, 1988). Presumably, these transcription factors are competing for a limited number of transcriptional coactivators in the cell. Squelching was initially demonstrated for NRs by Meyer *et al.* (1989) who showed that stimulating the transcription of reporter genes by the progesterone receptor (PR) can be reduced by coexpression of the estrogen receptor (ER) in the same cell. The study showed that overexpression of ER also inhibited transactivation by the glucocorticoid receptor (GR) and, as would be expected if there were common coactivators, overexpression of either GR or PR inhibited the transactivation activity of ER (Meyer *et al.*, 1989). Interestingly, squelching experiments suggested that different types of AFs exist. Two AFs that were identified in ER and GR differed in their ability to squelch one another (Hollenberg and Evans, 1988; Tasset *et al.*, 1990; Tora *et al.*, 1989). For example, the AFs of ER will squelch one another, but AF-1 of GR will not squelch AF-2 of ER (Tasset *et al.*, 1990). Both AFs of ER squelch the acidic activation domains of both VP16 and GAL4; the AFs of GR can both squelch GAL4 but not VP16. This suggests that there are multiple coactivator protein functioning with specificity for the AFs of various NRs.

The identity of coactivator proteins was pursued using several methods. Halachmi *et al.* (1994) at the Brown Laboratory predicted that since AF-2 of ER was dependent on a ligand, the interaction with a putative coactivator would also be dependent on a ligand. Hence, they used biochemical methods to isolate ERAP140 and ERAP160, which interacted with ER only in the presence of estradiol (Halachmi *et al.*, 1994). Consistent with a role as coactivators, ERAPs did not recognize mutant ERs that were transcriptionally deficient, and the estradiol-dependent ER/ERAP interaction was antagonized by antiestrogens. In addition, far-Western blotting was used to characterize ER interacting proteins (receptor interacting proteins or RIPs) that interacted in a hormone-dependent manner; the interaction decreased in the presence of antiestrogens and transcriptionally deficient ER mutants. Biochemical methods were also used to identify GR interacting protein 170 (GRIP170), which increased the transactivation activity of GR in a hormone-dependent manner in an *in vitro* assay system (Eggert *et al.*, 1995). Steroid receptor coactivator-1 (SRC-1) was the first NR coactivator cloned and identified in a yeast two-hybrid screening using PR as bait (Onate *et al.*, 1995). These researchers also used the premise that the coactivators would interact with AF-2 in a ligand-dependent manner. Thus, the researchers screened in the presence and absence of the PR agonist R5020 and focused on only those clones that interacted with PR in an R5020-dependent manner. In assays that have since become standard in identifying and characterizing coactivators, overexpression of SRC-1 in cotransfection experiments led to an increase in transactivation activity of PR and, in addition, overexpression of SRC-1 relieved the squelching effect induced by overexpressing only the AF of PR. SRC-1 also functioned as a coactivator for a variety of NRs, indicating that NRs share a common coactivator (predicted from the squelching experiments described previously). The initial identification and characterization of SRC-1 was only the first step in discovering a number of coactivator proteins, many of which are described in this chapter.

Corepressors, which are functionally analogous to coactivators except that they function in transcription silencing instead of activation, were also first suggested to exist based on the squelching activity of NRs. Several NRs display the ability to silence gene transcription, and TR is particularly effective in silencing transcription of its target genes in the absence of the thyroid hormone. The silencing activity is mediated by the LBD (Baniahmad *et al.*, 1992) and can be relieved by the addition of a ligand that results in the transition of the receptor to a transactivator. Interestingly, the silencing activity of TR can be squelched by overexpression of either the LBD of TR or an oncogenic form of TR (v-erbA), suggesting a soluble factor mediates the silencing activity that can be sequestered by a competitor (Baniahmad *et al.*, 1995). Subsequently, two related corepressor proteins, silencing mediator for retinoic acid and thyroid hormone receptors (SMRT) and nuclear receptor corepressor (NCoR), were identified using the yeast two-hybrid screening system. As one would expect, these two proteins interacted with the NR only in the absence of a hormone, which is consistent with the predicted role in silencing activity (Chen and Evans, 1995; Horlein *et al.*, 1995).

B. COACTIVATOR AND COREPRESSOR MECHANISMS OF ACTION

The first step toward transcriptional activation by many NRs involves the ligand-dependent release of corepressor proteins from the LBD (Fig. 1). The corepressor proteins are actively involved in silencing of gene transcription

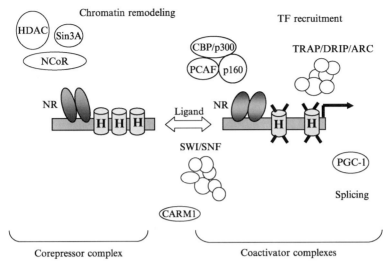

FIGURE 1. Two-step mechanism of coregulator action. Binding of a ligand to a nuclear receptor (NR) causes a conformational change in the LBD of the NR, resulting in the release of corepressor proteins that are involved in silencing of gene transcription (N-CoR, histone deacetylase, and Sin3A). Chromatin remodeling is mediated by the ATPase-coupled BRG/BRM proteins (SWI/SNF proteins), the methyltransferase harboring proteins (CARM1), and the p160 coactivator complexes that possess histone acetyltransferase activities (p160, PCAF, and CBP/p300). Coactivator complexes are recruited by an active receptor (TF recruitment), which localizes proteins that engage basal transcriptional machinery, such as the TRAP/DRIP/ARC complex. Coactivators like PGC-1 could serve as a link between the action of site-specific DNA-binding transcription factors and mRNA processing.

via histone deacetylation, and in the case of at least some orphan receptors, this may be their dominant function. Binding of corepressors to some NRs in the absence of a ligand may not be as relevant as they may be sequestered away from the chromatin, such as is the case for steroid receptors. Corepressors, however, may play a role in antagonist function at the steroid receptors as discussed later in this section. The ligand-induced conformational change in the LBD not only forces release of corepressors, but it allows association of coactivator proteins that mediate a series of events leading to transcriptional activation. For transactivation to be possible, chromatin remodeling must occur and several coactivator complexes harbor this class of activity, such as ATP-dependent chromatin remodeling complexes: the BRG/BRM (SWI/SNF) proteins and the p160 coactivator complexes that carry with them a number of proteins displaying histone acetyltransferase (HAT) activity. Coactivator complexes recruited by an active receptor may "shift," resulting in localization of proteins, including thyroid hormone receptor-associated protein/vitamin D receptor-interacting protein/SREBP-interacting complex (TRAP/DRIP/ARC), that engage the basal transcriptional machinery. Furthermore, with transcription being coupled to RNA processing, coactivators like PGC-1 serve as a link between the action of site-specific DNA binding transcription factors and mRNA processing.

II. COREGULATORS INVOLVED IN NUCLEAR RECEPTOR ACTION

A. COACTIVATORS

Based on their primary role in mediating transcriptional activation, coactivator proteins have been broadly classified into two types for the purposes of the descriptions in the upcoming sections: (1) those that are involved in chromatin remodeling and (2) those that have primary roles other than chromatin remodeling, such as interaction with basal transcriptional machinery or mRNA splicing. Several examples of coactivator structures are illustrated in Figure 2.

1. p160 Coactivators

The p160 steroid receptor coactivators (SRCs) are members of a protein family and include SRC-1 (NCoA-1) (Kamei et al., 1996; Onate et al., 1995; Torchia et al., 1997), SRC-2 (TIF2, GRIP1, NCoA-2) (Hong et al., 1996; Torchia et al., 1997; Voegel et al., 1996), and SRC-3 (ACTR, RAC3, AIB1, p/CIP, TRAM1) (Anzick et al., 1997; Chen et al., 1997; Li et al., 1997a; Suen et al., 1998; Takeshita et al., 1997; Torchia et al., 1997). The three members of the SRC family possess the ability to interact with several NRs in a ligand-dependent and ligand-independent manner and enhance NR-dependent transcription (McKenna et al., 1999).

The proteins encoded by the SRC gene family are ~160 kDa in size with a sequence similarity of 50 to 55% and a sequence identity of 43 to 48% between the three members. The p160/SRC proteins are about 1450 amino acids in length and contain distinct functional domains. The basic helix–loop–helix / Per-Arnt-Sim (bHLH/PAS) domain that is located in the N-terminal region of the protein plays a role in DNA binding and heterodimerization between proteins containing these motifs (Huang et al., 1993); however, the role of this domain in the p160s remains to be determined. The relatively conserved central region of the protein contains three LXXLL motifs called NR boxes (where L is leucine and X is any amino acid) that are responsible for interaction with NRs (Heery et al., 1997; Voegel et al., 1998) (Fig. 3A). The central and the carboxy-terminal regions contain two transcriptional activation domains (AD1 and AD2). The AD1 region is responsible for their interaction with the general transcriptional cointegrators—CREB binding protein (CBP) and p300—but does not interact with NRs (Onate et al., 1998;

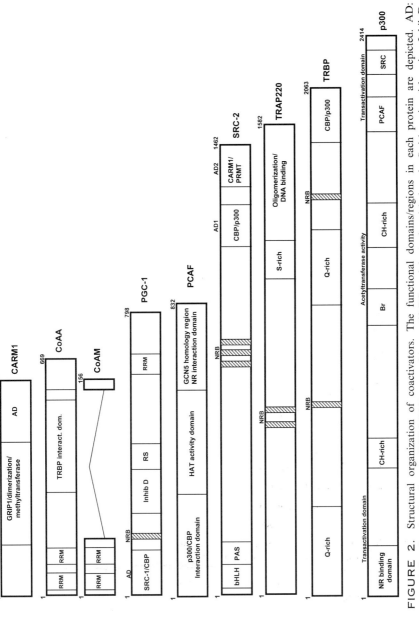

FIGURE 2. Structural organization of coactivators. The functional domains/regions in each protein are depicted. AD: transcriptional activation domain; RRM: RNA recognition motif; RS: arginine/serine-rich domain; S-rich: serine-rich region; Inhib D: inhibitory domain; NRB: nuclear receptor box (LXXLL motif); HAT: histone acetyltransferase; bHLH: basic helix–loop–helix motif; PAS: Per-Arnt-Sim region; Q-rich: glutamine-rich region; CH-rich: cysteine/histidine-rich domain; Br: bromodomain.

Voegel *et al.*, 1996). The AD1 also contains three LXXLL-like motifs that play a role in recruiting p300/CBP and the p300/CBP-associated factor (PCAF) for chromatin remodeling (Li *et al.*, 2000; Voegel *et al.*, 1998). The second activation domain (AD2) located in the C-terminal region of the SRC proteins is responsible for their interaction with the histone methyltransferases CARM1 and PRMT1 (Chen *et al.*, 1999a; Koh *et al.*, 2001). Recruitment of these histone methyltransferases to an enhancer or a promoter by binding to the AD2 domain of the SRC proteins would lead to remodeling of the chromatin and assembly of the transcriptional machinery around the promoter.

In addition to the ability of the SRC proteins to interact with histone acetyltransferases (HATs), such as p300/CBP and PCAF, recent studies

A

hSRC-1 NR1	Y	S	Q	T	S	H	K	**L**	V	Q	**L**	**L**	T	T	T	A	E	Q Q
hSRC-1 NR2	L	T	E	R	H	K	I	**L**	H	R	**L**	**L**	Q	E	G	S	P	S D
hSRC-1 NR3	E	S	K	D	H	Q	**L**	**L**	R	Y	**L**	**L**	D	K	D	E	K	D L
hSRC-1 NR4	Q	A	Q	Q	K	S	**L**	**L**	Q	Q	**L**	**L**	T	E				
hSRC-2 NR1	D	S	K	G	Q	T	K	**L**	**L**	Q	**L**	**L**	T	T	K	S	D	Q M
hSRC-2 NR2	L	K	E	K	H	K	I	**L**	H	R	**L**	**L**	Q	D	S	S	S	P V
hSRC-2 NR3	K	K	K	E	N	A	**L**	**L**	R	Y	**L**	**L**	D	K	D	D	T	K D
hSRC-3 NR1	E	S	K	G	H	K	K	**L**	**L**	Q	**L**	**L**	T	C	S	S	D	D R
hSRC-3 NR2	L	Q	E	K	H	R	I	**L**	H	K	**L**	**L**	Q	N	G	N	S	P A
hSRC-3 NR3	K	K	E	N	N	A	**L**	**L**	R	Y	**L**	**L**	D	R	D	D	P	S D
TRBP	V	T	L	T	S	P	**L**	**L**	V	N	**L**	**L**	Q	S	D	I	S	A G
TRAP220 NR1	K	V	S	Q	N	P	I	**L**	T	S	**L**	**L**	Q	I	T	G	N	G G
TRAP220 NR2	N	T	K	N	H	P	M	**L**	M	N	**L**	**L**	K	D	N	P	A	G D
PGC-1α	E	A	E	E	P	S	**L**	**L**	K	K	**L**	**L**	L	A	P	A	N	T Q

B

Class I NR box: (+)LXXLL

Class II NR box: PΦLXXLL

Class III NR box: (S/T)ΦLXXLL

Class IV NR box: (+)ΦLXXLL

FIGURE 3. (A) Amino acid sequence alignment of the NR boxes present in the NR coactivator proteins. A comparison of the NR boxes of the three p160 protein family (TRBP, TRAP220, and PGC-1α) reveals the presence of a conserved LXXLL motif. The first L of the LXXLL motif is referenced as position 1. (B) Classes of NR boxes. Four types of NR boxes can be classified according to residues present at the -1 and -2 positions (relative to the first L) of the LXXLL motif. The first three classes were identified using a phage display method to select for preferentially bound and randomly generated LXXLL motifs. The fourth class was identified after analyzing naturally occurring motifs in coactivators. The symbol $+$ represents a basic amino acid residue, and the symbol denotes a hydrophobic residue.

have also demonstrated that SRC-1 and SRC-3 (ACTR) possess an intrinsic acetyltransferase activity (Chen et al., 1997; Spencer et al., 1997). The HAT activity of SRC-1 and SRC-3 maps to the C-terminal domain of the protein and is specific toward histones H3 and H4. Thus, the synergistic acetylase activity of SRC-1 along with p300/CBP or PCAF may result in remodeling of the chromatin, permit the formation of the preinitiation complex, and enhance the transcription of the target genes.

2. p300/CBP Binding Protein

The p300/CBP proteins, which include the distinct but related proteins p300 and CBP and potentially other proteins such as p270, are part of a family that is involved in many physiologic processes, including cell proliferation, differentiation, and apoptosis (Giordano and Avantaggiati, 1999; Goodman and Smolik, 2000). These proteins function as transcriptional coactivators that are involved in multiple signal-dependent transcription events. The viral oncoproteins, such as adenoviral E1A and the SV40 large T antigen, specifically target these proteins (Arany et al., 1995; Eckner et al., 1996). The formation of viral oncoprotein complexes with p300/CBP causes a loss of cell growth control, enhances DNA synthesis, and blocks cellular differentiation (Goodman and Smolik, 2000). Evidence also suggests that the p300/CBP genes are altered in various human tumors (Gayther et al., 2000; Giles et al., 1998), which is consistent with the tumor suppressor activity revealed in studies of $CBP^{-/-}$ mice (Kung et al., 2000).

The p300/CBP proteins are 2414 and 2441 amino acids in length, respectively, and share several conserved regions that contain most of the known functional domains in the protein: (a) the bromodomain that is frequently found in mammalian histone acetyltransferases, (b) three cysteine/histidine (CH)-rich domains (CH1, CH2, and CH3), and (c) an Ada2 homology domain, which shares extensive homology with the yeast transcriptional coactivator, Ada2p (Chan and La Thangue, 2001). The CH1 and CH3 domains play a role in mediating protein–protein interactions with a number of cellular and viral proteins binding to these domains. The bromodomain that is located in the central region of the p300/CBP proteins harbors the acetyltransferase activity. Given that the p300/CBP proteins acetylate several cellular factors, the bromodomain plays a role in recognizing different acetylation motifs. Thus, both the N- and the C-terminal regions of the p300/CBP contain the transactivation domains, with the HAT domain residing in the central region of the protein molecule. Furthermore, the N-terminal region also harbors a region that interacts with the NRs, such as RXR, ER, retinoic acid receptor (RAR), GR, TR and PR; the SRC-1/PCAF interacting domain resides in the C-terminal region of the protein (Chan and La Thangue, 2001). This modular organization of the p300/CBP permits the proteins to provide a scaffold for the assembly of multicomponent transcription coactivator complexes.

3. p300/CBP-Associated Factor

The adenoviral E1A oncoprotein induces cellular transformation by interacting with proteins involved in cell growth and differentiation. A major target of the E1A oncoprotein is the p300/CBP protein family (Arany et al., 1995). The p300/CBP-associated factor (PCAF) was originally identified as competing with E1A for binding to p300/CBP (Yang et al., 1996). Initial characterization of PCAF suggested that the protein possesses intrinsic HAT activity. It has also been demonstrated, however, that the substrate for PCAF includes the general transcription factors TFIIEβ and TFIIF (Imhof et al., 1997), the sequence-specific transcription factor p53 (Gu and Roeder, 1997; Liu et al., 1999; Sakaguchi et al., 1998), and the myogenic transcription factor MyoD (Sartorelli et al., 1999). PCAF is an 832 amino acid protein that comprises three main domains. The C-terminal region of PCAF is homologous to the yeast histone acetylase, GCN5, with an amino acid identity of 64% between the two proteins. In addition, a PCAF-related gene has been discovered in humans (Smith et al., 1998). The human GCN5 protein lacks about 300 to 350 residues corresponding to the amino-terminal region of PCAF and is 62% similar to the yeast GCN5 throughout its length. The N-terminal region of PCAF is necessary for binding to p300/CBP (Yang et al., 1996). The absence of a similar N-terminal domain in yeast GCN5 and the lack of p300/CBP homologues in this organism suggests that the higher eukaryotes have acquired this domain to facilitate the interaction of GCN5 homologues with p300/CBP.

The central domain of PCAF (amino acids 352–658) contains the HAT activity, while the C-terminal region of PCAF harbors the NR binding domain (Blanco et al., 1998). Binding of PCAF to NRs occurs via the DBD of the NR, which is consistent with the observation that the interaction can be independent of ligand binding. However, PCAF may still contribute to ligand-activated NR action via recruitment by p300/CBP.

4. Coactivator-Associated Arginine Methyltransferase 1

Protein function is often modulated by post-translation modification. One such modification is the N-methylation of the side chain of guanidino arginine residues (Clarke, 1993; Gary and Clarke, 1998). Protein arginine N-methyltransferases (PRMT) catalyze the formation of asymmetric dimethyl arginine residues in proteins by transferring methyl groups from S-adenosyl methionine to the guanidino nitrogen atoms of arginine residues. These enzymes are generally methylate arginines found in RGG consensus sequences in the context of glycine arginine-rich (GAR) domains in proteins (Gary and Clarke, 1998; Liu and Dreyfuss, 1995).

Coactivator-associated arginine methyltransferase 1 (CARM1) is the only member of the PRMT family that possesses the ability to cooperate synergistically with p300, CBP, or PCAF to enhance transcriptional activation by NRs (Lee et al., 2002). CARM1 functions as a secondary coactivator through its association with the p160 coactivators and possesses the ability to methylate histones. Chromatin immunoprecipitation studies have demonstrated that steroid hormones stimulate the recruitment of CARM1 and methylation of histone H3 at promoters of stably integrated steroid responsive genes (Bauer et al., 2002; Chen et al., 1999a; Lee et al., 2002; Ma et al., 2001), indicating that transcriptional coactivation of CARM1 can be attributed at least partially to its methyltransferase activity.

CARM1 is a 608 amino acid protein and harbors several distinct domains. The central core region of about 330 amino acids that is highly conserved among other PRMT family members contains the methyltransferase activity. The N-terminal region of the central core, which is the most highly conserved in the primary amino acid sequence among family members, is composed of mixed α-helical and β-strands. The C-terminal part of the core forms an elongated nine-stranded β-barrel structure (Teyssier et al., 2002). The central core also harbors the homodimerization region, which is required for the enzyme activity. The dimer interface is formed by the contact between the $\alpha/\beta-$ region of one monomer and β-barrel structure of the other monomer. The S-adenosyl methionine binding pocket is formed by the $\alpha/\beta-$ region. The portion of the protein substrate surrounding the target arginine residues is predicted to fit in a groove between one monomer's $\alpha/\beta-$ region and the other monomer's β-barrel structure. The central core that contains the homodimerization region is responsible for its binding to GRIP1, a p160 coactivator. The GRIP1 binding activity is required for the coactivator function of CARM1 (Chen et al., 1999a, 2000). The C-terminal region of CARM1 also contains a strong autonomous activation domain, and deletion of this region abolishes the protein's transcriptional activation when fused with the GAL4 DBD (Teyssier et al., 2002).

CARM1, the unique member of the PRMT family with its functionally distinct domains, is thought to be recruited to the promoter through its interaction with the C-terminal domain of a p160 coactivator. The autonomous activation domain in the C-terminal region of CARM1 most likely synergistically acts with the methyltransferase activity in the central core region of the protein to mediate CARM1's coactivator function. Histone H3 is methylated, which results in chromatin remodeling and enhances transcription of the target genes.

5. Thyroid Hormone Receptor-Binding Protein

The thyroid hormone receptor-binding protein (TRBP) was originally identified based on its binding to the LBD of TRβ1 in a yeast two-hybrid screening system (Ko et al., 2000). TRBP (designated as NCoA6) has been shown to interact with several NRs in a ligand-dependent manner and to enhance NR-mediated transcriptional activation. TRBP, identified by several groups, was also designated as AIB3/ASC-2/RAP250/PRIP/NRC (Hermanson et al., 2002; McKenna and O'Malley, 2002). TRBP has been shown to associate with both the p300/CBP and the DRIP complexes (Ko et al., 2000; Lee et al., 1999; Mahajan and Samuels, 2000). TRBP coactivates multiple transcription factors, including AP-1 and NFκB (Ko et al., 2000), and its gene is amplified in human breast cancers (Guan et al., 1996; Lee et al., 1999). Furthermore, TRBP has been shown to interact with the RRM-containing coactivator activators CoAA and CoAM (Iwasaki et al., 2001).

TRBP, a ubiquitously expressed protein containing 2063 amino acids, has an apparent molecular mass of 210 kDa and harbors two distinct domains. The N-terminal region of the protein is rich in glutamines (Q-rich region). The central and the C-terminal regions of the protein contain two LXXLL motifs that are responsible for its binding to NRs. Studies have demonstrated that a single LXXLL motif is required for the interaction of TRBP with TR, ER, and several other NRs (Ko et al., 2002).

The C-terminal region allows TRBP to bind to a DNA-dependent protein kinase (DNA–PK) complex (Ko et al., 2000). The interaction of DNA–PK with TRBP is mediated by direct binding to the regulatory subunit of DNA–PK, Ku70 (Ko and Chin, 2003). The strong affinity of TRBP's C-terminal region with DNA–PK and poly(ADP-ribose) polymerase (PARP) suggests that TRBP may be associated with the DNA–PK complex *in vivo*. In addition to its interaction with the DNA–PK complex, TRBP was also found to be a potent substrate of DNA–PK *in vitro*. DNA–PK, a serine/threonine kinase, has been implicated as a factor in chromatin remodeling and transcriptional activation. The association of TRBP with DNA–PK suggests that TRBP may play an active role in transcription, especially in silencing of heterochromatin and in chromatin remodeling. In addition to binding DNA–PK, the C-terminal region of TRBP is also responsible for binding to DRIP and p300 (Ko et al., 2000). Because TRBP does not possess inherent HAT activity, the association with p300/CBP may permit TRBP to recruit HAT activity during the activation. Thus, in addition to binding to NRs, TRBP functions as a unique general coactivator by virtue of association with other proteins that possess chromatin remodeling and acetyltransferase activities.

6. TRAP/DRIP/ARC Complex

The thyroid hormone receptor-associated protein (TRAP) complex was first identified using biochemical methods based on its association with TRα in HeLa cells. The TRAP complex was shown to enhance the function of TRα in cell-free systems that are reconstituted with transcription factors and DNA templates containing TR-response elements (Fondell et al., 1996). Complexes related to TRAP that have been identified in other large transcriptional coregulatory complexes include the SRB/MED-containing coregulator complex (SMCC) (Gu et al., 1999; Ito et al., 1999), vitamin D receptor-interacting protein (DRIP) (Rachez et al., 1999), an SREBP-interacting complex activator recruited cofactor (ARC) (Naar et al., 1999), and an E1A-interacting complex (mediator) (Boyer et al., 1999).

The TRAP/DRIP/ARC, complex is a 1.5 to 2.0 MDa complex comprising more than 25 distinct polypeptides that range in size from 12 to 240 kDa (Ito et al., 1999; Malik and Roeder, 2000). Among the several subunits of the TRAP complex, the 220 kDa subunit (TRAP220/DRIP205/ARC205) is unique in that it demonstrates direct ligand-dependent binding to several nuclear receptors, including TRα, vitamin D receptor (VDR), RXRα, RARα, peroxisome proliferator-activated receptor α (PPARα), PPARγ, ERα, and GR (Hittelman et al., 1999; Yuan et al., 1998; Zhu et al., 1997). The TRAP220 protein spans 1582 amino acids in length with a predicted mass of 168 kDa and contains three distinct domains (Yuan et al., 1998). The central NR interaction domain contains two LXXLL motifs that are involved in the ligand-dependent binding to the various NRs. The two NR boxes of the TRAP220 protein show differential affinities for different receptors, with ligand-dependent interactions involving TRα, VDR, and PPARα being the strongest for NR box 2, and ligand-dependent interactions with their heterodimerization partner RXRα being strongest for NR box 1 (Rachez et al., 2000; Ren et al., 2000; Yuan et al., 1998). The C-terminal region of TRAP220 also includes a serine-rich domain followed by a cluster of mixed-charged amino acids. This charged cluster has been suggested to be responsible for oligomerization, DNA binding, and binding to the tumor suppressor protein p53 (Drane et al., 1997). TRAP220 protein has been proposed to target and possibly anchor the entire TRAP complex to a ligand-activated NR (Yuan et al., 1998). Based on the differential affinities of the two NR boxes for different receptors, it has been postulated that a single molecule of TRAP220 might interact simultaneously, through the two NR boxes, with both units of DNA-bound receptor heterodimers.

Of the other TRAP/DRIP/ARC subunits, TRAP170/DRIP150/ARC150 shows ligand-independent binding to GR (Hittelman et al., 1999) and is thought to facilitate the function of the complex by serving as a TRAP220-independent anchor or supplemental anchor for the complex. The TRAP150β/DRIP130/ARC130 component of the complex is thought to be involved in the E1A and RAS signaling pathways (Boyer et al., 1999; Malik and Roeder, 2000), whereas the TRAP80 subunit shows specific interactions with and is thought to mediate function of p53 and VP16 activation domains (Ito et al., 1999). The evidence that the TRAPs enhance the function of NRs in cell-free systems that are reconstituted with general transcription factors and naked DNA templates suggests that TRAPs are involved directly in the assembly or function of the preinitiation complex. This could involve either the interaction with initiation factors/coactivators that enhance the intrinsic activity of the factors or involve reversing the action of specific negatively acting coregulators. Thus, the general model for the mode of action is an initial ligand-mediated dissociation of the corepressors from promoter-bound receptors with the concomitant association of HAT-containing or HAT-interacting coactivators. This is followed by HAT-mediated acetylation of adjacent nucleosome histones and possibly other factors, which is in turn followed by the exchange of promoter-bound HATs with the TRAP complex and subsequent recruitment of transcription factors and RNA Pol. II for the formation of a functional preinitiation complex occurs (Ito and Roeder, 2001).

7. Coactivator Activator and Coactivator Modulator

The coactivator activator (CoAA) was first identified as associated with TRBP in a yeast two-hybrid screen (Iwasaki *et al.*, 2001a). CoAA functions as a general activator of transcription for several NRs and stimulates transcription through its interaction with TRBP and RNA-containing transcriptional complexes. Furthermore, CoAA is also associated with the DNA–PK/PARP complex and acts synergistically p300/CBP. Thus, CoAA appears to function as a mediator of coactivators.

CoAA and coactivator modulator (CoAM) are encoded by a single gene that is located on chromosome 11q13. Both proteins belong to a family of RNA binding proteins and harbor at least two distinct domains that are responsible for their coactivator function. CoAA protein is 669 amino acids in length with an estimated molecular mass of 69 kDa, and contains three domains. The N-terminal region of the protein contains two RNA recognition motifs (RRMs) at amino acid positions 3–68 and 81–144. Both RRMs are composed of two conserved ribonucleoprotein (RNP)—RNP1 and RNP2—consensus motifs (Iwasaki *et al.*, 2001). The RRM motif consists of four antiparallel β-strands and two α–helices that serve as an RNA binding platform (Nagai *et al.*, 1995; Shamoo *et al.*, 1997). The central region of CoAA harbors the TRBP interacting domain with an amino acid composition that is rich in serines, alanines, glycines, and tyrosines. This region also contains a unique XYXXQ motif that is present with more than 20 copies throughout the domain. The aromatic residues that are interspersed among the motifs play an important role in the protein–protein interaction between hnRNP proteins (Cartegni *et al.*, 1996).

CoAM is generated as a result of alternative splicing of exon 2 of the CoAA pre-mRNA (Iwasaki *et al.*, 2001). The CoAM protein is 156 amino acids in length and has a predicted molecular mass of 17 kDa. Similar to CoAA, the N-terminal region of CoAM contains two RRM that serve as RNA binding platforms. CoAM lacks the central TRBP-interacting domain due to the exclusion of exon 2. Furthermore, the C-terminal six amino acids of CoAM differ from those in CoAA due to a frame shift that occurs during the alternative splicing event. A frame shift at amino acid 150 with a subsequent premature termination codon gives rise to the truncated CoAM.

Although CoAA acts as an activator of transcription in conjunction with TRBP, the absence of the TRBP interaction domain in CoAM results in CoAM acting as a repressor of transcription. This is probably due to competition between CoAA and CoAM regarding the RRM function. Thus, CoAM would function as a dominant negative regulator of CoAA and may act to modulate CoAA function *in vivo*.

8. PPARγ Coactivator-1 and Related Coactivators

Peroxisome proliferator-activated receptorγ coactivator-1 (PGC-1) was initially identified as a coactivator for the nuclear receptor PPAR-γ. Expression analysis of PGC-1, designated PGC-1α, and its induction during exposure to lower temperatures have suggested that PGC-1α plays a role in energy metabolism and adaptive thermogenesis (Puigserver *et al.*, 1998). PGC-1α is preferentially expressed in brown adipose tissue and skeletal muscle and enhances transcriptional activation of the uncoupling promoter-1 (UCP-1) by TR and PPARγ. Overexpression of PGC-1α in white adipose tissue activates UCP-1 and other key mitochondrial enzymes. Thus, cell- and promoter-specific coactivator functions of PGC-1α are controlled by external environmental stimuli.

PGC-1α is 798 amino acids in length and is a protein that contains several distinct domains. The amino-terminal region of the protein has a transcriptional activation domain (AD) (Puigserver *et al.*, 1999; Vega *et al.*, 2000) and an LXXLL motif. The LXXLL motif of PGC-1α has been shown to mediate ligand-dependent interactions with NRs, such as ERα (Tcherepanova *et al.*, 2000), ERRα (Schreiber *et al.*, 2003), GR (Knutti *et al.*, 2001), PPARα (Vega *et al.*, 2000), RXR (Delerive *et al.*, 2002), PPARγ (Wu *et al.*, 2003), and TR (Wu *et al.*, 2002). The amino-terminal region also contains a domain that is responsible for PGC-1α's interaction with SRC-1 and CBP (Puigserver *et al.*, 1999). PGC-1α contains an "inhibitory" domain that represses its transcriptional activation function. Docking to recruiting transcriptional factors relieves the inhibitory activity, induces a conformational change in PGC-1α, and enables the interaction with effectors. The C-terminal region of PGC-1α contains a serine/arginine-rich domain called the RS domain and an RRM domain. The presence of these domains in PGC-1α implicates a role in the processing of pre-mRNA or in the export of mRNA from the nucleus to the cytoplasm (Graveley, 2000; Huang and Steitz, 2001). The putative role of PGC-1α in pre-mRNA processing has been corroborated by the observation that PGC-1α is involved in pre-mRNA splicing (Monsalve *et al.*, 2000). Furthermore, PGC-1α has been found to associate with RNA polymerase II and splice proteins like U1 snRNP and the SR proteins; the C-terminal region of PGC-1α is responsible for its association with these splicing factors.

PGC-1β, another isoform of PGC-1, has also been identified (Meirhaeghe *et al.*, 2003). The two variants of PGC-1β, PGC-1β-1a and PGC-1β-2a, are expressed predominantly in the heart, brown adipose tissue, skeletal muscle, and brain. Results, *in vitro* provide evidence that PGC-1β-1a functions as a coactivator for TRβ, PPARα, PPARγ, and GR (Meirhaeghe *et al.*, 2003). Furthermore, overexpression of PGC-1β in myoblasts suggests that the isoform could play a role in constitutive nonadrenergic-mediated mitochondrial biogenesis. Unlike PGC-1α, the expression of PGC-1β is not induced in response to cold, obesity, or exercise. PGC-1β is also not induced in the liver, suggesting that unlike PGC-1α, it may not be involved in hepatic gluconeogenesis.

PGC-1 related coactivator (PRC) is a 177 kDa protein, spanning 1664 amino acids, that was originally identified as a transcriptional coactivator for the nuclear respiratory factor-1 (NRF-1) (Andersson and Scarpulla, 2001). PRC is ubiquitously expressed in a cell cycle-dependent manner in a variety of tissues and cell lines. The overall sequence of PRC is very different from PGC-1; however, the spatial pattern for the specific regions of significant sequence similarity is conserved (Andersson and Scarpulla, 2001). These include the amino-terminal transcriptional activation domain followed by the LXXLL motif and the C-terminal region containing the RS domain and the RRM. The role of PRC as transcriptional coactivator has been suggested based on nuclear localization. This role is also suggested by PRC's ability to enhance transcription of target promoters, which is dependent on NRF-1, through its direct interaction with the DBD of NRF-1. Deletion analysis of PRC indicates that, similar to PGC-1, the N-terminal region of the protein functions in transcriptional activation, whereas the C-terminal region is implicated in the processing of pre-mRNA.

9. ATP-Dependent Nucleosome Remodeling Factors

A role for ATP-dependent chromatin remodeling in NR action was first suggested after observation that GR was nonfunctional in yeast strains lacking SWI proteins (Yoshinaga *et al.*, 1992). Chromatin remodeling proteins whose activities are dependent or associated with ATP hydrolysis have recently been reviewed (Peterson, 2002a,b; Vignali *et al.*, 2000), and the prototype of this class is the yeast SWI2/SNF2. SWI2/SNF2 contains sequence motifs closely related to those founding DNA-dependent ATPases/DNA helicases (Eisen *et al.*, 1995). The SWI2/SNF2 homologous protein present in *Drosophila* is called *brahma* (*brm*) (Tamkun *et al.*, 1992), and the homologues in humans are called BRG1 and hBRM (Chiba *et al.*, 1994). BRG1 and hBRM are 1647 and 1572 amino acids in length respectively, and have been shown to enhance transcriptional activation by GR and ER (DiRenzo *et al.*, 2000; Fryer and Archer, 1998; Muchardt and Yaniv, 1993). The BRG1 and hBRM proteins harbor several distinct domains. The N-terminal region is rich in glycine/proline/serine residues and is termed the

GPS domain. This domain is immediately followed by a glutamine repeat in hBRM and a glycine/proline repeat in BRG1 (Chiba et al., 1994). The central region contains the helicase-related motifs, and the C-terminal region contains charged amino acids and also harbors the bromodomain. The BRG-1/hBRM proteins 2 MDa and BRG-1-associated factors (BAFs) exist in a complex of (Wang et al., 1996) that regulates transcription through its ability to remodel chromatin. Both BRG-1 and hBRM display ATPase activity that is required for the complex to function in remodeling (Kingston and Narlikar, 1999). The BRG-1/hBRM complex activates transcription through ligand-dependent recruitment of the proteins to the receptors, thereby targeting BRG-1/hBRM to the promoters of the target genes and subsequently promoting the activity of other recruited factors that alter the acetylation state of chromatin.

10. Steroid Receptor RNA Activator

The majority of coactivators and cointegrators that play a role in mediating transcription of target genes are proteins that function through direct interactions with other proteins or via the formation of protein complexes. Steroid receptor RNA activator (SRA) is unusual and unique and differs from other coactivators in two main ways. First, SRA transcripts are not translated; hence, this coactivator acts as an RNA and not as a protein. Second, whereas most receptor-interacting factors act as general coactivators, SRA is specific for steroid-hormone receptors and mediates transactivation via the N-terminal AF-1.

SRA exists as several different isoforms (Emberley et al., 2003; Lanz et al., 1999) that can be grouped into three splice variant classes based on their sequences outside a common core region. The three splice classes of the SRA range in size from 0.7 to 1.5 kb, contain an identical core of 687 nt, and differ in their 5' and 3' extensions. The tissue expression profile analysis indicated that SRA was expressed in all tissues but was predominantly expressed in the liver and skeletal muscle.

Studies have demonstrated that SRA is a component of distinct RNP complexes and is associated with SRC-1. Furthermore, this RNP complex containing SRA and SRC-1 is recruited by steroid hormone receptors through the receptor's AF-1 and AF-2 domains, respectively. Secondary structure prediction of SRA has identified six substructures in the core region of SRA that are important for SRA coactivation (Lanz et al., 2002). The identification of SRA-interacting proteins, however, would be essential for understanding the molecular mechanism of SRA's coactivation of steroid receptors. Because no evidence has yet been obtained for the direct binding of SRA to steroid receptors, it has been proposed that SRA functions to confer specificity on RNP complexes and enable steroid receptor-mediated transcription. Studies have shown that SRA-containing RNP complexes enhance transcription by integrating coregulator activities on selective binding to steroid receptors (Shi et al., 2001; Watanabe et al., 2001). The specificity of the SRA–protein interactions would be dictated by spatial folding of the RNA as a function of its secondary and tertiary structure. Thus, unlike the well-defined domains in the coactivator proteins, the coactivation function of SRA is likely an integration of individual activities mediated through the secondary structure of the RNA, which is distributed over the entire region of the SRA RNA molecule.

B. COREPRESSORS

Observed inhibition of basal transcription by some unliganded NRs, such as TR and RAR, led to the discovery of nuclear receptor corepressor (NCoR) proteins that are directly involved in repression of target genes. NCoR and silencing mediator for retinoic acid and thyroid hormone receptors (SMRT) were first identified via specific protein–protein interaction with unliganded nuclear receptors in 1995 (Chen and Evans, 1995; Horlein et al., 1995). These newly identified corepressors were found in multiprotein repression complexes with unliganded members of the nuclear receptor superfamily, proteins containing histone deacetylase (HDAC), and components of the basal transcriptional machinery, such as TFIIB (Muscat et al., 1998; Wong and Privalsky, 1998a). NCoR and SMRT have been shown to interact with various transcription factors in a multiprotein complex and actively repress transcription by chromatin modification at the lysine residues of histone tails, maintaining them in a deacetylated state (Hassig and Schreiber, 1997; Hu et al., 2001; Yu et al., 2003). These actions directly oppose the histone acetylation function of many nuclear receptor coactivator proteins and, as such, provide for dynamic chromatin remodeling and an additional layer of transcriptional control by members of the NR superfamily.

NCoR and SMRT are 270 kDa proteins that contain distinct functional domains responsible for interaction with NRs, as well as interaction and activation of HDAC proteins, ultimately resulting in targeted repression of transcription. The C-terminal portions of NCoR and SMRT contain two receptor interaction domains (RIDs) that are responsible for NR binding (Fig. 4). Examination of these RIDs show conserved CoRNR motifs necessary for NR interaction. The CoRNR boxes closely resemble the previously described NR box used by NR coactivators to bind ligand-activated nuclear receptors (Hu and Lazar, 1999; Nagy et al., 1999; Perissi et al., 1999). The amino-terminal portions of NCoR and SMRT contain multiple distinct domains responsible for repression (Li et al., 1997b). The repression domains (RDs) are involved in recruitment and activation of HDAC enzymes, thus leading to targeted hypoacetylation of histones and subsequent repression of transcription. In the amino-terminal portion of the corepressor proteins reside two distinct SANT domains (a domain common

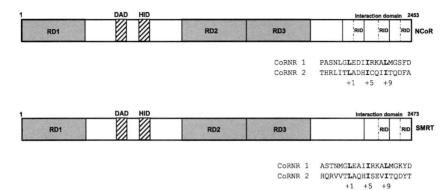

FIGURE 4. Structural organization of corepressors, NCoR and SMRT. The N-terminal region contains multiple repression domains shown in gray (RDs). The RIDs in the C-terminal region each contain a conserved CoRNR box sequence as denoted by the dashed lines in the RIDs. The specific amino acid sequences of the CoRNR boxes are shown with the conserved LXXXIXXXL sequence in bold. The N-terminal portion of the corepressors also harbors the two matched SANT domains designated by diagonal boxes in the DAD and the HID, respectively.

to SWI/SNF, Ada, N-CoR, and TFIIIB) that are integral to the repressive function of the NCoR and SMRT proteins. A conserved deacetylase activating domain (DAD) is present in both NCoR and SMRT and contains one of the two SANT domains. The DAD is not only critical in the interaction of corepressors with HDAC3 but is also necessary to create and activate the HDAC enzyme complex (Guenther et al., 2001). The second conserved SANT domain is part of a histone interaction domain (HID) and serves to increase the affinity of the active corepressor/HDAC complex for histones. The HID is also thought to have an inhibitory effect on HAT activity, which is a key mechanism involved in NR coactivation as described earlier (Yu et al., 2003).

The varied functional domains of NCoR and SMRT allow the proteins to use multiple complementary mechanisms to repress transcription to targeted promoters in DNA. Both NCoR and SMRT have not only directly associated with multiple members of the HDAC family, including classes I and II members HDACs 3, 4, 5, and 7, but also to interact with HDACs 1 and 2 via association with Sin3 (Thiagalingam et al., 2003). These interactions, combined with the HDAC activating and enhancing functions of the DAD and HID domains, cooperatively allow active transrepression through NR targeting to specific response elements in the promoter regions of target genes either by unliganded NRs or by antagonist-bound NRs.

III. NUCLEAR RECEPTOR–COREGULATOR RECOGNITION

A. COACTIVATORS

The recruitment of coregulators, either coactivators or corepressors, by NRs is essential for these transcription factors to regulate expression of target genes. Thus, the mechanism used by NRs to recognize coregulators is crucial for understanding how their transcriptional activity is regulated. Although there are examples of both ligand-dependent and ligand-independent interactions between NRs and coregulators, the mechanism by which ligands regulate the interaction is particularly intriguing since this function is modulated by both physiologic and pharmacological agents. Modulation of coactivator and corepressor recognition by the ligand-dependent AF-2 has been an area of intense investigation for the past several years.

Although there are likely several classes of recognition interfaces between NRs and coregulators, especially when one considers the AF-1 region, two motifs have been well characterized: the coactivator NR recognition motif (NR box) and the corepressor NR recognition motif (CoRNR box). The NR box was identified during examination of the NR interaction regions of p/CIP, TIF1 α, RIP140, GRIP1, and SRC-1 (Ding et al., 1998; Heery et al., 1997; Le Douarin et al., 1996; Torchia et al., 1997). The NR box is composed of a core α-helical LXXLL motif. In addition, residues flanking the core also play an important role in NR recognition. The NR box is found in a wide range of coactivators as indicated in Figures 2 and 3A and the NR interaction domain may be composed of just a single NR box or an array of repeated NR boxes. A single NR box, however, is sufficient for interaction although there may be cooperativeness displayed when more than one NR box is present in the coactivator's NR interaction domain. The amphipathic α-helical NR box displays the hydrophobic leucine side chains on one surface of the helix, which indicates that it most likely recognizes a hydrophobic surface on the LBD of the NR.

Feng et al. (1998) mapped the surface of the LBD of TR with point mutations using the X-ray structure as a guide. They identified a region that, when mutated, inhibited GRIP1 binding but did not affect ligand binding (Feng et al., 1998). This region was relatively conserved in the NR superfamily and was composed of charged and hydrophobic residues on the periphery, but only hydrophobic residues in the core surface and was predicted to be large enough to accommodate the LXXLL helix. The putative interaction surface was composed of residues contributed by helices 3 through 6 as well as helix 12 (H12) of the TR LBD. The potential contribution of H12 to the coactivator interaction surface is particularly interesting given that the position of H12 is dependent on ligand binding. Structures of several NR LBDs have been determined by X-ray crystallographic methods illustrating that the LBD is a globular structure composed of a three-layered α-helical sandwich (Fig. 5A, B, C). The carboxy-terminal H12, which had previously been identified as the activation domain AF-2 (Baniahmad *et al.*, 1995; Danielian *et al.*, 1992), appears to be highly mobile and is stabilized upon ligand binding into a position that completes the proposed recognition surface for coactivators described previously. This is consistent with the proposed "moustrap" model in which H12 in the apo-receptor is highly mobile and the coactivator binding surface is nonfunctional. Once the ligand is bound, H12 is stabilized and packs itself against the LBD, yielding the final components of a functional coactivator binding surface.

Crystal structures of the NR box complexed with several ligand-bound NRs confirmed these predictions. Nolte *et al.* (1998) solved the structure of liganded PPARγ complexed with an SRC-1 fragment containing two copies of an NR box, which illustrated several interesting features of the molecular basis for coactivator recognition by NRs. As had been demonstrated for other liganded NRs, H12 was packed against the core globular structure, and its positioning was stabilized by interactions with the ligand itself bound in the ligand binding pocket (Nolte *et al.*, 1998). This ligand-dependent positioning of H12 completed the formation of the recognition surface for the NR box on the LBD, which also included helices 3, 4, and 5 (H3–H5).

As shown in Fig. 5D, the leucine side chains displayed on one surface of the NR box helix face the hydrophobic surface recognition surface of the LBD. Side chains of the leucines at positions +1 and +5 are entrenched in a hydrophobic groove while the leucine side chain of the +4 position lies across the plane of the hydrophobic surface. The X residues of positions +2 and +3 are exposed to solvent which is consistent with their lack of conservation in a variety of coactivator NR boxes. Hydrophobic interactions provide one component of the recognition surface, whereas a series of hydrogen bonds provides yet another component.

The "charge clamp" motif consists of two charged residues of the NR LBD, a lysine residue (contributed by H3), and a glutamic acid residue (contributed by H12 (AF-2)) that effectively clamp the NR box α–helix in place by hydrogen bonding with the peptide backbone of the LXXLL motif. The γ-carboxylate of the glutamic acid residue (H12) hydrogen bonds with the backbone amides in the amino-terminal of the LXXLL motif, while the ε-amino group of the lysine residue (H3) hydrogen bonds with the carboxy-terminal backbone carbonyls of the LXXLL motif. This provides a capping interaction at both ends of the LXXLL motif that is believed to play an essential role in orientation and placement of the NR box in the coactivator binding groove of the NR (Nolte *et al.*, 1998). In addition, since H12 contributes the glutamic acid component of the charge clamp, correct positioning of this residue to provide a surface competent for coactivator binding requires ligand binding so that the positioning of this helix is stabilized. Hence, this provides a mechanism for ligand-dependent NR

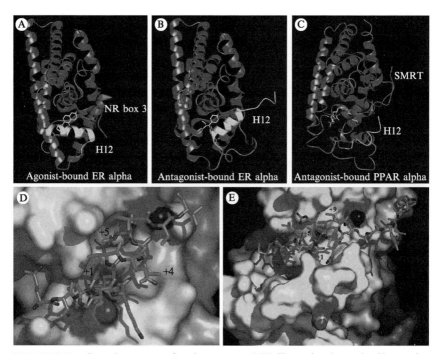

FIGURE 5. Crystal structures of nuclear receptor LBD illustrating the mode of interaction with coregulators. (A) The 2.45 Å structure of the human ER alpha (ERα) LBD, residues 306–549, in complex with the agonist 17β-estradiol and a TIF2 NR box 3 derived peptide, demonstrating the nuclear receptor–coactivator interface (1GWR.pdb; Warnmark et al., 2002). Helix 12 (H12) is highlighted in white, and the TIF2 NR box 3 peptide is as indicated. (B) The 1.90 Å structure of the human ERα LBD, residues 306–551, in complex with the antagonist 4-hydroxytamoxifen (3ERT.pdb; Shiau et al., 1998). Note the positioning of H12 in the identical groove that was occupied by NR box 3 in Figure 5A is highlighted in yellow. (C) The 3.00 Å structure of the human PPAR alpha (PPARα) LBD, residues 200–468, in complex with the antagonist GW6471 and a SMRT CoRNR box-derived peptide (1KKQ.pdb; Xu et al., 2002). Note the unique positioning of H12 that allows for the extended LXXXIXXXL CoRNR box peptide to bind. The LXXXIXXXL motif is longer than the LXXLL motif; thus, the LXXXIXXXL can only bind when H12 is not positioned in the agonist mode. H12 is highlighted in white, and the SMRT CoRNR box peptide is highlighted in white. (D) Surface representation of the interactions observed in the 2.30 Å structure of the human PPAR gamma (hPPARγ) LBD residues 207–477, in complex with the agonist rosiglitazone (BRL-49653) and a coactivator SRC-1-derived peptide is shown (2PRG.pdb; Nolte et al., 1998). The charge clamp residues K301 (dark grey) and E471 (light grey) are highlighted on the surface. The hydrophobic leucine residues of positions +1, +4, and +5 of the NR box are labeled to illustrate their recognition mode of the NR surface. (E) Surface representation of the interactions observed in the PPARα LBD–GW6471 complex and the SMRT CoRNR box peptide (1KKQ.pdb; Xu et al., 2002). PPARα charge clamp residues K292 (dark grey) and E462 (light grey) are highlighted on the surface. Note the displaced positioning of E462 from H12, allowing LXXXIXXXL motif binding. Positions +1, +5, and +9 of the SMRT CoRNR box peptide (LXXXIXXXL motif) are labeled.

recruitment of transcriptional coactivators. Structures of both ER and TR bound to NR box peptides confirmed these observations (Darimont *et al.*, 1998; Shiau *et al.*, 1998), but the studies with the longer SRC-1 fragment also suggest a mechanism for NR box cooperativeness in coactivators that contain multiple NR boxes. A single SRC-1 fragment can contain two NR boxes bound to each PPARγ homodimer, with one NR box associated with each monomer. This observation is consistent with the higher affinity noted for proteins containing multiple NR boxes versus single NR boxes (Gee *et al.*, 1999). This observation, however, does not rule out the possibility that an NR dimer could potentially simultaneously recognize NR boxes from distinct coactivator proteins.

Structures of a variety of NR LBDs show that the glutamic acid and lysine charge clamp residues of the coactivator LXXLL recognition surface are absolutely conserved with only a couple of exceptions in the superfamily. It is interesting to note, however, that there are numerous examples of coactivator specificity displayed by various NRs, even though the minimal recognition surface of the LBD and the LXXLL motif do not yield much diversity for providing specificity. For example, GR displays preference for the third NR box of GRIP1, whereas TR prefers the second box (Darimont *et al.*, 1998; Ding *et al.*, 1998; Leers *et al.*, 1998). ERα has been demonstrated to have a strong preference for the second NR box of SRC-1, while RAR prefers the third motif (Bramlett *et al.*, 2001; Ding *et al.*, 1998, Mak *et al.*, 1999; Torchia *et al.*, 1997). Even between subtypes of a particular receptor, such as ERα or ERβ, significant NR box preferences may be exhibited (Bramlett *et al.*, 2001). Although these studies focused on receptor selectivity for NR boxes, at a more macroscopic level receptors have demonstrated specificity for entire coactivator proteins. This specificity suggests that the selectivity may be physiologically relevant (Ding *et al.*, 1998; Kalkhoven *et al.*, 1998; Warnmark *et al.*, 2001; Zhou *et al.*, 1998). The specificity appears to be generated by unique amino acid residues flanking either side of the core LXXLL motif, which was originally suggested based on the observation with the crystal structures residues beyond the core motif made contacts with the LBD surface in regions that were relatively receptor specific. McInerney *et al.* (1998) examined the carboxy-terminal NR box residues for their role in receptor selectivity and determined that the +12 and +13 residues (relative to the first L) provide specificity for ER, whereas the +6 residue is critical for RAR functionality.

Numerous investigators have focused on the amino-terminal residues of the LXXLL motif, revealing its critical nature in receptor selectivity and pharmacology (Coulthard *et al.*, 2003; Heery *et al.*, 2001; Mak *et al.*, 1999; Needham *et al.*, 2000; Northrop *et al.*, 2000a). Using a phage display method to select for preferentially bound, randomly generated LXXLL motifs, Chang *et al.* (1999) defined three classes of motifs based on their residues at the −1 and −2 positions. An additional fourth class of LXXLL motif was later added after analysis of naturally occurring motifs in coactivators (Bramlett and Burris, 2002) (Fig. 3B). Nearly all of the natural LXXLL motifs or those selected from random generation by phage display contain a hydrophobic residue at the −1 position, which is consistent with the crystal structure illustrating that the hydrophobic side chain of the −1 amino acid makes contact with the nonpolar surface of the LBD at this position (Fig. 5D). A variety of hydrophobic amino acids appear to be tolerated at this position, including isoleucine, leucine, valine, and methionine (Fig. 3B). NR boxes from classes II, III, and IV display this conserved hydrophobic residue, and only class I boxes deviate from this pattern and contain a basic residue at the −1 position. At the −2 position, the class I box does not appear to display a preference for a particular residue, while the class II box contains a proline, the class III box contains either a serine or threonine, and the class IV box contains a basic residue (Fig. 3B).

Interestingly, these various classes of NR boxes provide patterns of receptor selectivity. For example, several laboratories have demonstrated that the class II box that contains a proline at the −2 position exhibit, preference for TR, ERβ, and GR over receptors such as ERα, AR, and PR (Chang et al., 1999; Ko et al., 2002; Northrop et al., 2000). Consistent with this observation, this particular class of NR box is found in two coactivators, TRAP220 and TRBP, that were initially identified and characterized based on their strong TR coactivation activity (Ko et al., 2000; Yuan et al., 1998). The receptor selectivity of some members of this class of NR box can also apparently be regulated by post-translational modification. The NR box of TRBP has a serine at the −3 position to yield the LXXLL motif SPLLVNLL, which contains a consensus MAP kinase phosphorylation site (Ko et al., 2002). Phosphorylation of the serine residue eliminates the ligand-dependent recruitment by ERβ but interestingly increases ligand-independent recruitment, suggesting that post-translational modification of a coactivator may directly regulate the transcriptional activity of an NR. Post-translational modification has also been demonstrated to affect the function of the coactivators containing a class I box. The p160s contain class I NR boxes, among others, and the lysine residue at the −1 and −2 positions has been shown to be a target of acetylation by CBP/p300, resulting in a loss of the coactivator's affinity for the receptor (Chen et al., 1999b). This was proposed as one potential mechanism to control coactivator dissociation from an NR after activation has occurred and may be associated with coregulator exchange at the promoter. The class III NR box exhibits particularly interesting characteristics concerning its mode of interaction with the coactivator binding surface of the NR, which appears to be dependent on the serine or threonine at the −2 position. Before it was characterized as a component of the charge clamp, the conserved glutamic acid residue contained in H12 of the LBD was shown to be essential for the transcriptional activity of all ligand-activated members of the NR superfamily. During an analysis of randomly generated LXXLL motifs, however, it was noted that class III NR boxes did not require the charge clamp glutamic acid residue of ER for binding activity (Chang et al., 1999). Although the charge clamp was not required, deletion of H12 resulted in loss of binding to the motif, indicating that some component of the helix was still required for recognition. This loss of binding was also consistent with the observation that the binding was still ligand-dependent. The class III NR box is found in several coactivators, including PGC-1 and it was demonstrated that in the context of a natural coactivator, neither TR nor PPARγ required an intact charge clamp to be coactivated by PGC-1 (Wu et al., 2002, 2003). The lack of requirement for the charge clamp was demonstrated to be a function of the unique serine residue at the -2 position, since mutation of this residue to an alanine resulted in recovery of the requirement (Wu et al., 2003).

B. COREPRESSORS

Corepressors have been demonstrated to interact with a subset of NR superfamily members in a reversed ligand-dependent manner relative to that of coactivators. For a limited number of NRs, corepressors are bound in the absence of a ligand and released upon ligand binding. Other NRs may be constitutively bound to corepressors, which is consistent with their unvarying transcriptional repressive activity, while others may only bind to corepressors in context of a pharmacological antagonist. Both NCoR and SMRT contain two independent NR interaction domains, RID-1 and RID-2, that display significant receptor preference (Hu et al., 2001; Wong and Privalsky, 1998b). When these interaction domains were closely examined, a motif similar to the coactivator NR box was identified (Hu and Lazar, 1999; Nagy et al., 1999; Perissi et al., 1999). This motif, termed the CoRNR box, contained an additional three amino-terminal amino acids relative to the NR box resulting in the following consensus sequence: LXXXIXXXL. This motif was predicted to be an amphipathic α–helix with an extra turn of the helix relative to the NR box. Mutation of the conserved residues in the context of the full-length corepressor resulted in the inability of the protein to interact with NRs. In addition, peptides containing the CoRNR box motifs interacted with NRs in the expected fashion in the absence of a ligand, which is consistent with their role in mediation of protein–protein recognition. The NR interaction preferences displayed by the RID-1 and RID-2 interaction domains were also reflected by the individual CoRNR box polypeptides. The preferences were mediated by these short motifs, and sequences flanking either end of a motif were responsible for the preferential binding pattern.

The similarity between the NR and CoRNR box motifs led to suggestions that they may recognize the same area of the NR LBD. The additional three amino acids in the CoRNR box yielding the extra turn of the helix, however, could not be accommodated by an NR with H12 positioned in the ligand-bound conformation due to steric hindrance. In the apo-receptor state, H12 is not folded against the LBD as a component of the recognition surface, and the longer CoRNR box helix may in fact recognize this surface, which significantly overlaps with the NR box recognition surface (Hu and Lazar, 1999; Marimuthu *et al.*, 2002; Nagy *et al.*, 1999; Perissi *et al.*, 1999; Renaud *et al.*, 2000). The role of H12 in displacing the corepressor was predicted prior to knowing the crystal structure of an NR or the cloning of a corepressor (Baniahmad *et al.*, 1995). The $\tau 4$ activation domain (H12) was shown to be required for the release of a proposed corepressor's silencing activity, and it was proposed that a conformational change induced by a ligand resulted in $\tau 4$ displacing the corepressor and allowing coactivator to bind.

Recently, a crystal structure of the CoRNR box peptide associated with an NR was solved, revealing the nature of the protein–protein interaction (Fig. 5C). The SMRT CoRNR box co-crystalized with PPARα bound to the antagonist GW6471, which demonstrated that indeed there was significant overlap of the CoRNR and NR box recognition surface (Xu *et al.*, 2002b). H12 was positioned against H3, leaving adequate space for the additional turn of the CoRNR box helix to occupy the space once held by H12 in the activated state of the NR LBD. Interestingly, although the glutamic acid charge clamp residue from H12 was relocated and not used for CoRNR box recognition, the remaining half of the charge clamp from H3, lysine 292, formed a series of hydrogen bonds with the carboxy-terminal carbonyls of the CoRNR box peptides, serving to cap the corepressor helix just as it did for the coactivator helix.

Although the CoRNR box helix is amphipathic, it diverges from the classic α-helical structure, allowing for the $L+1$, $I+5$, and $L+9$ residues to align on the same face of the helix to enable interaction with the hydrophobic recognition surface of the LBD. Comparison of the structures of the LXXLL domain to the LXXXIXXXL domain showed that the $L+1$ position of the NR box corresponds to the $I+5$ position of the CoRNR box. One key structural feature that allows distinction between the two motifs beyond the additional helical turn is the large leucine side chains of the NR box at the $+1$ and $+5$ positions, which prevent the motif from binding to the LBD surface in the absence of stabilization by H12.

IV. PHARMACOLOGY OF NUCLEAR RECEPTOR–COREGULATOR RECOGNITION

Receptor ligands modulate the activity of NRs by binding to the ligand binding pocket in the LBD and altering the conformation of the LBD to modulate the receptor's ability to interact with coregulator proteins. The conformational change induced by a ligand has been characterized using a variety of biophysical techniques; however, the molecular distinctions differentiating agonist action from antagonist action were not delineated until the crystal structures of the LBDs were determined. Positioning of H12 is critical for translation of the pharmacological character of a ligand to alteration of function of the NR. If a ligand correctly positions H12 in such a manner that the NR box can be recruited efficiently, it functions as an agonist. If the ligand binds to the LBD and does not position H12 in a productive manner, the ligand functions as an antagonist.

The crystal structure of ER bound to raloxifene, a ligand with antagonist character, revealed the flexible positioning of H12 and indicated that antagonists induce distinct positioning of this carboxy-terminal helix (Brzozowski et al., 1997). A second ER crystal structure with the antagonist tamoxifen demonstrated that H12 itself could mimic the NR box motif with the sequence LLEML instead of LXXLL, resulting in positioning of H12 in the coactivator groove and preventing coactivator binding (Shiau et al., 1998). Hence, in both cases, the raloxifene and tamoxifen molecules extend from the ligand binding pocket and sterically prevent productive positioning of H12. In the case of ICI 164384, a pure ER antagonist, an aliphatic extension from the steroidal core of the ligand also extends from the ligand binding pocket; however, the extension is significantly longer and lies along the surface of the LBD, resulting in instability of H12 positioning (Pike et al., 2001). The antagonist itself is not necessarily required to sterically hinder positioning of H12. Shiau et al. (2002) described an ERβ antagonist that appears to allosterically place H12 in an antagonist position. The structure of antagonist RU486 complexed with GR revealed yet another position of H12. In an unusual homodimer structure of each individual H12 subunit blocks the coactivator binding groove of its respective partner (Kauppi et al., 2003). Thus, in all these cases, the unique conformation induced by an antagonist positions H12 in such a manner that prevents access to the coactivator binding groove.

Interestingly, the structure of a genestein, a partial agonist, complexed with ERβ has also been solved, providing additional insight into H12 positioning (Pike et al., 1999). Although genestein clearly displays partial agonist activity, H12 is in a position that resembles more of the antagonist positioning indicated in the raloxifene or tamoxifen structures. Thus, H12 of a partial agonist-bound NR is likely in equilibrium between a state productive for coactivator binding and an antagonist state, and the degree to which it occupies a particular state would reflect its level of agonism. This supports the notion that H12 is very flexible and that the agonist/antagonist nature of a particular ligand is a function of how it modulates the position and stability of H12.

Although the agonist/antagonist nature of a particular ligand could be explained in terms of efficacy of coactivator recruitment, several researchers have demonstrated that corepressors also play an essential role in NR pharmacology. The ER agonist/antagonist character of tamoxifen was demonstrated to be a function of the relative levels of the coactivator SRC-1 and the corepressor SMRT (Smith *et al.*, 1997). Overexpression of the coactivator resulted in increased agonist character, whereas overexpression of the corepressor led to increased antagonist character. The relative expression ratios of coactivators to corepressors have been demonstrated to play an important function in determination of the agonist/antagonist character of ligands for several other NRs. Szapary *et al.* (1999) demonstrated that the relative coregulator levels not only modulate the level of agonism of a GR ligand, but also affect the potency of a particular compound.

The coactivator: corepressor ratio also determines the pharmacological character of PR ligands, such as RU486, dexamethasone mesylate, and dexamethasone oxetanone, in a variety of cell contexts (Giannoukos *et al.*, 2001; Liu *et al.*, 2002). ER has been shown to interact with SMRT and NCoR, suggesting that antagonists may be able to stabilize an NR–corepressor complex (Lavinsky *et al.*, 1998; Shang *et al.*, 2000; Smith *et al.*, 1997). Recruitment of a corepressor by ER has been characterized as either antagonist-dependent (Lavinsky *et al.*, 1998) or antagonist-independent (Smith *et al.*, 1997), and the observation that corepressors play a role in antagonist action was quickly extended to PR and RU486 (Zhang *et al.*, 1998). The role of NCoR was confirmed in a more physiologic model in which it was noted that tamoxifen displayed considerably higher agonist character in fibroblasts isolated from $NCoR^{-/-}$ mice (Jepsen *et al.*, 2000). These findings are of particular interest because they demonstrate that NR antagonism is not necessarily a passive event preventing coactivator recruitment and that it may also result from active recruitment of corepressors playing a direct role in target gene silencing. At least in some circumstances, antagonist-mediated recruitment of corepressors is mediated by the CoRNR box motif (Xu *et al.*, 2002a). The PPARα antagonist GW6471 was shown to disrupt interactions of the receptor with NR box motifs and promote interaction with the CoRNR box motif. Xu *et al.* (2002a) solved the crystal structure of PPARα bound to the antagonist GW6471, illustrating the role that an antagonist can play by causing a stable interaction with a corepressor via the CoRNR box motif (Fig. 5C). GW6471 cannot stabilize positioning of H12 in an agonist position like structurally related PPARα agonists and, via an extension, displaces H12. This displacement allows for a productive binding groove to be formed for the CoRNR box motif binding.

Beyond the unique pharmacological profiles that can be derived from coactivator versus corepressor preferences, yet another area of active investigation exists in which the relative preference for a particular NR's coregulator binding is regulated by the identity of the ligand to which it is bound. As previously described, NRs often display relative preferences for specific coactivators and corepressors; however, it appears that this preference is regulated by the nature of the bound ligand. This area has been heavily investigated because of the character of a certain NR ligands class known as *selective modulators*. The prototypic selective modulator is the selective estrogen receptor modulator (SERM), exemplified by tamoxifen and raloxifene. These ligands display tissue and promoter/target gene context dependent agonist/antagonist character. For example, tamoxifen acts as a strong partial agonist in uterine tissue but as an antagonist in breast tissue (Bryant, 2001). This is a complex profile because even in the same cell type, SERMs may display a differential pharmacological character depending on the target gene examined (Bramlett and Burris, 2003). Interestingly, the relative expression levels of coregulators overlaid with ligand-specific preferences for the coregulators appear to be important for this unique pharmacological profile of the selective modulator (Shang and Brown, 2002). Distinct preferences for particular coactivators specified by ligands have been shown for a number of NRs. Kodera *et al.* (2000) noted that while the natural PPARγ agonist 15-deoxy-Δ12,14 prostaglandin J2 was able to mediate recruitment of a variety of NR box-containing coactivators, including the p160s and TRAP220 to the receptor, the synthetic agonist troglitazone did not recruit the same array of coactivators. Similar observations have been made with VDR: while the natural hormone 1α,25-dihydroxyvitamin D_3 mediates recruitment of all three p160 to the receptor, the synthetic VDR agonist 22-oxa-1α,25 dihydroxyvitamin D_3 is only able to mediate recruitment of SRC-2 (Takeyama *et al.*, 1999). Distinct ligands for ER also appear to result in specific affinities for various coactivators (Kraichely *et al.*, 2000; Wong *et al.*, 2001), and this is apparently driven by distinct affinities for the NR box motifs specified by the conformation induced by a particular ligand (Bramlett *et al.*, 2001).

V. CONCLUSION

The observed pharmacological character of an NR ligand is a combination of a variety of factors. Much of the character is a function of the specific conformation induced by the ligand because this drives the relative affinity for coactivators. At a macroscopic level, the ligand specifies whether H12 will adopt a position competent for coactivator binding or for defining the agonist/antagonist character of the ligand. If the conformation is competent for coactivator binding, the ligand determines the relative equilibrium state of the helix, specifying the level of agonist character. As discussed previously in this chapter, simply preventing coactivator binding does not completely characterize the array of NR antagonists that have been described because they may also be involved in active stabilization of an NR–corepressor complex. At a more microscopic level, each ligand defines a specific conformation in the coactivator recognition surface that is displayed to the cellular milieu, leading to a unique pattern of coregulator recruitment and thus potentially novel pharmacological pattern of activity. Finally, since each tissue or cell type may differentially express coregulators with distinct coactivator: corepressor ratios, novel pharmacology can also be driven in this manner.

REFERENCES

Andersson, U., and Scarpulla, R. C. (2001). PGC-1-related coactivator, a novel, serum-inducible coactivator of nuclear respiratory factor 1-dependent transcription in mammalian cells. *Mol. Cell. Biol.* **21**, 3738–3749.

Anzick, S. L., Kononen, J., Walker, R. L., Azorsa, D. O., Tanner, M. M., Guan, X. Y., Sauter, G., Kallioniemi, O. P., Trent, J. M., and Meltzer, P. S. (1997). AIB1, a steroid receptor coactivator amplified in breast and ovarian cancer. *Science* **277**, 965–968.

Arany, Z., Newsome, D., Oldread, E., Livingston, D. M., and Eckner, R. (1995). A family of transcriptional adaptor proteins targeted by the E1A oncoprotein. *Nature* **374**, 81–84.

Baniahmad, A., Kohne, A. C., and Renkawitz, R. (1992). A transferable silencing domain is present in the thyroid hormone receptor, in the v-erbA oncogene product, and in the retinoic acid receptor. *EMBO J.* **11**, 1015–1023.

Baniahmad, A., Leng, X. H., Burris, T. P., Tsai, S. Y., Tsai, M. J., and O'Malley, B. W. (1995). The Tau-4 activation domain of the thyroid hormone receptor is required for release of a putative corepressor(s) necessary for transcriptional silencing. *Mol. Cell. Biol.* **15**, 76–86.

Bauer, U. M., Daujat, S., Nielsen, S. J., Nightingale, K., and Kouzarides, T. (2002). Methylation at arginine 17 of histone H3 is linked to gene activation. *EMBO Rep.* **3**, 39–44.

Blanco, J. C., Minucci, S., Lu, J., Yang, X. J., Walker, K. K., Chen, H., Evans, R. M., Nakatani, Y., and Ozato, K. (1998). The histone acetylase PCAF is a nuclear receptor coactivator. *Genes Develop.* **12**, 1638–1651.

Boyer, T. G., Martin, M. E., Lees, E., Ricciardi, R. P., and Berk, A. J. (1999). Mammalian Srb/mediator complex is targeted by adenovirus E1A protein. *Nature* **399**, 276–279.

Bramlett, K. S., and Burris, T. P. (2002). Effects of selective estrogen receptor modulators (SERMs) on coactivator nuclear receptor (NR) box binding to estrogen receptors. *Mol. Gen. Metabol.* **76**, 225–233.

Bramlett, K. S., and Burris, T. P. (2003). Target specificity of selective estrogen receptor modulators within human endometrial cancer cells. *J. Steroid Biochem. Mol. Biol.* **86**, 27–34.

Bramlett, K. S., Wu, Y. F., and Burris, T. P. (2001). Ligands specify coactivator nuclear receptor (NR) box affinity for estrogen receptor subtypes. *Mol. Endocrinol.* **15**, 909–922.

Bryant, H. U. (2001). Mechanism of action and preclinical profile of raloxifene, a selective estrogen receptor modulation. *Rev. Endocr. Metab. Disord.* **2**, 129–138.

Brzozowski, A. M., Pike, A. C., Dauter, Z., Hubbard, R. E., Bonn, T., Engstrom, O., Ohman, L., Greene, G. L., Gustafsson, J. A., and Carlquist, M. (1997). Molecular basis of agonism and antagonism in the oestrogen receptor. *Nature* **389**, 753–758.

Cartegni, L., Maconi, M., Morandi, E., Cobianchi, F., Riva, S., and Biamonti, G. (1996). hnRNP A1 selectively interacts through its Gly-rich domain with different RNA-binding proteins. *J. Mol. Biol.* **259**, 337–348.

Chan, H. M., and La Thangue, N. B. (2001). p300/CBP proteins: HATs for transcriptional bridges and scaffolds. *J. Cell. Sci.* **114**, 2363–2373.

Chang, C., Norris, J. D., Gron, H., Paige, L. A., Hamilton, P. T., Kenan, D. J., Fowlkes, D., and McDonnell, D. P. (1999). Dissection of the LXXLL nuclear receptor–coactivator interaction motif using combinatorial peptide libraries: Discovery of peptide antagonists of estrogen receptors alpha and beta. *Mol. Cell. Biol.* **19**, 8226–8239.

Chen, D., Huang, S. M., and Stallcup, M. R. (2000). Synergistic, p160 coactivator-dependent enhancement of estrogen receptor function by CARM1 and p300. *J. Biol. Chem.* **275**, 40810–40816.

Chen, D., Ma, H., Hong, H., Koh, S. S., Huang, S. M., Schurter, B. T., Aswad, D. W., and Stallcup, M. R. (1999a). Regulation of transcription by a protein methyltransferase. *Science* **284**, 2174–2177.

Chen, H., Lin, R. J., Schiltz, R. L., Chakravarti, D., Nash, A., Nagy, L., Privalsky, M. L., Nakatani, Y., and Evans, R. M. (1997). Nuclear receptor coactivator ACTR is a novel histone acetyltransferase and forms a multimeric activation complex with PCAF and CBP/p300. *Cell* **90**, 569–580.

Chen, H., Lin, R. J., Xie, W., Wilpitz, D., and Evans, R. M. (1999b). Regulation of hormone-induced histone hyperacetylation and gene activation via acetylation of an acetylase. *Cell* **98**, 675–686.

Chen, J. D., and Evans, R. M. (1995). A transcriptional corepressor that interacts with nuclear hormone receptors. *Nature* **377**, 454–457.

Chiba, H., Muramatsu, M., Nomoto, A., and Kato, H. (1994). Two human homologues of *Saccharomyces cerevisiae* SWI2/SNF2 and *Drosophila brahma* are transcriptional coactivators cooperating with the estrogen receptor and the retinoic acid receptor. *Nucleic Acids Res.* **22**, 1815–1820.

Clarke, S. (1993). Protein methylation. *Curr. Opin. in Cell. Biol.* **5**, 977–983.

Coulthard, V. H., Matsuda, S., and Heery, D. M. (2003). An extended LXXLL motif sequence determines the nuclear receptor binding specificity of TRAP220. *J. Biol. Chem.* **278**, 10942–10951.

Danielian, P. S., White, R., Lees, J. A., and Parker, M. G. (1992). Identification of a conserved region required for hormone dependent transcriptional activation by steroid hormone receptors. *EMBO J.* **11**, 1025–1033.

Darimont, B. D., Wagner, R. L., Apriletti, J. W., Stallcup, M. R., Kushner, P. J., Baxter, J. D., Fletterick, R. J., and Yamamoto, K. R. (1998). Structure and specificity of nuclear receptor–coactivator interactions. *Genes Develop.* **12**, 3343–3356.

Delerive, P., Wu, Y. F., Burris, T. P., Chin, W. W., and Suen, C. S. (2002). PGC-1 functions as a transcriptional coactivator for the retinoid X receptors. *J. Biol. Chem.* **277**, 3913–3917.

Ding, X. F., Anderson, C. M., Ma, H., Hong, H., Uht, R. M., Kushner, P. J., and Stallcup, M. R. (1998). Nuclear receptor-binding sites of coactivators glucocorticoid receptor interacting protein 1 (GRIP1) and steroid receptor coactivator-1 (SRC-1): Multiple motifs with different binding specificities. *Mol. Endocrinol.* **12**, 302–313.

DiRenzo, J., Shang, Y., Phelan, M., Sif, S., Myers, M., Kingston, R., and Brown, M. (2000). BRG-1 is recruited to estrogen-responsive promoters and cooperates with factors involved in histone acetylation. *Mol. Cell. Biol.* **20**, 7541–7549.

Drane, P., Barel, M., Balbo, M., and Frade, R. (1997). Identification of RB18A, a 205-kDa new p53 regulatory protein which shares antigenic and functional properties with p53. *Oncogene* **15**, 3013–3024.

Eckner, R., Ludlow, J. W., Lill, N. L., Oldread, E., Arany, Z., Modjtahedi, N., DeCaprio, J. A., Livingston, D. M., and Morgan, J. A. (1996). Association of p300 and CBP with simian virus 40 large T antigen. *Mol. Cell. Biol.* **16**, 3454–3464.

Eggert, M., Mows, C. C., Tripier, D., Arnold, R., Michel, J., Nickel, J., Schmidt, S., Beato, M., and Renkawitz, R. (1995). A fraction enriched in a novel glucocorticoid receptor-interacting protein stimulates receptor-dependent transcription *in vitro*. *J. Biol. Chem.* **270**, 30755–30759.

Eisen, J. A., Sweder, K. S., and Hanawalt, P. C. (1995). Evolution of the SNF2 family of proteins: Subfamilies with distinct sequences and functions. *Nucleic Acids Res.* **23**, 2715–2723.

Emberley, E., Huang, G. J., Hamedani, M. K., Czosnek, A., Ali, D., Grolla, A., Lu, B., Watson, P. H., Murphy, L. C., and Leygue, E. (2003). Identification of new human coding steroid receptor RNA activator isoforms. *Biochem. Biophys. Res. Comm.* **301**, 509–515.

Feng, W., Ribeiro, R. C., Wagner, R. L., Nguyen, H., Apriletti, J. W., Fletterick, R. J., Baxter, J. D., Kushner, P. J., and West, B. L. (1998). Hormone-dependent coactivator binding to a hydrophobic cleft on nuclear receptors. *Science* **280**, 1747–1749.

Fondell, J. D., Ge, H., and Roeder, R. G. (1996). Ligand induction of a transcriptionally active thyroid hormone receptor coactivator complex. *Proc. Natl. Acad. Sci. USA* **93**, 8329–8333.

Fryer, C. J., and Archer, T. K. (1998). Chromatin remodelling by the glucocorticoid receptor requires the BRG1 complex. *Nature* **393**, 88–91.

Gary, J. D., and Clarke, S. (1998). RNA and protein interactions modulated by protein arginine methylation. *Prog. Nucleic Acid Res. Mol. Biol.* **61**, 65–131.

Gayther, S. A., Batley, S. J., Linger, L., Bannister, A., Thorpe, K., Chin, S. F., Daigo, Y., Russell, P., Wilson, A., Sowter, H. M., Delhanty, J. D., Ponder, B. A., Kouzarides, T., and Caldas, C. (2000). Mutations truncating the EP300 acetylase in human cancers. *Nature Gen.* **24**, 300–303.

Gee, A. C., Carlson, K. E., Martini, P. G., Katzenellenbogen, B. S., and Katzenellenbogen, J. A. (1999). Coactivator peptides have a differential stabilizing effect on the binding of estrogens and antiestrogens with the estrogen receptor. *Mol. Endocrinol.* **13**, 1912–1923.

Giannoukos, G., Szapary, D., Smith, C. L., Meeker, J. E., and Simons, S. S. (2001). New antiprogestins with partial agonist activity: Potential selective progesterone receptor modulators (SPRMs) and probes for receptor- and coregulator-induced changes in progesterone receptor induction properties. *Mol. Endocrinol.* **15**, 255–270.

Giles, R. H., Peters, D. J., and Breuning, M. H. (1998). Conjunction dysfunction: CBP/p300 in human disease. *Trends Gen.* **14**, 178–183.

Gill, G., and Ptashne, M. (1988). Negative effect of the transcriptional activator GAL4. *Nature* **334**, 721–724.

Giordano, A., and Avantaggiati, M. L. (1999). p300 and CBP: Partners for life and death. *J. Cell. Phys.* **181**, 218–230.

Goodman, R. H., and Smolik, S. (2000). CBP/p300 in cell growth, transformation, and development. *Genes Dev.* **14**, 1553–1577.

Graveley, B. R. (2000). Sorting out the complexity of SR protein functions. *RNA* **6**, 1197–1211.

Gu, W., Malik, S., Ito, M., Yuan, C. X., Fondell, J. D., Zhang, X., Martinez, E., Qin, J., and Roeder, R. G. (1999). A novel human SRB/MED-containing cofactor complex, SMCC, involved in transcription regulation. *Mol. Cell* **3**, 97–108.

Gu, W., and Roeder, R. G. (1997). Activation of p53 sequence-specific DNA binding by acetylation of the p53 C-terminal domain. *Cell* **90**, 595–606.

Guan, X. Y., Xu, J., Anzick, S. L., Zhang, H., Trent, J. M., and Meltzer, P. S. (1996). Hybrid selection of transcribed sequences from microdissected DNA: Isolation of genes within amplified region at 20q11-q13.2 in breast cancer. *Cancer Res.* **56**, 3446–3450.

Guenther, M. G., Barak, O., and Lazar, M. A. (2001). The SMRT and NCoR corepressors are activating cofactors for histone deacetylase 3. *Mol. Cell. Biol.* **21**, 6091–6101.

Halachmi, S., Marden, E., Martin, G., MacKay, H., Abbondanza, C., and Brown, M. (1994). Estrogen receptor-associated proteins: Possible mediators of hormone-induced transcription. *Science* **264**, 1455–1458.

Hassig, C. A., and Schreiber, S. L. (1997). Nuclear histone acetylases and deacetylases and transcriptional regulation: HATs off to HDACs. *Curr. Opin. Chem. Biol.* **1**, 300–308.

Heery, D. M., Hoare, S., Hussain, S., Parker, M. G., and Sheppard, H. (2001). Core LXXLL motif sequences in CREB-binding protein SRC-1 and RIP140 define affinity and selectivity for steroid and retinoid receptors. *J. Biol. Chem.* **276**, 6695–6702.

Heery, D. M., Kalkhoven, E., Hoare, S., and Parker, M. G. (1997). A signature motif in transcriptional coactivators mediates binding to nuclear receptors. *Nature* **387,** 733–736.

Hermanson, O., Glass, C. K., and Rosenfeld, M. G. (2002). Nuclear receptor coregulators: Multiple modes of modification. *Trends Endocrinol. Metabol.* **13,** 55–60.

Hittelman, A. B., Burakov, D., Iniguez-Lluhi, J. A., Freedman, L. P., and Garabedian, M. J. (1999). Differential regulation of glucocorticoid receptor transcriptional activation via AF-1-associated proteins. *EMBO J.* **18,** 5380–5388.

Hollenberg, S. M., and Evans, R. M. (1988). Multiple and cooperative transactivation domains of the human glucocorticoid receptor. *Cell* **55,** 899–906.

Hong, H., Kohli, K., Trivedi, A., Johnson, D. L., and Stallcup, M. R. (1996). GRIP1, a novel mouse protein that serves as a transcriptional coactivator in yeast for the hormone binding domains of steroid receptors. *Proc. Natl. Acad. Sci. USA* **93,** 4948–4952.

Horlein, A. J., Naar, A. M., Heinzel, T., Torchia, J., Gloss, B., Kurokawa, R., Ryan, A., Kamei, Y., Soderstrom, M., and Glass, C. K. (1995). Ligand-independent repression by the thyroid hormone receptor mediated by a nuclear receptor corepressor. *Nature* **377,** 397–404.

Hu, X., and Lazar, M. A. (1999). The CoRNR motif controls the recruitment of corepressors by nuclear hormone receptors. *Nature* **402,** 93–96.

Hu, X., Li, Y., and Lazar, M. A. (2001). Determinants of CoRNR-dependent repression complex assembly on nuclear hormone receptors. *Mol. Cell. Biol.* **21,** 1747–1758.

Huang, Y., and Steitz, J. A. (2001). Splicing factors SRp20 and 9G8 promote the nucleocytoplasmic export of mRNA. *Mol. Cell* **7,** 899–905.

Huang, Z. J., Edery, I., and Rosbash, M. (1993). PAS is a dimerization domain common to *Drosophila* period and several transcription factors. *Nature* **364,** 259–262.

Imhof, A., Yang, X. J., Ogryzko, V. V., Nakatani, Y., Wolffe, A. P., and Ge, H. (1997). Acetylation of general transcription factors by histone acetyltransferases. *Curr. Biol.* **7,** 689–692.

Ito, M., and Roeder, R. G. (2001). The TRAP/SMCC/mediator complex and thyroid hormone receptor function. *Trends Endocrinol. Metabol.* **12,** 127–134.

Ito, M., Yuan, C. X., Malik, S., Gu, W., Fondell, J. D., Yamamura, S., Fu, Z. Y., Zhang, X., Qin, J., and Roeder, R. G. (1999). Identity between TRAP and SMCC complexes indicates novel pathways for the function of nuclear receptors and diverse mammalian activators. *Mol. Cell* **3,** 361–370.

Iwasaki, T., Chin, W. W., and Ko, L. (2001). Identification and characterization of RRM-containing coactivator activator (CoAA) as TRBP-interacting protein and its splice variant as a coactivator modulator (CoAM). *J. Biol. Chem.* **276,** 33375–33383.

Jepsen, K., Hermanson, O., Onami, T. M., Gleiberman, A. S., Lunyak, V., McEvilly, R. J., Kurokawa, R., Kumar, V., Liu, F., Seto, E., Hedrick, S. M., Mandel, G., Glass, C. K., Rose, D. W., and Rosenfeld, M. G. (2000). Combinatorial roles of the nuclear receptor corepressor in transcription and development. *Cell* **102,** 753–763.

Kalkhoven, E., Valentine, J. E., Heery, D. M., and Parker, M. G. (1998). Isoforms of steroid receptor coactivator-1 differ in their ability to potentiate transcription by the oestrogen receptor. *EMBO J.* **17,** 232–243.

Kamei, Y., Xu, L., Heinzel, T., Torchia, J., Kurokawa, R., Gloss, B., Lin, S. C., Heyman, R. A., Rose, D. W., Glass, C. K., and Rosenfeld, M. G. (1996). A CBP integrator complex mediates transcriptional activation and AP-1 inhibition by nuclear receptors. *Cell* **85,** 403–414.

Kauppi, B., Jakob, C., Farnegardh, M., Yang, J., Ahola, H., Alarcon, M., Calles, K., Engstrom, O., Harlan, J., Muchmore, S., Ramqvist, A. K., Thorell, S., Ohman, L., Greer, J., Gustafsson, J. A., Carlstedt_Duke, J., and Carlquist, M. (2003). The three-dimensional structures of antagonistic and agonistic forms of the glucocorticoid receptor ligand binding domain: RU486 induces a transconformation that leads to active antagonism. *J. Biol. Chem.* **278,** 22748–22754.

Kingston, R. E., and Narlikar, G. J. (1999). ATP-dependent remodeling and acetylation as regulators of chromatin fluidity. *Genes Develop.* **13,** 2339–2352.

Knutti, D., Kressler, D., and Kralli, A. (2001). Regulation of the transcriptional coactivator PGC-1 via MAPK-sensitive interaction with a repressor. *Proc. Natl. Acad. Sci. USA* **98,** 9713–9718.

Ko, L., Cardona, G. R., and Chin, W. W. (2000). Thyroid hormone receptor-binding protein, an LXXLL motif-containing protein, functions as a general coactivator. *Proc. Natl. Acad. Sci. USA* **97,** 6212–6217.

Ko, L., Cardona, G. R., Iwasaki, T., Bramlett, K. S., Burris, T. P., and Chin, W. W. (2002). Ser-884 adjacent to the LXXLL motif of coactivator TRBP defines selectivity for ERs and TRs. *Mol. Endocrinol.* **16,** 128–140.

Ko, L., and Chin, W. W. (2003). Nuclear receptor coactivator thyroid hormone receptor binding protein (TRBP) interacts with and stimulates its associated DNA-dependent protein kinase. *J. Biol. Chem.* **278,** 11471–11479.

Kodera, Y., Takeyama, K., Murayama, A., Suzawa, M., Masuhiro, Y., and Kato, S. (2000). Ligand type-specific interactions of peroxisome proliferator-activated receptor gamma with transcriptional coactivators. *J. Biol. Chem.* **275,** 33201–33204.

Koh, S. S., Chen, D., Lee, Y. H., and Stallcup, M. R. (2001). Synergistic enhancement of nuclear receptor function by p160 coactivators and two coactivators with protein methyltransferase activities. *J. Biol. Chem.* **276,** 1089–1098.

Kraichely, D. M., Sun, J., Katzenellenbogen, J. A., and Katzenellenbogen, B. S. (2000). Conformational changes and coactivator recruitment by novel ligands for estrogen receptor alpha and estrogen receptor beta: Correlations with biological character and distinct differences among SRC coactivator family members. *Endocrinology* **141,** 3534–3545.

Kung, A. L., Rebel, V. I., Bronson, R. T., Ch'ng, L. E., Sieff, C. A., Livingston, D. M., and Yao, T. P. (2000). Gene dose-dependent control of hematopoiesis and hematologic tumor suppression by CBP. *Genes Develop.* **14,** 272–277.

Lanz, R. B., McKenna, N. J., Onate, S. A., Albrecht, U., Wong, J., Tsai, S. Y., Tsai, M. J., and O'Malley, B. W. (1999). A steroid receptor coactivator, SRA, functions as an RNA and is present in an SRC-1 complex. *Cell* **97,** 17–27.

Lanz, R. B., Razani, B., Goldberg, A. D., and O'Malley, B. W. (2002). Distinct RNA motifs are important for coactivation of steroid hormone receptors by steroid receptor RNA activator (SRA). *Proc. Natl. Acad. Sci. USA* **99,** 16081–16086.

Lavinsky, R. M., Jepsen, K., Heinzel, T., Torchia, J., Mullen, T. M., Schiff, R., Del-Rio, A. L., Ricote, M., Ngo, S., Gemsch, J., Hilsenbeck, S. G., Osborne, C. K., Glass, C. K., Rosenfeld, M. G., and Rose, D. W. (1998). Diverse signaling pathways modulate nuclear receptor recruitment of NCoR and SMRT complexes. *Proc. Natl. Acad. Sci. USA* **95,** 2920–2925.

Le Douarin, B., Nielsen, A. L., Garnier, J. M., Ichinose, H., Jeanmougin, F., Losson, R., and Chambon, P. (1996). A possible involvement of TIF1 alpha and TIF1 beta in the epigenetic control of transcription by nuclear receptors. *EMBO J.* **15,** 6701–6715.

Lee, S. K., Anzick, S. L., Choi, J. E., Bubendorf, L., Guan, X. Y., Jung, Y. K., Kallioniemi, O. P., Kononen, J., Trent, J. M., Azorsa, D., Jhun, B. H., Cheong, J. H., Lee, Y. C., Meltzer, P. S., and Lee, J. W. (1999). A nuclear factor, ASC-2, as a cancer-amplified transcriptional coactivator essential for ligand-dependent transactivation by nuclear receptors *in vivo*. *J. Biol. Chem.* **274,** 34283–34293.

Lee, Y. H., Koh, S. S., Zhang, X., Cheng, X., and Stallcup, M. R. (2002). Synergy among nuclear receptor coactivators: Selective requirement for protein methyltransferase and acetyltransferase activities. *Mol. Cell. Biol.* **22,** 3621–3632.

Leers, J., Treuter, E., and Gustafsson, J. A. (1998). Mechanistic principles in NR box-dependent interaction between nuclear hormone receptors and the coactivator TIF2. *Mol. Cell. Biol.* **18,** 6001–6013.

Li, H., Gomes, P. J., and Chen, J. D. (1997a). RAC3, a steroid/nuclear receptor-associated coactivator that is related to SRC-1 and TIF2. *Proc. Nat. Acad. Sci. USA* **94**, 8479–8484.

Li, H., Leo, C., Schroen, D. J., and Chen, J. D. (1997b). Characterization of receptor interaction and transcriptional repression by the corepressor SMRT. *Mol. Endocrinol.* **11**, 2025–2037.

Li, J., O'Malley, B. W., and Wong, J. (2000). p300 requires its histone acetyltransferase activity and SRC-1 interaction domain to facilitate thyroid hormone receptor activation in chromatin. *Mol. Cell. Biol.* **20**, 2031–2042.

Liu, L., Scolnick, D. M., Trievel, R. C., Zhang, H. B., Marmorstein, R., Halazonetis, T. D., and Berger, S. L. (1999). p53 sites acetylated *in vitro* by PCAF and p300 are acetylated *in vivo* in response to DNA damage. *Mol. Cell. Biol.* **19**, 1202–1209.

Liu, Q., and Dreyfuss, G. (1995). In vivo and in vitro arginine methylation of RNA-binding proteins. *Mol. Cell. Biol.* **15**, 2800–2808.

Liu, Z., Auboeuf, D., Wong, J., Chen, J. D., Tsai, S. Y., Tsai, M. J., and O'Malley, B. W. (2002). Coactivator/corepressor ratios modulate PR-mediated transcription by the selective receptor modulator RU486. *Proc. Natl. Acad. Sci. USA* **99**, 7940–7944.

Ma, H., Baumann, C. T., Li, H., Strahl, B. D., Rice, R., Jelinek, M. A., Aswad, D. W., Allis, C. D., Hager, G. L., and Stallcup, M. R. (2001). Hormone-dependent, CARM1-directed, arginine-specific methylation of histone H3 on a steroid-regulated promoter. *Curr. Biol.* **11**, 1981–1985.

Mahajan, M. A., and Samuels, H. H. (2000). A new family of nuclear receptor coregulators that integrate nuclear receptor signaling through CREB-binding protein. *Mol. Cell. Biol.* **20**, 5048–5063.

Mak, H. Y., Hoare, S., Henttu, P. M., and Parker, M. G. (1999). Molecular determinants of the estrogen receptor–coactivator interface. *Mol. Cell. Biol.* **19**, 3895–3903.

Malik, S., and Roeder, R. G. (2000). Transcriptional regulation through mediator-like coactivators in yeast and metazoan cells. *Trends Biochem. Sci.* **25**, 277–283.

Marimuthu, A., Feng, W., Tagami, T., Nguyen, H., Jameson, J. L., Fletterick, R. J., Baxter, J. D., and West, B. L. (2002). TR surfaces and conformations required to bind nuclear receptor corepressor. *Mol. Endocrinol.* **16**, 271–286.

McInerney, E. M., Rose, D. W., Flynn, S. E., Westin, S., Mullen, T. M., Krones, A., Inostroza, J., Torchia, J., Nolte, R. T., Assa-Munt, N., Milburn, M. V., Glass, C. K., and Rosenfeld, M. G. (1998). Determinants of coactivator LXXLL motif specificity in nuclear receptor transcriptional activation. *Genes Develop.* **12**, 3357–3368.

McKenna, N. J., Lanz, R. B., and O'Malley, B. W. (1999). Nuclear receptor coregulators: Cellular and molecular biology. *Endocr. Rev.* **20**, 321–344.

McKenna, N. J., and O'Malley, B. W. (2002). Combinatorial control of gene expression by nuclear receptors and coregulators. *Cell* **108**, 465–474.

Meirhaeghe, A., Crowley, V., Lenaghan, C., Lelliott, C., Green, K., Stewart, A., Hart, K., Schinner, S., Sethi, J. K., Yeo, G., Brand, M. D., Cortright, R. N., O'Rahilly, S., Montague, C., and Vidal-Puig, A. J. (2003). Characterization of the human, mouse, and rat PGC1 beta (peroxisome proliferator-activated receptor gamma coactivator 1 beta) gene *in vitro* and *in vivo*. *Biochem. J.* **373**, 155–165.

Meyer, M. E., Gronemeyer, H., Turcotte, B., Bocquel, M. T., Tasset, D., and Chambon, P. (1989). Steroid hormone receptors compete for factors that mediate their enhancer function. *Cell* **57**, 433–442.

Monsalve, M., Wu, Z., Adelmant, G., Puigserver, P., Fan, M., and Spiegelman, B. M. (2000). Direct coupling of transcription and mRNA processing through the thermogenic coactivator PGC-1. *Mol. Cell* **6**, 307–316.

Muchardt, C., and Yaniv, M. (1993). A human homologue of *Saccharomyces cerevisiae* SNF2/SWI2 and *Drosophila brm* genes potentiate transcriptional activation by the glucocorticoid receptor. *EMBO J.* **12**, 4279–4290.

Muscat, G. E., Burke, L. J., and Downes, M. (1998). The corepressor NCoR and its variants RIP13a and RIP13Delta1 directly interact with the basal transcription factors TFIIB, TAFII32, and TAFII70. *Nucleic Acids Res.* **26**, 2899–2907.

Naar, A. M., Beaurang, P. A., Zhou, S., Abraham, S., Solomon, W., and Tjian, R. (1999). Composite coactivator ARC mediates chromatin-directed transcriptional activation. *Nature* **398**, 828–832.

Nagai, K., Oubridge, C., Ito, N., Avis, J., and Evans, P. (1995). The RNP domain: A sequence-specific RNA binding domain involved in processing and transport of RNA. *Trends Biochem. Sci.* **20**, 235–240.

Nagy, L., Kao, H. Y., Love, J. D., Li, C., Banayo, E., Gooch, J. T., Krishna, V., Chatterjee, K., Evans, R. M., and Schwabe, J. W. (1999). Mechanism of corepressor binding and release from nuclear hormone receptors. *Genes Develop.* **13**, 3209–3216.

Needham, M., Raines, S., McPheat, J., Stacey, C., Ellston, J., Hoare, S., and Parker, M. (2000). Differential interaction of steroid hormone receptors with LXXLL motifs in SRC-1a depends on residues flanking the motif. *J. Ster. Biochem. Molec. Biol.* **72**, 35–46.

Nolte, R. T., Wisely, G. B., Westin, S., Cobb, J. E., Lambert, M. H., Kurokawa, R., Rosenfeld, M. G., Willson, T. M., Glass, C. K., and Milburn, M. V. (1998). Ligand binding and coactivator assembly of the peroxisome proliferator-activated receptor gamma. *Nature* **395**, 137–143.

Northrop, J. P., Nguyen, D., Piplani, S., Olivan, S. E., Kwan, S. T., Go, N. F., Hart, C. P., and Schatz, P. J. (2000). Selection of estrogen receptor beta- and thyroid hormone receptor beta-specific coactivator-mimetic peptides using recombinant peptide libraries. *Mol. Endocrinol.* **14**, 605–622.

Onate, S. A., Boonyaratanakornkit, V., Spencer, T. E., Tsai, S. Y., Tsai, M. J., Edwards, D. P., and O'Malley, B. W. (1998). The steroid receptor coactivator-1 contains multiple receptor interacting and activation domains that cooperatively enhance the activation function-1 (AF-1) and AF-2 domains of steroid receptors. *J. Biol. Chem.* **273**, 12101–12108.

Onate, S. A., Tsai, S. Y., Tsai, M. J., and O'Malley, B. W. (1995). Sequence and characterization of a coactivator for the steroid hormone receptor superfamily. *Science* **270**, 1354–1357.

Perissi, V., Staszewski, L. M., McInerney, E. M., Kurokawa, R., Krones, A., Rose, D. W., Lambert, M. H., Milburn, M. V., Glass, C. K., and Rosenfeld, M. G. (1999). Molecular determinants of nuclear receptor–corepressor interaction. *Genes Develop.* **13**, 3198–3208.

Peterson, C. L. (2002a). Chromatin remodeling enzymes: Taming the machines. Third in review series on chromatin dynamics. *EMBO Rep.* **3**, 319–322.

Peterson, C. L. (2002b). Chromatin remodeling: Nucleosomes bulging at the seams. *Curr. Biol.* **12**, R245–R247.

Pike, A. C., Brzozowski, A. M., Hubbard, R. E., Bonn, T., Thorsell, A. G., Engstrom, O., Ljunggren, J., Gustafsson, J. A., and Carlquist, M. (1999). Structure of the ligand binding domain of oestrogen receptor beta in the presence of a partial agonist and a full antagonist. *EMBO J.* **18**, 4608–4618.

Pike, A. C., Brzozowski, A. M., Walton, J., Hubbard, R. E., Thorsell, A. G., Li, Y. L., Gustafsson, J. A., and Carlquist, M. (2001). Structural insights into the mode of action of a pure antiestrogen. *Structure* **9**, 145–153.

Puigserver, P., Adelmant, G., Wu, Z., Fan, M., Xu, J., O'Malley, B., and Spiegelman, B. M. (1999). Activation of PPAR gamma coactivator-1 through transcription factor docking. *Science* **286**, 1368–1371.

Puigserver, P., Wu, Z., Park, C. W., Graves, R., Wright, M., and Spiegelman, B. M. (1998). A cold-inducible coactivator of nuclear receptors linked to adaptive thermogenesis. *Cell* **92**, 829–839.

Rachez, C., Gamble, M., Chang, C. P., Atkins, G. B., Lazar, M. A., and Freedman, L. P. (2000). The DRIP complex and SRC-1/p160 coactivators share similar nuclear receptor binding determinants but constitute functionally distinct complexes. *Mol. Cell. Biol.* **20**, 2718–2726.

Rachez, C., Lemon, B. D., Suldan, Z., Bromleigh, V., Gamble, M., Naar, A. M., Erdjument-Bromage, H., Tempst, P., and Freedman, L. P. (1999). Ligand-dependent transcription activation by nuclear receptors requires the DRIP complex. *Nature* **398**, 824–828.

Ren, Y., Behre, E., Ren, Z., Zhang, J., Wang, Q., and Fondell, J. D. (2000). Specific structural motifs determine TRAP220 interactions with nuclear hormone receptors. *Mol. Cell. Biol.* **20**, 5433–5446.

Renaud, J. P., Harris, J. M., Downes, M., Burke, L. J., and Muscat, G. E. (2000). Structure–function analysis of the Rev-erbA and RVR ligand binding domains reveals a large hydrophobic surface that mediates corepressor binding and a ligand cavity occupied by side chains. *Mol. Endocrinol.* **14**, 700–717.

Sadowski, I., Ma, J., Triezenberg, S., and Ptashne, M. (1988). GAL4-VP16 is an unusually potent transcriptional activator. *Nature* **335**, 563–564.

Sakaguchi, K., Herrera, J. E., Saito, S., Miki, T., Bustin, M., Vassilev, A., Anderson, C. W., and Appella, E. (1998). DNA damage activates p53 through a phosphorylation-acetylation cascade. *Genes Develop.* **12**, 2831–2841.

Sartorelli, V., Puri, P. L., Hamamori, Y., Ogryzko, V., Chung, G., Nakatani, Y., Wang, J. Y., and Kedes, L. (1999). Acetylation of MyoD directed by PCAF is necessary for the execution of the muscle program. *Mol. Cell* **4**, 725–734.

Schreiber, S. N., Knutti, D., Brogli, K., Uhlmann, T., and Kralli, A. (2003). The transcriptional coactivator PGC-1 regulates the expression and activity of the orphan nuclear receptor estrogen-related receptor alpha (ERR alpha). *J. Biol. Chem.* **278**, 9013–9018.

Shamoo, Y., Krueger, U., Rice, L. M., Williams, K. R., and Steitz, T. A. (1997). Crystal structure of the two RNA binding domains of human hnRNP A1 at 1.75 Å resolution. *Nat. Struct. Biol.* **4**, 215–222.

Shang, Y., and Brown, M. (2002). Molecular determinants for the tissue specificity of SERMs. *Science* **295**, 2465–2468.

Shang, Y., Hu, X., DiRenzo, J., Lazar, M. A., and Brown, M. (2000). Cofactor dynamics and sufficiency in estrogen receptor-regulated transcription. *Cell* **103**, 843–852.

Shi, Y., Downes, M., Xie, W., Kao, H. Y., Ordentlich, P., Tsai, C. C., Hon, M., and Evans, R. M. (2001). SHARP, an inducible cofactor that integrates nuclear receptor repression and activation. *Genes Develop.* **15**, 1140–1151.

Shiau, A. K., Barstad, D., Loria, P. M., Cheng, L., Kushner, P. J., Agard, D. A., and Greene, G. L. (1998). The structural basis of estrogen receptor/coactivator recognition and the antagonism of this interaction by tamoxifen. *Cell* **95**, 927–937.

Shiau, A. K., Barstad, D., Radek, J. T., Meyers, M. J., Nettles, K. W., Katzenellenbogen, B. S., Katzenellenbogen, J. A., Agard, D. A., and Greene, G. L. (2002). Structural characterization of a subtype-selective ligand reveals a novel mode of estrogen receptor antagonism. *Nat. Struct. Biol.* **9**, 359–364.

Smith, C. L., Nawaz, Z., and O'Malley, B. W. (1997). Coactivator and corepressor regulation of the agonist/antagonist activity of the mixed antiestrogen, 4-hydroxytamoxifen. *Mol. Endocrinol.* **11**, 657–666.

Smith, E. R., Belote, J. M., Schiltz, R. L., Yang, X. J., Moore, P. A., Berger, S. L., Nakatani, Y., and Allis, C. D. (1998). Cloning of *Drosophila* GCN5: Conserved features among metazoan GCN5 family members. *Nucleic Acids Res. (Online)* **26**, 2948–2954.

Spencer, T. E., Jenster, G., Burcin, M. M., Allis, C. D., Zhou, J., Mizzen, C. A., McKenna, N. J., Onate, S. A., Tsai, S. Y., Tsai, M. J., and O'Malley, B. W. (1997). Steroid receptor coactivator-1 is a histone acetyltransferase. *Nature* **389**, 194–198.

Suen, C. S., Berrodin, T. J., Mastroeni, R., Cheskis, B. J., Lyttle, C. R., and Frail, D. E. (1998). A transcriptional coactivator, steroid receptor coactivator-3, selectively augments steroid receptor transcriptional activity. *J. Biol. Chem.* **273**, 27645–27653.

Szapary, D., Huang, Y., and Simons, S. S. (1999). Opposing effects of corepressor and coactivators in determining the dose-response curve of agonists, and residual agonist activity of antagonists, for glucocorticoid receptor-regulated gene expression. *Mol. Endocrinol.* **13**, 2108–2121.

Takeshita, A., Cardona, G. R., Koibuchi, N., Suen, C. S., and Chin, W. W. (1997). TRAM-1, a novel 160-kDa thyroid hormone receptor activator molecule, exhibits distinct properties from steroid receptor coactivator-1. *J. Biol. Chem.* **272**, 27629–27634.

Takeyama, K., Masuhiro, Y., Fuse, H., Endoh, H., Murayama, A., Kitanaka, S., Suzawa, M., Yanagisawa, J., and Kato, S. (1999). Selective interaction of vitamin D receptor with transcriptional coactivators by a vitamin D analog. *Mol. Cell. Biol.* **19**, 1049–1055.

Tamkun, J. W., Deuring, R., Scott, M. P., Kissinger, M., Pattatucci, A. M., Kaufman, T. C., and Kennison, J. A. (1992). Brahma: A regulator of *Drosophila* homeotic genes structurally related to the yeast transcriptional activator SNF2/SWI2. *Cell* **68**, 561–572.

Tasset, D., Tora, L., Fromental, C., Scheer, E., and Chambon, P. (1990). Distinct classes of transcriptional activating domains function by different mechanisms. *Cell* **62**, 1177–1187.

Tcherepanova, I., Puigserver, P., Norris, J. D., Spiegelman, B. M., and McDonnell, D. P. (2000). Modulation of estrogen receptor-alpha transcriptional activity by the coactivator PGC-1. *J. Biol. Chem.* **275**, 16302–16308.

Teyssier, C., Chen, D., and Stallcup, M. R. (2002). Requirement for multiple domains of the protein arginine methyltransferase CARM1 in its transcriptional coactivator function. *J. Biol. Chem.* **277**, 46066–46072.

Thiagalingam, S., Cheng, K. H., Lee, H. J., Mineva, N., Thiagalingam, A., and Ponte, J. F. (2003). Histone deacetylases: Unique players in shaping the epigenetic histone code. *Ann. NY Acad. Sci.* **983**, 84–100.

Tora, L., White, J., Brou, C., Tasset, D., Webster, N., Scheer, E., and Chambon, P. (1989). The human estrogen receptor has two independent nonacidic transcriptional activation functions. *Cell* **59**, 477–487.

Torchia, J., Rose, D. W., Inostroza, J., Kamei, Y., Westin, S., Glass, C. K., and Rosenfeld, M. G. (1997). The transcriptional coactivator p/CIP binds CBP and mediates nuclear receptor function. *Nature* **387**, 677–684.

Triezenberg, S. J., Kingsbury, R. C., and McKnight, S. L. (1988). Functional dissection of VP16, the transactivator of herpes simplex virus immediate early gene expression. *Genes Develop.* **2**, 718–729.

Vega, R. B., Huss, J. M., and Kelly, D. P. (2000). The coactivator PGC-1 cooperates with peroxisome proliferator-activated receptor alpha in transcriptional control of nuclear genes encoding mitochondrial fatty acid oxidation enzymes. *Mol. Cell. Biol.* **20**, 1868–1876.

Vignali, M., Hassan, A. H., Neely, K. E., and Workman, J. L. (2000). ATP-dependent chromatin-remodeling complexes. *Mol. Cell. Biol.* **20**, 1899–1910.

Voegel, J. J., Heine, M. J., Tini, M., Vivat, V., Chambon, P., and Gronemeyer, H. (1998). The coactivator TIF2 contains three nuclear receptor-binding motifs and mediates transactivation through CBP binding-dependent and -independent pathways. *EMBO J.* **17**, 507–519.

Voegel, J. J., Heine, M. J., Zechel, C., Chambon, P., and Gronemeyer, H. (1996). TIF2, a 160-kDa transcriptional mediator for the ligand-dependent activation function AF-2 of nuclear receptors. *EMBO J.* **15**, 3667–3675.

Wang, W., Xue, Y., Zhou, S., Kuo, A., Cairns, B. R., and Crabtree, G. R. (1996). Diversity and specialization of mammalian SWI/SNF complexes. *Genes Develop.* **10**, 2117–2130.

Warnmark, A., Almlof, T., Leers, J., Gustafsson, J. A., and Treuter, E. (2001). Differential recruitment of the mammalian mediator subunit TRAP220 by estrogen receptors ER alpha and ER beta. *J. Biol. Chem.* **276**, 23397–23404.

Warnmark, A., Treuter, E., Gustafsson, J.-A., Hubbard, R. E., Brzozowski, A. M., and Pike, A. C. W. (2002). Interaction of TIF2 NR Box Peptides with the coactivator binding site of ERalpha. *J. Biol. Chem.* **277**, 21862–21868.

Watanabe, M., Yanagisawa, J., Kitagawa, H., Takeyama, K., Ogawa, S., Arao, Y., Suzawa, M., Kobayashi, Y., Yano, T., Yoshikawa, H., Masuhiro, Y., and Kato, S. (2001). A subfamily of RNA-binding DEAD-box proteins acts as an estrogen receptor alpha coactivator through the N-terminal activation domain (AF-1) with an RNA coactivator, SRA. *EMBO J.* **20**, 1341–1352.

Wong, C. W., Komm, B., and Cheskis, B. J. (2001). Structure–function evaluation of ER alpha and beta interplay with SRC family coactivators. ER selective ligands. *Biochemistry* **40**, 6756–6765.

Wong, C. W., and Privalsky, M. L. (1998a). Transcriptional repression by the SMRT–mSin3 corepressor: Multiple interactions, multiple mechanisms, and a potential role for TFIIB. *Mol. Cell. Biol.* **18**, 5500–5510.

Wong, C. W., and Privalsky, M. L. (1998b). Transcriptional silencing is defined by isoform- and heterodimer-specific interactions between nuclear hormone receptors and corepressors. *Mol. Cell. Biol.* **18**, 5724–5733.

Wu, Y., Chin, W. W., Wang, Y., and Burris, T. P. (2003). Ligand and coactivator identity determines the requirement of the charge clamp for coactivation of the peroxisome proliferator-activated receptor gamma. *J. Biol. Chem.* **278**, 8637–8644.

Wu, Y. F., Delerive, P., Chin, W. W., and Burris, T. P. (2002). Requirement of helix 1 and the AF-2 domain of the thyroid hormone receptor for coactivation by PGC-1. *J. Biol. Chem.* **277**, 8898–8905.

Xu, H. E., Stanley, T. B., Montana, V. G., Lambert, M. H., Shearer, B. G., Cobb, J. E., McKee, D. D., Galardi, C. M., Plunket, K. D., Nolte, R. T., Parks, D. J., Moore, J. T., Kliewer, S. A., Willson, T. M., and Stimmel, J. B. (2002a). Structural basis for antagonist-mediated recruitment of nuclear corepressors by PPAR alpha. *Nature* **415**, 813–817.

Xu, H. E., Stanley, T. B., Montana, V. G., Lambert, M. H., Shearer, B. G., Cobb, J. E., McKee, D. D., Galardi, C. M., Plunket, K. D., Nolte, R. T., Parks, D. J., Moore, J. T., Kliewer, S. A., Willson, T. M., and Stimmel, J. B. (2002b). Structural basis for antagonist-mediated recruitment of nuclear corepressors by PPAR alpha. *Nature* **415**, 813–817.

Yang, X. J., Ogryzko, V. V., Nishikawa, J., Howard, B. H., and Nakatani, Y. (1996). A p300/CBP-associated factor that competes with the adenoviral oncoprotein E1A. *Nature* **382**, 319–324.

Yoshinaga, S. K., Peterson, C. L., Herskowitz, I., and Yamamoto, K. R. (1992). Roles of SWI1, SWI2, and SWI3 proteins for transcriptional enhancement by steroid receptors. *Science* **258**, 1598–1604.

Yu, J., Li, Y., Ishizuka, T., Guenther, M. G., and Lazar, M. A. (2003). A SANT motif in the SMRT corepressor interprets the histone code and promotes histone deacetylation. *EMBO J.* **22**, 3403–3410.

Yuan, C. X., Ito, M., Fondell, J. D., Fu, Z. Y., and Roeder, R. G. (1998). The TRAP220 component of a thyroid hormone receptor-associated protein (TRAP) coactivator complex interacts directly with nuclear receptors in a ligand-dependent fashion. *Proc. Natl. Acad. Sci. USA* **95**, 7939–7944.

Zhang, X., Jeyakumar, M., Petukhov, S., and Bagchi, M. K. (1998). A nuclear receptor corepressor modulates transcriptional activity of antagonist-occupied steroid hormone receptor. *Mol. Endocrinol.* **12**, 513–524.

Zhou, G., Cummings, R., Li, Y., Mitra, S., Wilkinson, H. A., Elbrecht, A., Hermes, J. D., Schaeffer, J. M., Smith, R. G., and Moller, D. E. (1998). Nuclear receptors have distinct affinities for coactivators: Characterization by fluorescence resonance energy transfer. *Mol. Endocrinol.* **12**, 1594–1604.

Zhu, Y., Qi, C., Jain, S., Rao, M. S., and Reddy, J. K. (1997). Isolation and characterization of PBP, a protein that interacts with peroxisome proliferator-activated receptor. *J. Biol. Chem.* **272**, 25500–25506.

6

Thyroid Hormone Receptor Subtypes and Their Interaction with Steroid Receptor Coactivators

Roy E. Weiss and Helton E. Ramos

*University of Chicago, Thyroid Study Unit
Chicago, Illinois 60637*

I. Introduction
   A. Thyroid Hormone Receptors and
      Their Organization
   B. Thyroid Response Elements and Gene Regulation
   C. Corepressors: Transcription Silencing Cofactors
   D. Coactivators: Transcription
      Amplifying Cofactors
II. Physiology of TR Function
   A. Thyroid Hormone Regulation
   B. Mice with TR Gene Knockouts: A Model to
      Understand TR Isoform Action
III. Physiology of SRC Function (SRC-1, SRC-2, and SRC-3)
IV. Interaction of TRs and SRC in the Pituitary
   A. SRC-1 Aiding TRβ in Mediation of TSH
      Gene Expression
   B. Effect of SRC-1 on Other p160 Coactivators
   C. Interaction of SRC-1 and TRα
V. SRC-1 and TRs in Peripheral Tissues

VI. Conclusions
References

Thyroid hormone (TH) is required for normal growth, development, and metabolism in metazoans. To influence this broad range of physiologic actions, TH is necessarily involved in the regulation of a multitude of genes in virtually every tissue. The diversity of gene expression regulation in response to TH is mediated through specific intranuclear TH receptors (TRs) and other nuclear coregulators. This chapter reviews TRs and nuclear coregulators, specifically coactivators, based on *in vivo* data from knockout (KO) mouse models. © 2004 Elsevier Inc.

I. INTRODUCTION

A. THYROID HORMONE RECEPTORS AND THEIR ORGANIZATION

The intranuclear localization of 3,5,3'-triiodothyronine (T_3) (Oppenheimer *et al.*, 1972) and the demonstration of a nuclear protein that fulfills the criteria of a hormone receptor (Oppenheimer, 1983) were followed by the cloning of two thyroid hormone (TH) receptors alpha and beta (TRα and TRβ) (Sap *et al.*, 1986; Weinberger *et al.*, 1986) mapped to human chromosomes 17 and 3, respectively. These receptors share structural and functional similarities with other members of the nuclear receptor (NR) superfamily such as those for adrenal steroids, sex hormones, vitamin D, and retinoic acids (Evans, 1988). NRs possess a well-conserved DNA binding domain (DBD) separated from a carboxy-terminal ligand binding domain (LBD) by a short segment of amino acids that constitutes the "hinge" region. The DBD of TRs contains two stretches of 13 and 12 amino acids separated by pairs of cysteines that interact with zinc to create two peptide loops (Evans, 1988). These "zinc fingers," projecting from the surface of the protein, interact with specific DNA sequences known as TH response elements (TREs) located usually near the transcription start point of genes regulated by TH. Transactivation of these target genes requires activation of the receptor by hormone binding to the LBD and the presence of additional cofactors. A highly conserved region in the distal carboxy-terminal of the LBD, termed activation function-2 (AF-2), has little effect on ligand binding or dimerization. AF-2, however, is necessary for nuclear coactivator (NCoA) transcriptional activation because it is composed of an amphipathic alpha–helix that interacts with NCoAs (Feng *et al.*, 1998; Tone *et al.*, 1994).

TRα and TRβ genes have substantial structural and sequence similarities. Each generates at least two TR proteins (α1 and α2; β1 and β2) from alternative transcription start sites and by splicing. TRα2 binds to TREs but, because of a sequence difference at the LBD, it does not bind TH and thus does not function as a TR proper (Mitsuhashi et al., 1988). Data from mice with a deletion of the TRα, however, suggest that this unliganded isoform of TRα may have a physiologic function (see the later section about TRα knockout mice). The relative expression of two TR genes and the distribution of the products vary among tissues and during different stages of development (Hodin et al., 1990; Macchia et al., 1990; Strait et al., 1990). Furthermore, an internal promoter is located in intron 7 of the TRα gene. In mice, this promoter is responsible for the expression of truncated isoforms of the TRα1 and TRα2 (TRΔα1 and TRΔα2) containing the carboxy-terminal segment of the molecule. In the case of TRα1, the segment is part of the LBD and contains AF-2. These partial products of the TRα gene may play a role in down-regulation of transcriptional activity (Chassande et al., 1997). In addition, rodents have a TRβ3 isoform that has a unique 23 amino acid N-terminal and acts as a functional receptor (Williams, 2000).

B. THYROID RESPONSE ELEMENTS AND GENE REGULATION

TREs, located in TH-regulated genes, consist of half-sites having the consensus sequence of AGGTCA and vary in number, spacing, and orientation (Forman et al., 1992; Glass, 1994). Each half-site usually binds a single TR molecule (monomer), and two half-sites bind two TRs, called a homodimer, or one TR and a retinoid X receptor (RXR), called a heterodimer (Lazar, 1993). The latter belongs to the NR superfamily and when not engaged in a heterodimer interaction with TR, binds 9-cis retinoic acid (Darling et al., 1991; Zhang et al., 1992). Heterodimer formation is facilitated by the presence of an intact "leucine zipper" motif located in the middle of the TR LBD.

C. COREPRESSORS: TRANSCRIPTION SILENCING COFACTORS

Occupation of TREs by unliganded TRs inhibits the constitutive expression of genes that are positively regulated by TH (Brent et al., 1989) through association with corepressors, such as the nuclear receptor corepressor (NCoR) or the silencing mediator for retinoic acid and thyroid hormone receptors (SMRT) (Koenig, 1998). Five NCoRs have already been identified (McKenna et al., 1999). Transcriptional repression is mediated through the recruitment of mSin3A and histone deacetylases (HDAC) (Pazin and Kadonaga, 1997). The repression compacts nucleosomes into a tight and inaccessible structure, effectively shutting down gene expression.

With regard to the **estrogen receptor**, of the presence of nuclear receptor antagonist, such as Tamoxifen or RU486, corepressor recruitment occurs. This effect on the TR is relieved by TH, which, by binding to TR, releases the corepressor, dissociates TR dimers bound to TRE, enhances the occupation of TREs by TR: RXR heterodimers (Yen *et al.*, 1992), and recruits NCoAs such as the CREB binding protein-associated factor (p/CAF) and SRC-1 with histone acetyltransferase (HAT) activity (see following sections).

D. COACTIVATORS: TRANSCRIPTION AMPLIFYING COFACTORS

More than 15 different nuclear coactivators (NCoAs) have been identified, some of which appear be involved in TH action; each coactivator has several different names (McKenna *et al.*, 1999; Xu *et al.*, 1999). NCoAs are proteins that can remodel chromatin via enzymatic modification of histone tails by histone acetylase or via regulation of transcription at the promoter by interaction with RNA polymerase II (Pol II) and general transcription factors (GTFs) (Fondell *et al.*, 1999) (Fig. 1). Actually, the ligand-dependent association with 10 to 20 polypeptides, known as TR-associated proteins or TRAPs, in conjunction with the general coactivators PC2 and PC4 act to mediate transcription by RNA Pol II and general initiation factors.

TR dimerization is not required for hormone binding, and binding does not induce dimerization. Furthermore, it is believed that T_3 exerts its effects by inducing conformational changes of the TR molecule and that RXR stabilizes the association of TR with TRE. Recently, Kim *et al.* (1999) demonstrated that distant regulatory elements, such as an enhancer and a promoter, come in close proximity when the TR binds T_3. The histones maintain the promoter of a responsive gene in a "closed" state by catalyzing the acetylation of the histone lysines, the NCoAs, and specifically the SRC and CBP, which makes the promoter 'open and ready for business.'

Among the many NCoAs, three ubiquitous coactivators of nuclear hormone receptors contain one or several copies of the amino acid sequence $L_1XXL_2L_3$ that is an amphipathic α–helix necessary for binding to the receptor (McInerney *et al.*, 1998). The number of copies and the spacing of these sequences may confer specific NR specificity (Coulthard *et al.*, 2003). These three coactivators are SRC-1 (NCoA-1/ERAP), SRC-2 (TIF2/NCoA-2/GRIP1), and SRC-3 (AIB1/NCoA-3/RAC-3/ACTR/TRAM1) (Koenig, 1998; McKenna *et al.*, 1999). SRC-1, SRC-2, and SRC-3 share a 40% sequence homology primarily in the N-terminal, which contains the bHLH/PAS domain. All three SRCs have been shown to be involved in the activation of TRs through interaction with several activation domains located in the LBD, the most important being the AF-2 domain (Benecke *et al.*, 2000). The interaction between TR and NCoA is facilitated by the $L_1XXL_2L_3$ sites. The designation of these factors as coactivators was

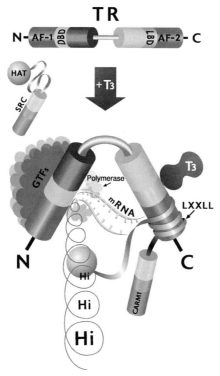

FIGURE 1. Model of SRC and TR interaction. The upper part of the figure represents the unliganded TR. In the presence of T3, recruitment of SRC and its interaction with the AF-2 domain of the TR results in a conformational change of the TR (lower panel). This allows the HAT activity of the SRC to act upon the thyroid hormone responsive gene allowing for transcription. Hi, histone; GTFs, general trascription factors; DBD, DNA binding domain; LBD, ligand binding domain; AF-2, activating function 2; HAT, histone acetyltransferase; CARM 1, coactivator arginine methyl transferase.

initially predicated on their nuclear localization, their ability to interact with and amplify ligand-dependent functions of NRs on TRE-linked reporter genes, and their capacity to relieve the nuclear receptor (gene repression) (Onate et al., 1995). The observation that mice in which the coactivator gene had been deleted had partial resistance to steroid and thyroid hormones confirmed that at least SRC-1 functions physiologically as a coactivator. Although cellular transfection assays have suggested a functional redundancy among NCoAs (McKenna et al., 1999), biologic specificity lies largely in spatial and temporal control of gene expression and probably in the regulation of NCoA expression. Even though mice with deficiency in SRC-1 have several steroid hormone abnormalities, their progesterone and glucocorticoid receptors recruit distinct coactivator complexes and promote distinct patterns of local chromatin modification (Li et al., 2003). The SRC binding pockets of NR bind $L_1XXL_2L_3$ consensus motif while relying on the difference in SRC structure flanking the binding region to convey selectivity. This has been proven by the development of the novel selective inhibitor 1(3,37,37) of TRβ with SRC-2 (Geistlinger and Guy, 2003).

II. PHYSIOLOGY OF TR FUNCTION

A. THYROID HORMONE REGULATION

The response of a tissue to TH is directly dependent on (1) the amount of TH available to the tissue, (2) the transport of TH into the cells, (3) the conversion of thyroxine (T_4) into the active hormone T_3 enabled by deiodinases, (4) the number of TH receptors (TRs) that are available to interact with T_3, and (5) the quantity and nature of transcriptional cofactors that are available in the cell. The main subject of this review is the latter point, namely how TRs and NCoAs interact and result in regulation of TH action.

The availability of TH to the tissue is dependent on the TH released from the gland. Thyroid hormone synthesis occurs in the thyroid follicular cell under the stimulation of thyrotropin. Several steps are necessary for the synthesis of TH, starting with active transport of iodine via a Sodium/Iodide (Na^+/I^-) symporter, oxidation of the iodine by peroxidase to form monotyrosine or dityrosine that is coupled with the thyroglobulin molecule to produce the iodothyronines T_4 and T_3. Thyroglobulin containing TH is stored in the eosinophilic, acellular colloid, where it remains until the thyroid stimulating hormone (TSH) activates internalization and proteolysis of thyroglobulin and the release of T_4 and T_3 from the gland (Fig. 2).

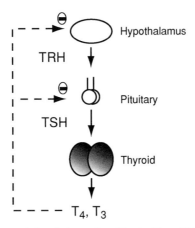

FIGURE 2. Physiology of the pituitary thyroid axis. Thyroid hormone (TH) is released from the thyroid gland under the influence of TSH from the pituitary. In addtion, the hypothalamus regulates TSH release. TH events negative feedback at the pituitary and hypothalamus (dashed line, negative arrows).

T_4 is released into the blood and circulates both free or bound to one of several binding proteins. Cell uptake of TH requires the presence of a thyroid hormone transporter, and the uptake process is possibly also aided by passive diffusion. Once inside the cell, T_4 is converted into the active hormone T_3 by deiodination. T_3 is then shuttled into the nucleus where it binds to the TRs. This complex together with nuclear coregulators control TR responsive gene transcription. Although TH affects all tissues, its effect on the pituitary thyrotroph is repressive and a component of the negative feedback loop.

The ultimate effect of TH on gene transcription relies on the availability of TRs and cofactors. It appears that TH regulates the expression of TRs as well as of NCoAs. The importance of TR "autoregulation" has been demonstrated in GH1 cells treated with T_3, where a decrease in the number of TRs leads to a decrease in T_3 responsiveness (Samuels et al., 1977) or desensitization to TH. Autoregulation of TRs has been described in amphibian metamorphosis (Tata, 2000).

B. MICE WITH TR GENE KNOCKOUTS: A MODEL TO UNDERSTAND TR ISOFORM ACTION

1. TRβ Knockout Mice: Nutley and Lyon

The importance of TRs in TR action is demonstrated by mice that lack the TRs (TR knockout mice) and subsequently are insensitive to TH or, in other words, resistant to TH (RTH). These mice have been extensively investigated as models to study TR isoform action. The first report of a mouse with a deletion of one of the TRs was the TRβ1 and 2 knockout mouse of Forrest et al. (1996a). Currently, there are two different lines of mice with similar but not identical deletions of the TRβ gene and with a specific deletion of TRβ2 (Abel et al., 1999). There are also three lines of mice with the TRα gene deletions (TRα1 or combined TRα1 and 2). In this chapter's discussion, they are distinguished by their city of origin: the Nutley TRβ knockout, the Lyon TRβ knockout, and the Stockholm knockout mice. For the Nutley mice, disruption of TRβ1 and TRβ2 expression associated with elevated serum TSH levels resulted in elevated thyroid hormone levels. Other abnormalities reported were goiter and sensorineural deafness (Forrest, 1996; Forrest et al., 1996b). Therefore, the phenotype is almost identical to the one observed more than 30 years earlier in humans: the mice had resistance to thyroid hormone (RTH) due to deletion of the TRβ gene (Refetoff et al., 1967; Takeda et al., 1992). The Nutley TRβ$^{-/-}$ mice did not display overt behavioral or neuroanatomical abnormalities and did not have fertility problems. Mice that were heterozygous for the gene disruption (Nutley TRβ$^{+/-}$) had a phenotype similar to the Nutley wild-type mice (TRβ$^{+/+}$).

The regulation of the thyroid axis of the Nutley TRβ$^{-/-}$ mice demonstrated that while TRβ is not required for the up-regulation of TSH induced by TH deprivation, TRβ enhances the sensitivity of TSH down-regulation by TH and may be essential for the complete suppression of TSH (Weiss et al., 1998). Thus, the unliganded TRβ is not required for the up-regulation of TSH. Human fibroblasts from subjects with a TR deletion (Hayashi et al., 1993) and from experiments using tissue from the ear, brain, and liver, respectively, of the TR$\beta^{-/-}$ mouse (Forrest et al., 1996a; Sandhofer et al., 1998; Weiss et al., 1998) show that the expression of TRα1 and 2 at the mRNA is not altered by the lack of TRβ1 and 2. Serum cholesterol (CHOL), alkaline phosphatase (AP), and the mRNAs of liver S14 and ME show no response or attenuated response to TH in the absence of TRβ, indicating TRβ dependence for these actions (Weiss et al., 1998). Although regulation of heart rate (HR) and energy expenditure (EE) by TH are independent of TRβ (Weiss et al., 1998) as shown later in Vennström's laboratory (Wikström et al., 1998), they are primarily dependent on TRα1 (see upcoming discussion).

Samarut and colleagues in Lyon, France, produced another TRβ knockout strain of mice (Lyon TR$\beta^{-/-}$) (Gauthier et al., 1999). Similar to the Nutley TR$\beta^{-/-}$ mice, the Lyon TR$\beta^{-/-}$ mice also expressed an amino-terminal fragment of the TRβ1 and TRβ2, though the Lyon mice had 10 more amino acids than the Nutley mice. In both the Nutley and Lyon mice, a recombination cassette containing the selection markers was inserted either in exon 3 (Nutley) or downstream from exon 3 with deletion of exons 4 and 5 (Lyon) precluding the expression of functional TRβ1 and TRβ2 proteins. The Lyon TR$\beta^{-/-}$ displayed an elevation of serum T$_4$ and TSH concentrations similar to those of the Nutley TR$\beta^{-/-}$ (Gauthier et al., 1999). Detailed physiologic studies with the Lyon TR$\beta^{-/-}$ mice revealed that they were similar to the Nutley TR$\beta^{-/-}$ mice. The mice were fertile and had normal growth and development.

2. TRα KO Mice: Lyon and Stockholm

The first TR$\alpha^{-/-}$ mouse was reported by Fraichard et al. (1997). This mouse was generated by placing the recombination DNA cassette in the first coding exon (exon 2) immediately downstream of the TRα promoter. This disrupted both TRα1 and TRα2 mRNAs, but did not affect expression of the carboxy-terminal fragment transcribed from an internal promoter located in intron 7 of the TRα gene. These $\Delta\alpha$1 and $\Delta\alpha$2 fragments have 169 and 249 amino acids, respectively. After 2 weeks of life, the homozygous mice (Lyon TR$\alpha^{-/-}$) had hypothyroidism that progressively worsened and exhibited a growth arrest. Marked delay in maturation of small intestine and bones was noted and all mice died by 5 weeks of age. When treated for 1 week with L-T$_3$ starting on week 3 of postnatal life, they survived as did the wild-type mice (Fraichard et al., 1997). The serum concentrations of T$_3$ and T$_4$ in 3-week-old Lyon TR$\alpha^{-/-}$ mice were reduced to 60% of the concentrations for the wild type. The degree of hypothyroninemia increased with age. These animals demonstrate the functional importance of TRα1 and/or TRα2. Because of the expression of both TRαΔ1 and/or TRαΔ2, however, it was difficult at that time to ascertain if the hypothyroid effects resulted from the unopposed action of these truncated receptors interfering with TH action or from the lack of TRα. Subsequently, the Lyon group produced a mouse that also lacked the alternative promoter in intron 7, eliminating the TRαΔ mRNAs. These mice, termed TRα$^{0/0}$, displayed TH hypersensitivity (Macchia et al., 2001).

Another TRα knockout mouse was produced by Wikström et al. (1998). These mice lacked the functional TRα1 and TRΔα1, but still expressed the TRα2 splice variant and presumably also TRαΔ2 (Stockholm TRα1$^{-/-}$ mice). This was achieved by insertion of a DNA cassette at the end of exon 7 of TRα1. The Stockholm TRα1$^{-/-}$ male mice had lower levels of free T$_4$ than the corresponding TRα1$^{+/+}$ mice 8.5 ± 1.7 versus 12.5 ± 4.3 pmol/l. The Stockholm TRα1$^{-/-}$ mice had normal growth, fertility, and development. Their body temperature, however, was lower by 0.5 °C than the corresponding wild-type mice, and their heart rate (HR) was slower by 20% with prolonged QRS and QT durations (Johansson et al., 1998). These results together with those obtained in my laboratory with the TRβ$^{-/-}$ mice (Nutley and Lyon) indicate that TRα1 plays an important role in TH-dependent regulation of HR, energy expenditure (EE), and body temperature (Weiss et al., 1998).

3. Combined TRα/TRβ Knockout Mice: Lyon and Stockholm

Production of combined TRα and TRβ knockout mice was achieved by crossing the Lyon TRα$^{-/-}$ with the TRβ$^{-/-}$ mice (Gauthier et al., 1999). In the double knockout mice, the serum T$_4$ and T$_3$ concentration were dramatically elevated, with serum TSH values greater than 300-fold the mean levels of the +/+ animals. These mice had goiters and delayed growth. The ablation of the TRα gene exaggerated the phenotype of the TRβ-only knockout. Therefore, although the mice were not robust, they survived in the absence of all TR isoforms except for the TRαΔ fragments.

Vennström and colleagues in Stockholm also produced viable combined TRα1$^{-/-}$/TRβ$^{-/-}$ mice, but these mice still contained the TRα2 transcript (Göther et al., 1999). They demonstrated that there was no detectable T$_3$-binding to nuclear extracts from liver or lung, eliminating the possibility that there were other TH binding proteins. However, these animals were not normal in the absence of TRα1 and TRβ because they demonstrated a slight reduction in growth. As did the Lyon double knockout mice, they had markedly elevated TSH levels that could not be further increased by the reduction of serum TH levels or reduced by treatment with L-T$_3$. The female mice also had reduced fertility.

A third type of double knockout mouse has been produced in France, the Lyon TRα$^{0/0}$ TRβ$^{-/-}$ that lacks TRα1 and 2 as well as TRαΔ1 and TRαΔ2 (Gauthier et al., 2001). Males and females survive until at least 6 months of age. The females are completely sterile, and the males have atrophied testes and show markedly reduced fertility. Although a goiter is present in these mice, it is about one third the size described for the Stockholm TRα TR$\beta^{-/-}$ mice.

III. PHYSIOLOGY OF SRC FUNCTION (SRC-1, SRC-2, AND SRC-3)

As previously explained in this discussion, the disruption of the NCoA-1 gene in mice results in partial resistance to several hormones (Weiss et al., 1999; Xu et al., 1998). SRC-1 knockout mice were shown to have a resistance to sex hormones (Xu et al., 1998) as well as to TH (Weiss et al., 1999). Serum total and free T_4 and T_3 levels, as well as TSH concentrations, were significantly higher compared to wild-type mice by approximately 1.5-fold (T_4 and T_3) and 2.5-fold (TSH) (Table 1). In addition to baseline data, decreased suppression of TSH and T_4 with L-T_3 confirmed the resistance to TH in the SRC-1$^{-/-}$ mice. Similarly, deletion of the coregulator RXRγ in mice has been shown to cause pituitary RTH associated with increased oxygen consumption (Brown et al., 2000).

SRC-2 knockout mice have defective spermatogenesis, testicular degeneration, placental hypoplasia, and decreased male and female fertility (Gehin et al., 2002), as well as higher energy expenditure in brown fat, higher lipolysis in white fat, and resistance to obesity (Picard et al., 2002). NCoA-3 has been recently shown to be required for normal growth, puberty, female reproductive function, and mammary gland development in knockout mice (Xu et al., 2000). A useful review has been published on the *in vivo* actions of the SRCs (Xu and Li, 2003). SRC-2 interacts with the estrogen receptor (Eng et al., 1998) *in vitro* and competes with TRAP 220 for binding to nuclear receptors (Treuter et al., 1999). A recent study has demonstrated (based on immunohistochemistry) that although SRC-2 mRNA is expressed at very low levels in brain, it is expressed in the anterior pituitary at a level comparable to that of SRC-1 (Meijer et al., 2000). Therefore, it is not surprising that these coactivators may play a role in TH action in health and in disease. The evidence of the coactivators' role in human disease is that NCoA-3 is highly amplified in 10% of primary breast cancers and its mRNA is overexpressed in 64% of primary breast tumors (Anzick et al., 1997). In addition, NCoA-2 fused with the mesenchyme homeobox (MOX) genes has been identified in acute myeloid leukemia (Carapeti et al., 1998).

Previous data from Northern blot analysis (Chen et al., 1997; Misiti et al., 1998; Takeshita et al., 1997; Torchia et al., 1997; Voegel et al., 1996; Xu et al., 1998) suggest that although NCoAs are widely expressed, their relative expression levels are dependent on tissue types. Furthermore, a single tissue may express all of the NCoAs. This implies that there may be some redundancy in the NCoA roles. To investigate redundancy, thyroid

TABLE 1. Baseline Thyroid Function Tests in Male Mice of Different Genotypes

Genotype	β1,2	α1,2	SRC-1	TT$_4$ μg/dl	FT$_4$I units	TT$_3$ ng/dl	FT$_3$I units	TSH mU/l
Wild type	+	+	+	3.7 ± 0.1 (53)***	4.9 ± 0.2 (23)**	73 ± 5 (21)***	124 ± 8 (11)***	30 ± 5 (65)**
TRβ$^{-/-}$	−	+	+	7.6 ± 0.3 (53)^{c,***}	12.7 ± 1.5 (19)^{c,**}	139 ± 9 (23)^c	267 ± 42 (9)^a	258 ± 44 (57)^{c,*}
TRα$^{0/0}$	+	−	+	3.2 ± 0.1 (33)^{b,***}	4.9 ± 0.2 (23)**	96 ± 6 (12)**	134 ± 28 (9)***	23 ± 4 (34)**
SRC-1$^{-/-}$	+	+	−	4.7 ± 0.2 (19)^c	8.8 ± 0.6 (12)^b	130 ± 4 (12)^c	235 ± 13 (12)^b	74 ± 19 (19)^b
TRβ$^{-/-}$/SRC-1$^{-/-}$	−	+	−	12.6 ± 2.6 (11)$^{c,***,\triangledown}$	25.7 ± 2.2 (6)$^{c,***,\triangledown}$	404 ± 39 (7)$^{c,***,\triangledown}$	624 ± 70 (8)$^{c,***,\triangledown}$	733 ± 152 (19)$^{c,***,\triangledown}$
TRα$^{0/0}$/SRC-1$^{-/-}$	+	−	−	6.5 ± 0.3 (18)$^{c,***,\triangledown}$	14.0 ± 1.1 (16)$^{c,***,\triangledown}$	159 ± 8 (14)$^{c,***,\triangledown}$	333 ± 25 (12)$^{c,***,\triangledown}$	52 ± 12 (18)$^{a,\triangledown}$
Wild type	+	+	+	3.9 ± 0.3 (10)		133 ± 5 (10)		43.4 ± 5.2 (12)
SRC-1$^{-/-}$	+	+	−	8.3 ± 0.5 (18)		210 ± 10 (15)		Not given^1
TRβ$^{PV/+}$	PV/+	+	+	11.0 ± 0.8 (13)		230 ± 10 (17)		56.0 ± 5.2 (12)
TRβ$^{PV/+}$/SRC-1$^{-/-}$	PV/+	+	−	23.4 ± 0.8 (20)		450 ± 30 (15)		104.9 ± 10.1 (29)

Notes: Mean ± standard error, numbers in parentheses = number of mice. Statistics compared to wild type: $^a p < 0.05$, $^b p < 0.005$, $^c p < 0.0005$; Statistics compared to SRC-1$^{-/-}$: *$p < 0.05$, **$p < 0.005$, ***$p < 0.0005$; Statistics compared to either TRβ$^{-/-}$ or TRα$^{0/0}$: $^\triangledown p < 0.0005$.
^{1}Reported as 1.9 times the value of the wild type.
Sources: For upper half of table, the source is Sadow et al. (2003c). For the lower half, the source is Kamiya (2003).

function was measured in compound SRC-1 and SRC-2 KO mice (Table 1) (Weiss et al., 2002). Whereas SRC-1 KO (SRC-1$^{-/-}$) mice are resistant to TH and SRC-1$^{+/-}$ mice are not, SRC-2 KO (SRC-2$^{-/-}$) mice had normal thyroid function. Yet double heterozygous SRC-1$^{+/-}$/SRC-2$^{+/-}$ mice manifested resistance to TH of a similar degree as mice completely deficient in SRC-1. KO mice of both SRC-1 and SRC-2 resulted in marked increases of serum TH and TSH concentrations. The increases demonstrated gene dosage effect in nuclear coactivators manifesting as haploinsufficiency and functional redundancy of SRC-1 and SRC-2.

IV. INTERACTION OF TRs AND SRC IN THE PITUITARY

A. SRC-1 AIDING TRβ IN MEDIATION OF TSH GENE EXPRESSION

Different tissues may express different arrays of coactivators. In this section we explore the specific interaction between TRB and SRC-1 in the pituitary. During TH deprivation, the greatest increase in serum TSH was in the TR$\beta^{-/-}$/SRC-1$^{-/-}$ mice, in which only TRα is present (Sadow et al., 2003a). This indicates the importance of TRα and SRC-1 interaction in the absence of TH. Similar to what happens due to TH withdrawal, during TH treatment there is a strong interaction between TRα and SRC-1 as the TR$\beta^{-/-}$/SRC-1$^{-/-}$ mouse experiences mild suppression of TSH. The term '*interaction*' implies not only a direct physical interaction between TR and SRC-1, but also refers to the interaction among SRC-1 and other transcription factors, which may influence the transcriptional activity of TR. These findings are further supported in a similar model showing that the lack of SRC-1 intensified the pituitary dysfunction in a mouse that was not a knockout but had a PV mutation (TRβPV) (Kamiya et al., 2003).

The interaction of TRβ and SRC-1 can be assessed by comparing the TR$\alpha^{0/0}$ and TR$\alpha^{0/0}$/SRC-1$^{-/-}$ mice. During TH withdrawal, the absence of SRC-1 does not influence release of TH. During TH treatment, however, the absence of SRC-1 impairs the ability of TRβ to suppress TSH, indicating that SRC-1 mediates TRβ suppression by TH and not activation (Sadow et al., 2003c) (Fig. 3).

The mechanism of down-regulation of TSH expression in the presence of TH is not well understood (Yen, 2001). TH, similar to steroid hormones and retinoids, regulates gene expression by binding NRs and by recruiting coactivators that posses intrinsic histone acetyltransferase activity (HAT) (Xu et al., 1999). The HAT activity is essential for RNA polymerase initiation of gene transcription. Activation of the TR with TH, however, turns off TSH transcription. Xu et al. (2001) have shown that coactivator-associated arginine methyltransferase 1 (CARM1) can methylate the class

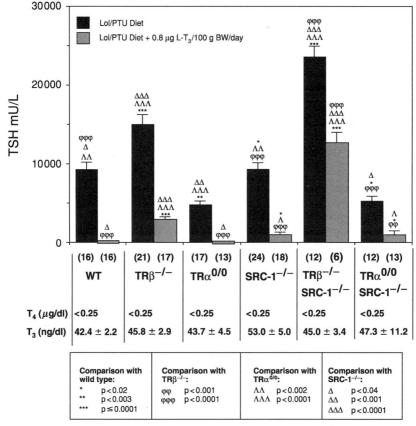

FIGURE 3. Effects of TH deprivation and T_3 treatment on serum TSH concentrations. Black bars represent data after 14 days on LoI/PTU diet, and gray bars represent data after 4 days of 0.8 μg L-T_3/100 g body weight per day while receiving the LoI/PTU diet. Each point is the mean ± SE, and the number of mice tested (N) is shown below each genotype. Relevant table of p values determined by Fisher PLSD is below the graph. Serum T_4 and T_3 concentration (mean ± SE) are shown for mice after 14 days of LoI/PTU diet.

of p160 coactivators. This results in loss of HAT activity and conversion of the SRC into a corepressor. The methylation of SRC-1 by CARM1 may thereby be a mechanism for TRβ/SRC-1 interaction during negative regulation of TSH with TH. In fact, it may be the interaction of SRC-1 with other transcription factors regulating TSH expression that contributes to the physiologic events observed. For example, one cannot overlook that RXR and other transcription factors may be involved in TH action mediated by SRC-1.

The interaction between SRC-1, TRβ, and TRα in regulation of thyrotroph TSH expression is summarized in Fig. 4. The interaction of

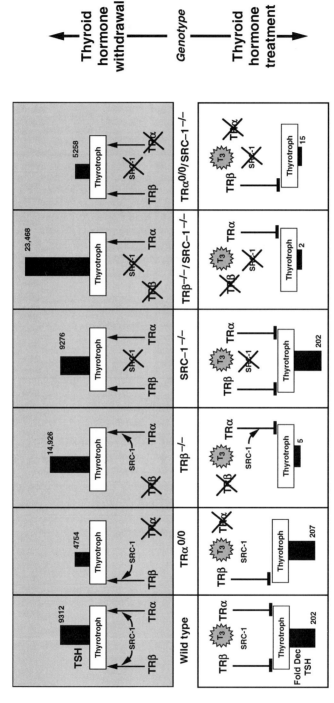

FIGURE 4. Model of TH/SRC interaction in the pituitary. The upper half of the diagram illustrates changes observed during thyroid hormone withdrawal. The lower half of the diagram illustrates the effect of treatment with L-T$_3$. The genotype for each case is in the middle.

SRC-1 and TRβ is stronger in the presence of TH, and the functional interaction of SRC-1 with TRα is stronger in the absence of TH. It should be noted that inherent in this model is that the absence of SRC-1 and TRs is not tissue selective and that one is unable to determine what adaptive processes have occurred during pituitary development that contribute to the final phenotype observed.

B. EFFECT OF SRC-1 ON OTHER p160 COACTIVATORS

Many SRC coactivators are expressed in multiple tissues, suggesting that there is either redundancy in their function or a receptor complex specificity. Deletion of SRC-1 resulted in a 2-fold increase in mRNA expression of the coactivator SRC-2 as measured by Northern blot in brain and testes tissues (Xu *et al.*, 1998). The same 2-fold up-regulation of SRC-2 in the pituitary of SRC-1$^{-/-}$ mice has been demonstrated by quantitative real time polymerase chain reaction (PCR) (Sadow *et al.*, 2003b). Furthermore, a 3.5-fold SRC-2 expression increase in pituitaries of TR$\beta^{-/-}$/SRC-1$^{-/-}$ mice over the wild type was shown. In addition, pituitary expression of the p160 coactivator SRC-3 was significantly increased in SRC-1$^{-/-}$, TR$\alpha^{0/0}$/SRC-1$^{-/-}$, and TR$\beta^{-/-}$/SRC-1$^{-/-}$ mice (2.6-, 2.1-, and 1.6-fold, respectively). The increase in SRC-3 expression is independent of receptor subtype and is only affected by the loss of SRC-1. Moreover, SRC-3 expression is increased in response to T_3, which in part may be responsible for SRC-3's increased expression in mice deficient in SRC-1 because the SRC-1$^{-/-}$ mice have higher TH levels. These data demonstrate that in the absence of SRC-1, coactivators of the p160 family increase expression in the pituitary, specifically mRNA expression of SRC-2 and SRC-3, yet this increased expression does not obviate the RTH seen in TR$\beta^{-/-}$, SRC-1$^{-/-}$, and TR$\beta^{-/-}$/SRC-1$^{-/-}$ mice.

These data, however, cannot rule out the possibility that the degree of resistance to TH would not be more severe in the absence of a combination of these coactivators (Takeuchi *et al.*, 2002). Though these coactivators have been shown to have interactions with individual TRs *in vitro* (McKenna *et al.*, 1999), these data are the first to demonstrate such an interaction between distinct coactivators and TRs *in vivo*. An important next step would be to investigate direct interaction between coactivators and TR subtypes *in vitro*. Isolation of a cell line devoid of all coactivators to test these specific interactions is difficult because these coactivators are ubiquitously expressed, as mentioned previously, and they interact with many molecules. To use an *in vitro* system would require not only isolation of individual coactivators, but also require depletion of other receptors that interact with them. Therefore, the *in vivo* approach, which acknowledges the additional interactions of other factors in the cell, seems to be the less biased approach as indicated by this discussion.

C. INTERACTION OF SRC-1 AND TRα

SRC-1 mRNA expression levels in pituitaries of TR$\alpha^{0/0}$ mice increased 4-fold over those of the wild type. SRC-1a compared to the SRC-1e isoform was previously reported to be higher in rat pituitary by *in situ* hybridization (Meijer *et al.*, 2000); however, distinguishing between the isoforms was not possible (Hayashi *et al.*, 1997) due to the design of the primers for the PCR reaction. Since the hypersensitivity of the TR$\alpha^{0/0}$ mice was abrogated by simultaneously removing SRC-1 (TR$\alpha^{0/0}$/SRC-1$^{-/-}$ mice) as determined by baseline TH levels, then the increase in SRC-1 mRNA levels (if accompanied by an increase in pituitary SRC-1 protein levels) rather than the removal of a constitutive inhibition by TRα2 may account for the hypersensitivity phenotype previously reported in TR$\alpha^{0/0}$ mice (Macchia *et al.*, 2001). Increased TH sensitivity would be caused by greater availability of the SRC-1 to interact with the TRβ and facilitate TH action, as has been shown previously *in vitro*.

Although the current data do not completely exclude a contribution by TRα2, they suggest that the contribution may be less important than the effect of increased SRC-1 levels: the TR$\alpha^{0/0}$/SRC-1$^{-/-}$ mice have moderate resistance to TH action, abrogating the T$_3$ hypersensitivity seen in TR$\alpha^{0/0}$ mice discussed previously. Furthermore, it is possible that the expression of SRC-1 is directly regulated by TRα2. One could answer this question by studying mice with a TRα2 specific deletion. Unfortunately, such mice have a mixed phenotype of hypothyroidism and hyperthyroidism. Because of the overexpression of TRα1, the inability to produce a mouse that normally expresses TRα1 but does not have TRα2 precludes the ability to test this hypothesis. Therefore, in the pituitary, SRC-1 action mediating TH-regulated transcription appears to be TR isoform specific. Whether the role SRC-1 plays in other tissues to mediate TH action is also TR isoform specific is under investigation.

From these studies, it was concluded that (1) TR$\beta^{-/-}$/SRC-1$^{-/-}$ mice demonstrate more severe TH resistance than either the SRC-1$^{-/-}$ or TR$\beta^{-/-}$ mice and the additive effect indicates that SRC-1 has an independent role in TH action over that of TRβ, possibly enhancing the activity of TRα; (2) SRC-1 facilitates TRβ and TRα-mediated down-regulation of TSH because TR$\alpha^{0/0}$/SRC-1$^{-/-}$ mice demonstrate TH resistance rather than the hypersensitivity of TR$\alpha^{0/0}$ mice; and (3) a compensatory increase in SRC-1 expression is associated with the TH hypersensitivity of TRα-deficient animals.

V. SRC-1 AND TRs IN PERIPHERAL TISSUES

Studies have examined the *in vivo* effect of TR subtypes in peripheral tissues, specifically by focusing on animal growth, heart rates, energy expenditure, and markers of TH action in the liver and heart (Sadow *et al.*,

2003a; Takeuchi *et al.*, 2002). Loss of SRC-1 in combination with either TR subtype (TR$\alpha^{0/0}$/SRC-1$^{-/-}$ and TR$\beta^{-/-}$/SRC-1$^{-/-}$) results in linear and weight growth retardation compared to the wild type both at 5 weeks (86.7/ 92.2% and 69.0/76.9%, respectively) and at 9 weeks (77.7/73.0% and 90.8/ 90.0%, respectively). Taken together, these data indicate that both TR subtypes play important roles in governing linear and body weight growth. In addition, SRC-1 plays a cooperative role in growth with each TR subtype in the absence of the other. In the absence of SRC-1, the presence of other cofactors may compensate for its loss, but they compensate insufficiently when there is additional loss of either TR. Although this chapter's study did not investigate expression of growth hormone in TR subtype/SRC-1 deficient mice, there is speculation that these levels may be normal, as has been shown in TR$\alpha^{0/0}$ (Gauthier *et al.*, 2001), TRα1$^{-/-}$, and TR$\beta^{-/-}$ with SRC-1$^{-/-}$ mice (Takeuchi *et al.*, 2002). Moreover, animals deficient in both TRβ and SRC-1 showed increased energy expenditure over all other genotypes on a normal diet ($p < 0.001$). This result corresponds with much higher baseline levels of thyroid hormone and progressive failure to gain weight in these mice at 9 weeks. The use of a TRβ isoform specific ligand, GC-1, failed to maintain core body temperature and reduced stimulation of the uncoupling protein in brown adipose tissue of hypothyroid mice (Ribeiro *et al.*, 2001). Taken together, these data indicate that TRα mediates the increased energy expenditure observed in response to TH, supporting previous observations (Wikström *et al.*, 1998), and that SRC-1 is not necessary for this action of TH. The increased energy expenditure and weight change in these animals do not correspond with illness. These animals demonstrated fertility and litter size at homozygosity, similar to those of the wild type. There was no increased mortality in adult animals at least through 60 weeks, which is the longest these animals have been maintained with respect to this chapter's study.

It has previously been shown that TRα is required for normal heart rate (Gloss *et al.*, 2001; Johansson *et al.*, 1998, 2000; Macchia *et al.*, 2001; Saltó *et al.*, 2001). In addition, the authors of this chapter as well as other researchers have reported that absence of TRβ does not cause such a decline in basal heart rates and, in fact, results in a slight increase in heart rate attributed to increased circulating levels of TH in these mice. There appear, however, to be a number of differences in baseline heart rates of wild-type mice. These differences seem dependent on strain and are even apparent in studies conducted by the same laboratories. SRC-1 appears to be essential for maintaining heart rate because in its absence, alone and in combination with each of the receptor subtypes, heart rate is decreased. In the absence of SRC-1, TRα and TRβ are unable to maintain normal heart rates, presumably due to competition for a limited supply of cofactors or an affinity of the TRα2, which does not bind TH, for other cofactors. In the absence of TRβ alone, TRα and SRC-1 are unable to maintain the wild-type's heart rate. In the absence of both TRβ and SRC-1, however, there is a marked increase in serum TH levels (Table I) over those of TR$\beta^{-/-}$ mice available to bind to TRα1, which does not happen in the absence of SRC-1 alone. This finding seems likely because TR$\beta^{-/-}$/SRC-1$^{-/-}$ mice achieve wild-type mouse heart rates when treated with exogenous TH. Despite the proposed role for SRC-1 in regulating heart rate the authors of this chapter have previously shown that other TH-dependent genetic markers in heart, specifically SERCA2, MHCα, and MHCβ, were unaffected by the absence of SRC-1 (Takeuchi et al., 2002). This was likely due to the low level of SRC-1 expression in the liver (unpublished results). In another model with the PV mutation present the TRβ mouse, the absence of SRC-1 did not have any effect on the liver's TH responsiveness (Kamiya et al., 2003).

VI. CONCLUSIONS

TH mediates a myriad of cellular responses in almost every tissue in the body. To achieve this function, each tissue has specific thyroid hormone receptors and nuclear coregulatory proteins. Together, these serve as gene transcription regulation factors that can lend to the specificity of TH action. Furthermore, developmental expression of the different TRs and SRCs allows for regulation of TH action at different stages of development. Specifically, SRC-1 can modulate TH action via the TRβ isotype in the pituitary. Modulation is achieved via a direct and an indirect interaction in that there is autoregulation of the expression of TRβ and SRC-1. In the liver, however, the interaction of SRC-1 and TRβ seems less important.

The response of TH responsive genes, therefore, are dependent on: (1) the total amount of TR expressed in any one tissue, (2) the relative amount of TRα versus TRβ, (3) cell versus tissue expression data, (4) amount of other cofactor types, and (5) unique differences among TREs.

An understanding of the interactions between TRs and SRCs and other nuclear coregulators will enable researchers to better understand the molecular basis of thyroid action and the factors that regulate processes dependent on thyroid hormone, such as normal neurological development, metabolism, and growth.

ACKNOWLEDGMENTS

We thank Professor Samuel Refetoff for his critical review of this manuscript. This work was supported in part by grants DK 58281, RR 000055, and RR 18372 from the National Institutes of Health and by the Seymour J. Abrams Thyroid Research Center.

REFERENCES

Abel, E. D., Boers, M. E., Pazos-Moura, C., Moura, E., Kaulbach, H., Zakaria, M., Lowell, B., Radovick, S., Liberman, M. C., and Wondisford, F. (1999). Divergent roles for thyroid hormone receptor beta isoforms in the endocrine axis and auditory system. *J. Clin. Invest.* **104,** 291–300.

Anzick, S. L., Kononen, J., Walker, R. L., Azorsa, D. O., Tanner, M. M., Guan, X. Y., Sauter, G., Kallioniemi, O. P., Trent, J. M., and Meltzer, P. S. (1997). A1B1, a steroid receptor coactivator amplified in breast and ovarian cancer. *Science* **277,** 965–968.

Benecke, A., Chambon, P., and Gronemeyer, H. (2000). Synergy between estrogen receptor alpha activation functions AF-1 and AF-2 mediated by transcription intermediatry factor TIF2. *EMBO Rep.* **1,** 151–157.

Brent, G. A., Dunn, M. K., Harney, J. W., Gulick, T., Larsen, P. R., and Moore, D. D. (1989). Thyroid hormone aporeceptor represses T_3-inducible promoters and blocks activity of the retinoic acid receptor. *New Biol.* **1,** 329–336.

Brown, N. S., Smart, A., Sharma, V., Brinkmeirer, M. L., Greenlee, L., Camper, S. A., Jensen, D. R., Eckel, R. H., Krezel, W., Chambon, P., and Haugen, B. R. (2000). Thyroid hormone resistance and increased metabolic rate in the RXR gamma deficient mouse. *J. Clin. Invest.* **106,** 73–79.

Carapeti, M., Aguiar, R. C., Goldman, J. M., and Cross, N. C. (1998). A novel fusion between MOZ and the nuclear receptor coactivator TIF2 in acute myeloid leukemia. *Blood* **91,** 3127–3133.

Chassande, O., Fraichard, A., Gauthier, K., Flamant, F., Legrand, C., Savatier, P., Laudet, V., and Samarut, J. (1997). Identification of transcripts initiated from an internal promoter in the c-erbAα locus that encode inhibitors of retinoic acid receptor α and triiodothyronine receptor activities. *Mol. Endocrinol.* **11,** 1278–1290.

Chen, H., Lin, R. J., Schiltz, R. L., Chacravarti, D., Hash, A., Nagy, L., Privalsky, M. L., Nakatani, Y., and Evans, R. M. (1997). Nuclear receptor coactivator ACTR is a novel histone acetyltransferase and forms a multimeric activation complex with P/CAF and CPB/p300. *Cell* **90,** 569–580.

Coulthard, V. H., Matsuda, S., and Heery, D. M. (2003). An extended LXXLL motif sequence determines the nuclear receptor binding specificity of TRAP220. *J. Biol. Chem.* **278,** 10942–10951.

Darling, D. S., Beebe, J. S., Burnside, J., Winslow, E. R., and Chin, W. W. (1991). 3,5,3′-triiodothyronine (T_3) receptor-auxiliary protein (TRAP) binds DNA and forms heterodimers with the T_3 receptor. *Mol. Endocrinol.* **5,** 73–84.

Eng, F. C., Barsalou, A., Akutsu, N., Mercier, I., Zechel, C., Mader, S., and White, J. H. (1998). Different classes of coactivators recognize distinct but overlapping binding sites on the estrogen receptor ligand binding domain. *J. Biol. Chem.* **273,** 28371–28377.

Evans, R. M. (1988). The steroid and thyroid hormone receptor superfamily. *Science* **240,** 889–895.

Feng, W., Ribeiro, R. C. J., Wagner, R. L., Nguyen, H., Apriletti, J. W., Fletterick, R. J., Baxter, J. D., Kushner, P. J., and West, B. L. (1998). Hormone-dependent coactivator binding to a hydrophobic cleft on nuclear receptors. *Science* **280,** 1747–1749.

Fondell, J. D., Guermah, M., Mali, S., and Roeder, R. G. (1999). Thyroid hormone receptor-associated proteins and general positive cofactors mediate thyroid hormone receptor function in the absence of the TATA box-binding protein associated factors of TFIID. *Proc. Natl. Acad. Sci. USA* **96,** 1959–1964.

Forman, B. M., Casanova, J., Raaka, B. M., Ghysdael, J., and Samuels, H. H. (1992). Half-site spacing and orientation determines whether thyroid hormone and retinoic acid receptors and related factors bind to DNA response elements as monomers, homodimers, or heterodimers. *Mol. Endocrinol.* **6,** 429–442.

Forrest, D. (1996). Deafness and Goiter: Molecular genetic considerations. *J. Clin. Endocrinol. Metab.* **81**, 2764–2767.

Forrest, D., Erway, L. C., Ng, L., Altschuler, R., and Curran, T. (1996a). Thyroid hormone receptor β is essential for development of auditory function. *Nat. Genet.* **13**, 354–357.

Forrest, D., Hanebuth, E., Smeyne, R. J., Evereds, N., Stewart, C. L., Wehner, J. M., and Curran, T. (1996b). Recessive resistance to thyroid hormone in mice lacking thyroid hormone receptor β: Evidence for tissue-specific modulation of receptor function. *EMBO J.* **15**, 3006–3015.

Fraichard, A., Chassande, O., Plateroti, M., Roux, J. P., Trouillas, J., Dehay, C., Legrand, C., Gauthier, K., Kedinger, M., Malaval, L., Rousset, B., and Samarut, J. (1997). The $T_3R\alpha$ gene encoding a thyroid hormone receptor is essential for postnatal development and thyroid hormone production. *EMBO J.* **16**, 4412–4420.

Gauthier, K., Chassande, O., Platerotti, M., Roux, J. P., Legrand, C., Rousset, B., Weiss, R., Trouillas, J., and Samarut, J. (1999). Different functions for the thyroid hormone receptors TRα and TRβ in the control of thyroid hormone production and postnatal development. *EMBO J.* **18**, 623–631.

Gauthier, K., Plateroti, M., Harvey, C. B., Williams, G. R., Weiss, R. E., Refetoff, S., Willott, J. F., Sundin, V., Roux, J. P., Malaval, L., Hara, M., Samarut, J., and Chassande, O. (2001). Genetic analysis reveals different functions for the products of the thyroid hormone receptor alpha locus. *Mol. Cell. Biol.* **21**, 4748–4760.

Gehin, M., Mark, M., Dennefeld, C., Dierich, A., Gronemeyer, H., and Chambon, P. (2002). The function of TIF2/GRIP1 in mouse reproduction is distinct from those of SRC-1 and p/CIP. *Mol. Cell. Biol.* **22**, 5923–5937.

Geistlinger, T. R., and Guy, R. K. (2003). Novel selective inhibitors of the interaction of individual nuclear hormone receptors with a mutually shared steroid receptor coactivator 2. *J. Am. Chem. Soc.* **125**, 6852–6853.

Glass, C. K. (1994). Differential recognition of target genes by nuclear receptor monomers, dimers, and heterodimers. *Endocr. Rev.* **15**, 391–407.

Gloss, B., Trost, S. U., Gluhm, W. F., Swanson, E. A., Clark, R., Winkfein, R., Janzen, K. M., Giles, W., Chassande, O., Samarut, J., and Dillmann, W. H. (2001). Cardiac ion channel expression and contractile function in mice with deletion of thyroid hormone receptor alpha or beta. *Endocrinology* **142**, 544–550.

Göther, S., Wang, Z., Ng, L., Kindblom, J. M., Campos Barros, A., Ohlsson, C., Vennstrom, B., and Forrest, D. (1999). Mice devoid of all known thyroid hormone receptors are viable but exhibit disorders of the pituitary–thyroid axis, growth, and bone maturation. *Genes Develop.* **13**, 1329–1341.

Hayashi, Y., Janssen, O. E., Weiss, R. E., Murata, Y., Seo, H., and Refetoff, S. (1993). The relative expression of mutant and normal thyroid hormone receptor genes in patients with generalized resistance to thyroid hormone determined by estimation of their specific messenger ribonucleic acid products. *J. Clin. Endocrinol. Metab.* **76**, 64–69.

Hayashi, Y., Ohmori, S., Ito, T., and Seo, H. (1997). A splicing variant of steroid receptor coactivator-1 (SRC-1E): The major isoform of SCR-1 to mediate thyroid hormone action. *Biochem. Biophys. Res. Commun.* **236**, 83–87.

Hodin, R. A., Lazar, M. A., and Chin, W. W. (1990). Differential and tissue-specific regulation of the multiple rat c-erbA messenger RNA species by thyroid hormone. *J. Clin. Invest.* **85**, 101–105.

Johansson, C., Lannergeren, J., Lunde, P. K., Vennstrom, B., Thoren, P., and Westerblad, H. (2000). Isometric force and endurance in soleus muscle of thyroid hormone receptor alpha 1 or beta-deficient mice. *Amer. J. Physiol. Reg. Integ. Comp. Phys.* **278**, R598–R603.

Johansson, C., Vennstrom, B., and Thoren, P. (1998). Evidence that decreased heart rate in thyroid hormone receptor alpha 1 deficient mice is an intrinsic defect. *Am. J. Physiol.* **275**, R640–R646.

Kamiya, Y., Zhang, X. K., Ying, H., Kato, Y., Willingham, M. C., Xu, J., O'Malley, B. W., and Cheng, S. Y. (2003). Modulation by steroid receptor coactivator-1 of target tissue responsiveness in resistance to thyroid hormone. *Endocrinology* **144,** 4144–4153.

Kim, M. K., Lee, J. S., and Chung, J. H. (1999). In vivo transcription factor recruitment during thyroid hormone receptor mediated activation. *Proc. Natl. Acad. Sci. USA* **96,** 10092–10097.

Koenig, R. J. (1998). Thyroid hormone receptor coactivators and corepressors. *Thyroid* **8,** 703–713.

Lazar, M. A. (1993). Thyroid hormone receptors: Multiple forms, multiple possibilities. *Endocr. Rev.* **14,** 184–193.

Li, X., Wong, J., Tsai, S. Y., Tsai, M. J., and O'Malley, B. (2003). Progesterone and glucocorticoid receptors recruit distinct coactivator complexes and promote distinct patterns of local chromatin modification. *Mol. Cell. Biol.* **23,** 3763–3773.

Macchia, E., Nakai, A., Janiga, A., Sakurai, A., Fisfalen, M. E., Gardner, P., Soltani, K., and DeGroot, L. J. (1990). Characterization of site-specific polyclonal antibodies to c-erbA peptides recognizing human thyroid hormone receptors α_1, α_2, and β and native 3,5,3'-triiodothyronine receptor and study of tissue distribution of the antigen. *Endocrinology* **126,** 3232–3239.

Macchia, P. E., Takeuchi, Y., Kawai, T., Cua, K., Gauthier, K., Chassande, O., Seo, H., Hayashi, Y., Samarut, J., Murata, Y., Weiss, R. E., and Refetoff, S. (2001). Increased sensitivity to thyroid hormone in mice deficient in thyroid hormone receptor alpha. *Proc. Natl. Acad. Sci. USA* **98,** 349–354.

McInerney, E. M., Rose, D. W., Flynn, S. E., Westin, S., Mullen, T. M., Krones, A., Inostroza, J., Torchia, J., Nolte, R. T., Assa-Munt, N., Milburn, M. V., Glass, C. K., and Rosenfeld, M. G. (1998). Determinants of coactivator LXXLL motif specificity in nuclear receptor transcriptional activation. *Genes Dev.* **12,** 3357–3368.

McKenna, N. J., Lanz, R. B., and O'Malley, B. W. (1999). Nuclear receptor coregulators: Cellular and molecular biology. *Endocr. Rev.* **20,** 321–344.

Meijer, O. J., Steenbergen, P. J., and De Kloet, E. R. (2000). Differential expression and regional distribution of steroid receptor coactivators SRC-1 and SRC-2 in brain and pituitary. *Endocrinology* **141,** 2192–2199.

Misiti, S., Schomburg, L., Yen, P. M., and Chin, W. W. (1998). Expression and hormonal regulation of coactivator and corepressor genes. *Endocrinology* **139,** 2493–2500.

Mitsuhashi, T., Tennyson, G. E., and Nikodem, V. M. (1988). Alternative splicing generates messages encoding rat c-erbA proteins that do not bind thyroid hormones. *Proc. Natl. Acad. Sci. USA* **85,** 5804–5805.

Onate, S. A., Tsai, S. Y., Tsai, M. J., and O'Malley, B. W. (1995). Sequence and characterization of a coactivator for the steroid hormone receptor superfamily. *Science* **270,** 1354–1357.

Oppenheimer, J. H. (1983). The nuclear receptor–triiodothyronine complex: Relationship to thyroid hormone distribution, metabolism, and biological action. In "Molecular Basis of Thyroid Hormone Action" (I. H. Oppenheimer and H. H. Samuels, Eds.). Academic Press, New York.

Oppenheimer, J. H., Koerner, D., Schwartz, H. L., and Surks, M. I. (1972). Specific nuclear triiodothyronine binding sites in rat liver and kidney. *J. Clin. Endocrinol. Metab.* **35,** 330–333.

Pazin, M. J., and Kadonaga, J. T. (1997). What's up and down with histone deacetylation and transcription? *Cell* **89,** 325–328.

Picard, F., Gehin, M., Annicotte, J., Rocchi, S., Champy, M. F., O'Malley, B. W., Chambon, P., and Auwerx, J. (2002). SRC-1 and TIF2 control energy balance between white and brown adipose tissues. *Cell* **111,** 931–941.

Refetoff, S., DeWind, L. T., and DeGroot, L. J. (1967). Familial syndrome combining deafmutism, stippled epiphyses, goiter, and abnormally high PBI: Possible target organ refractoriness to thyroid hormone. *J. Clin. Endocrinol. Metab.* **27**, 279–294.

Ribeiro, R. C., Feng, W., Wagner, R. L., Costa, C. H., Pereira, A. C., Apriletti, J. W., Fletterick, R. J., and Baxter, J. D. (2001). Definition of the surface in the thyroid hormone receptor ligand binding domain for association as homodimers and heterodimers with retinoid X receptor. *J. Biol. Chem.* **276**, 14987–14995.

Sadow, P. M., Chassande, O., Gauthier, K., Samarut, J., Xu, J., O'Malley, B. W., and Weiss, R. E. (2003a). Specificity of thyroid hormone receptor subtype and steroid receptor coactivator-1 on thyroid hormone action. *Am. J. Physiol. Endocrinol. Metab.* **284**, E36–46.

Sadow, P. M., Chassande, O., Koo, E. K., Gauthier, K., Samarut, J., Xu, J., O'Malley, B. W., and Weiss, R. E. (2003b). Regulation of expression of thyroid hormone receptor isoforms and coactivators in liver and heart by thyroid hormone. *Mol. Cell. Endocrinol.* **30**, 65–75.

Sadow, P. M., Koo, E., Chassande, O., Gauthier, K., Samarut, J., Xu, J., O'Malley, B., Seo, H., Murata, Y., and Weiss, R. E. (2003c). Thyroid hormone receptor-specific interactions with steroid receptor coactivator-1 in the pituitary. *Mol. Endocrinol.* **17**, 889–894.

Saltó, C., Kindblom, J. M., Johansson, C., Wang, Z., Gullberg, H., Norström, K., Mansén, A., Ohlsson, C., Thorén, P., Forrest, D., and Vennström, B. (2001). Ablation of TRα2 and a concomitant overexpression of α1 yields a mixed hypo- and hyperthyroid phenotype in mice. *Mol. Endocrinol.* **15**, 2115–2128.

Samuels, H. H., Stanley, F., and Shapiro, L. E. (1977). Modulation of thyroid hormone nuclear receptor levels by 3,5,3'-triiodo-L-thyronine in GH1 cells. Evidence for two functional components of nuclear-bound receptor and relationship to the induction of growth hormone synthesis. *J. Biol. Chem.* **252**, 6052–6060.

Sandhofer, C., Schwartz, H. L., Mariash, C. M., Forrest, D., and Oppenheimer, J. H. (1998). Beta receptor isoforms are not essential for thyroid hormone-dependent acceleration of PCP-2 and myelin basic protein gene expression in the developing brains of neonatal mice. *Mol. Cell. Endocrinol.* **137**, 109–115.

Sap, J., Muñoz, A., Damm, K., Goldberg, Y., Ghysdael, J., Lentz, A., Beng, H., and Vennström, B. (1986). The c-erbA protein is a high-affinity receptor for thyroid hormone. *Nature* **324**, 635–640.

Strait, K. A., Schwartz, H. L., Perez-Castillo, A., and Oppenheimer, J. H. (1990). Relationship of c-erbA mRNA content to tissue triiodothyronine nuclear binding capacity and function in developing and adult rats. *J. Biol. Chem.* **265**, 10514–10521.

Takeda, K., Sakurai, A., DeGroot, L. J., and Refetoff, S. (1992). Recessive inheritance of thyroid hormone resistance by complete deletion of the protein-coding region of the thyroid hormone receptor-β gene. *J. Clin. Endocrinol. Metab.* **74**, 49–55.

Takeshita, A., Cardona, G. R., Koibuchi, N., Suen, C. S., and Chin, W. W. (1997). TRAM-1, a novel 160-kDa thyroid hormone receptor activator molecule, exhibits distinct properties from steroid receptor coactivator-1. *J. Biol. Chem.* **272**, 27629–27634.

Takeuchi, Y., Murata, Y., Sadow, P. M., Hayashi, Y., Seo, H., Xu, J., O'Malley, B. W., Weiss, R. E., and Refetoff, S. (2002). Steroid receptor coactivator-1 deficiency causes variable alterations in the modulation of T_3-regulated transcription of genes *in vivo*. *Endocrinology* **143**, 1346–1352.

Tata, J. R. (2000). Autoinduction of nuclear hormone receptors during metamorphosis and its significance. *Insect Biochem. Molec. Biol.* **30**, 645–651.

Tone, Y., Collingwood, T. N., Adams, M., and Chatterjee, V. K. (1994). Functional analysis of a transactivation domain in the thyroid hormone β receptor. *J. Biol. Chem.* **269**, 31157–31161.

Torchia, J., Rose, D. W., Inostroza, J., Kamei, Y., Westin, S., Glass, C. K., and Rosenfeld, M. G. (1997). The transcriptional coactivator p/CIP binds CBP and mediates nuclear receptor function. *Nature* **387**, 677–684.

Treuter, E., Johansson, L., Thomsen, J. S., Warnmark, A., Leers, J., Pelto-Huikko, M., Sjoberg, M., Wright, A. P., Spyrou, G., and Gustafsson, J. A. (1999). Competition between thyroid hormone receptor-associated protein (TRAP) 220 and transcriptional intermediary factor (TIF) 2 for binding to nuclear receptors. Implications for the recruitment of TRAP and p160 coactivator complexes. *J. Biol. Chem.* **274**, 6667–6677.

Voegel, J. J., Heine, M. J. S., Zechel, C., Chambon, P., and Gronemeyer, H. (1996). TIF2, a 160-kD transcriptional mediator for the ligand-dependent activation function AF-2 nuclear receptor. *EMBO J.* **15**, 3667–3675.

Weinberger, C., Thompson, C. C., Ong, E. S., Lebo, R., Gruol, D. J., and Evans, R. M. (1986). The c-erb A gene encodes a thyroid hormone receptor. *Nature* **324**, 641–646.

Weiss, R. E., Gehin, M., Xu, J., Sadow, P. M., O'Malley, B. W., Chambon, P., and Refetoff, S. (2002). Thyroid function in mice with compound heterozygous and homozygous disruptions of SRC-1 and TIF2 coactivators: Evidence for haploinsufficiency. *Endocrinology* **143**, 1554–1557.

Weiss, R. E., Murata, Y., Cua, K., Hayashi, Y., Forrest, D., Seo, H., and Refetoff, S. (1998). Thyroid hormone action on liver, heart, and energy expenditure in thyroid hormone receptor β deficient mice. *Endocrinology* **139**, 4945–4952.

Weiss, R. E., Xu, J., Ning, G. J. P., O'Malley, B., and Refetoff, S. (1999). Mice deficient in the steroid receptor coactivator 1 (SRC-1) are resistant to thyroid hormone. *EMBO J.* **18**, 1900–1904.

Wikström, L., Johansson, C., Salto, C., Barlow, C., Campos Barros, A., Baas, F., Forrest, D., Thorén, P., and Vennström, B. (1998). Abnormal heart rate and body temperature in mice lacking thyroid hormone receptor $\alpha 1$. *EMBO J.* **17**, 455–461.

Williams, G. R. (2000). Cloning and characterization of two novel thyroid hormone receptor beta isoforms. *Mol. Cell. Biol.* **20**, 8329–8342.

Xu, J., and Li, Q. (2003). Review of the *in vivo* functions of the p160 steroid receptor coactivator family. *Mol. Endocrinol.* **17**, 1681–1692.

Xu, J., Liao, L., Ning, G., Yoshinda-Komiya, H., Deng, C., and O'Malley, B. W. (2000). The steroid receptor coactivator SRC-3 (p/CIP/RAC3/AiB1/ACTR/TRAM1) is required for normal growth, puberty, female reproductive function, and mammary gland development. *Proc. Natl. Acad. Sci. USA* **97**, 6379–6384.

Xu, J., Qui, Y., DeMayo, F. J., Tsai, S. Y., Tsai, M. J., and O'Malley, B. W. (1998). Partial hormone resistance in mice with disruption of the steroid receptor coactivator-1 (SRC-1) gene. *Science* **279**, 1922–1925.

Xu, L., Glass, C. K., and Rosenfeld, M. G. (1999). Coactivator and corepressor complexes in nuclear receptor function. *Curr. Opin. Genet. Dev.* **9**, 140–147.

Xu, W., Chen, H., Du, K., Asahara, H., Tini, M., Emerson, B. M., Montminy, M., and Evans, R. M. (2001). A transcription switch mediated b cofactor methylation. *Science* **294**, 2507–2511.

Yen, P. M. (2001). Physiological and molecular basis of thyroid hormone action. *Physiol. Rev.* **81**, 1097–1142.

Yen, P. M., Sunday, M. E., Darling, D. S., and Chin, W. W. (1992). Isoform-specific thyroid hormone receptor antibodies detect multiple thyroid hormone receptors in rat and human pituitaries. *Endocrinology* **130**, 1539–1546.

Zhang, X. K., Hoffmann, B., Tran, P. B. V., Graupner, G., and Pfahl, M. (1992). Retinoid X receptor is an auxiliary protein for thyroid hormone and retinoic acid receptors. *Nature* **355**, 441–446.

7

COREPRESSOR REQUIREMENT AND THYROID HORMONE RECEPTOR FUNCTION DURING XENOPUS DEVELOPMENT

LAURENT M. SACHS

*Département Régulations, Développement et Diversité Moléculaire
USM 501 Muséum National d'Histoire Naturelle
UMR-5166 CNRS, 75231 Paris cedex 05, France*

I. Introduction
II. Mechanism of TR Action
 A. TR Corepressor in Xenopus
 B. Mechanisms of Corepressor and Coactivator Action: Insight to Chromatin Modification
 C. Model of TR Action
III. Dual Role of TRs During Amphibian Development
 A. TRs and T_3 Response Genes: Expression Profiles During Amphibian Development
 B. Embryonic Development as a Model
 C. Hypothesis: The Dual Role of TRs During Xenopus Development
IV. Corepressor Function During Amphibian Development

    A. Corepressor Recruitment by TR: Repression of
        T_3-Response Genes During Larval Growth
    B. Corepressor Expression During Metamorphosis
    C. HDAC Involvement During Metamorphosis
    D. TR Related Function: The Feedback Hypothesis
V. Conclusion and Perspectives
    References

The biologic role of hormonal activation of nuclear receptors is well established. Only recently, however, has the biologic significance of repression begun to be appreciated. Amphibian metamorphosis is marked by dramatic thyroid hormone induced changes, including *de novo* morphogenesis, tissue remodeling, and organ resorption through programmed cell death. These changes involve cascades of gene regulation initiated by 3,5,3′-triiodothyronine (T_3). T_3 functions by regulating gene expression through thyroid hormone receptor (TR). TRs are DNA-binding transcription factors that belong to the steroid hormone receptor superfamily. In the absence of a ligand, TRs can repress gene expression by recruiting corepressor complexes, whereas liganded TRs recruit coactivator complexes for gene activation. Corepressor and coactivator complexes induce chromatin remodeling to mediate TR regulation of transcription. The mechanisms of TR action permit a dual function for TRs during development. In premetamorphic tadpoles, when TRs are expressed and T_3 levels are barely detectable, unliganded TRs repress transcription through corepressor recruitment. This TR-mediated repression of target genes is critical for proper larval development, allowing tadpole growth and acquisition of metamorphic competence. In contrast, during metamorphosis, endogenous T_3 causes TRs to activate gene expression, leading to tadpole transformation. Several results also support a role for corepressors during metamorphosis. Corepressor targeted functions, however, are still speculative but may again involve TRs. The requirement of active gene repression at different stages during amphibian development establishes an important biologic role for corepressors. © 2004 Elsevier Inc.

I. INTRODUCTION

Development and cell homeostasis are hormone-dependent processes. Amphibian development is indirect in that embryogenesis and adult growth are separated by a larval period that ends with metamorphosis. Although complex, thyroid hormones (TH) and, more precisely, the most active TH form termed 3,5,3′-triiodothyronine (T_3) initiate metamorphosis

(Shi, 1999). The biologic action of T_3 is mediated mainly through thyroid hormone receptors (TRs), which belong to the nuclear receptor (NR) superfamily (Tata, 2002). Thus, TRs are transcription factors that regulate gene expression by binding to DNA at specific sites referred to as T_3-response elements (T_3REs). As in mammals, there are two types of TRs: TRα and TRβ. In *Xenopus laevis*, due to pseudo-tetraploidy, there are two TRα (A and B) and two TRβ (A and B) (Yaoita *et al.*, 1990). The amino acid sequences of the *Xenopus* TRs are conserved in evolution as seen when compared with their counterparts described for mammals, fish, and chickens. The *Xenopus* TRs behave similarly to the mammalian and avian TRs in terms of their secondary structure organization, their ligand and DNA binding properties, and their requirement to heterodimerize with 9-cis retinoid X receptor (RXR) (Wong and Shi, 1995). As in mammals, three heterodimeric partners (RXRα, RXRβ, and RXRγ) have been cloned in *Xenopus* (Blumberg *et al.*, 1992; Marklew *et al.*, 1994).

The majority of the T_3-response genes are upregulated by liganded TRs; most studies so far on TR mechanisms of action have been carried out on this category of genes. Interestingly, it is well understood that for these liganded TR positively-induced genes, unliganded TRs repress basal transcription in the absence of T_3 and TRs activate gene transcription in the presence of T_3 (Damm *et al.*, 1989; Wong *et al.*, 1995). In mammals, a direct interaction of TRs with transcriptional machinery was first proposed to explain this basic switch from repression to activation (Baniahmad *et al.*, 1993; Fondell *et al.*, 1993). Use of the two-hybrid screening method, however, led to the discovery of intermediate proteins, now distinguished as corepressors and coactivators (Glass and Rosenfeld, 2000). Further work has led to the identification of three major categories of proteins involved in transcriptional regulation: transcription factors, multiprotein complexes, and chromatin remodeling and modification complexes. (Levine and Tjian, 2003). First, transcription factors, the sequence-specific DNA binding proteins, such as TRs, mediate gene-selective transcriptional repression and activation. Second, multiprotein complexes of transcriptional machinery with RNA polymerase are required for promoter recognition and RNA synthesis. Finally, chromatin remodeling and modification complexes contain coactivators and corepressors and assist the first categories to transduce the signal. In the context of TR action, unliganded TRs recruit corepressors to repress transcription, and liganded TRs recruit coactivators to activate transcription. This chapter's review concentrates on transcriptional repression and analyzes how the unliganded TRs work as active repressors through corepressors. These corepressors are involved in several processes during amphibian development, such as the formation of the head during embryogenesis (Koide *et al.*, 2001; Weston *et al.*, 2003) and the formation of embryonic musculature (Steinbac *et al.*,

2000). Given that the physiologic relevance of this phenomenon is largely unknown, this chapter's discussion presents the author's use of the TR-dependent amphibian development model to assess the role of unliganded TRs and corepressors.

II. MECHANISM OF TR ACTION

Several corepressors have been shown to interact with TRs in mammals (Jepsen and Rosenfeld, 2002). The nuclear receptor corepressor (NCoR) and the silencing mediator for retinoic acid and thyroid hormone receptors (SMRT) have emerged as important players for unliganded TR-mediated gene repression.

A. TR COREPRESSOR IN *XENOPUS*

The SMRT (Koide *et al.*, 2001) and the NCoR (Sachs *et al.*, 2002) complementary DNAs of *Xenopus laevis* have been isolated recently. *Xenopus* SMRT is only partially cloned, which does not allow domain analysis but has provided tools for developmental studies. Comparative analysis of full length *Xenopus* NCoR reveals strong identity (68%) with vertebrate NCoR (Hörlein *et al.*, 1995). This similarity suggests that *Xenopus* NCoR exerts similar functions and has a conserved domain organization. In the N-terminal, the presence of three independent repression domains (RD1, RD2, and RD3) is an indication that there are several mechanisms for repression through NCoR. Indeed, the different RDs have been shown to interact with different repressor proteins (Jepsen and Rosenfeld, 2002). In the C-terminal, two defined receptor interaction domains are the site of three conserved hydrophobic cores I/LXXI/VI referred to as the corepressor nuclear receptor recognition (CoRNR) boxes (Hu and Lazar, 1999; Webb *et al.*, 2000). CoRNR boxes are comparable with the LXXLL motif found in coactivators (Glass and Rosenfeld, 2000).

NCoR does not act alone and participates in multiprotein complexes recruited by TR in mammals and *Xenopus* (Jepsen and Rosenfeld, 2002). First, a high-speed fractionation of an extract using a conventional biochemical approach yielded three purified NCoR containing complexes (Jones *et al.*, 2001). Two of them carry histone deacetylase activity (HDAC). Complex NCoR-1 is the NCoR/HDAC1(Rpd3)/Sin3/Rabp48 complex predicted by the early *in vitro* studies (Alland *et al.*, 1997; Heinzel *et al.*, 1997). In addition to its four known components, five are unknown. Two of them are likely the *Xenopus* homologues to the mammalian Sin3-associated proteins SAP30 (Laherty *et al.*, 1998) and SAP18 (Zhang *et al.*, 1997). Complex NCoR-2 is composed of five unknown proteins in addition to NCoR. One of these carries robust but unidentified HDAC activity. Western blot analysis using antibodies known to be reactive against the *Xenopus* proteins has ruled out HDAC1, 2, or 3 presence in NCoR-2. Unfortunately, none of the other *Xenopus* HDACs are characterized. NCoR-3 lacks detectable HDAC activity, has been purified, and contains four polypeptides of which only NCoR has been identified. Biochemical purification of corepressor complexes from oocytes led to characterization of two complexes in which both NCoR and SMRT are associated with HDAC3 (Li *et al.*, 2000a; Urnov *et al.*, 2000). Such complexes have also been isolated in mammals (Jepsen and Rosenfeld, 2002).

B. MECHANISMS OF COREPRESSOR AND COACTIVATOR ACTION: INSIGHT TO CHROMATIN MODIFICATION

Transcriptional regulation has to overcome the chromatin barrier. Genomic DNA in eukaryotic cells is associated with histones and other nuclear proteins. Transcriptionally active DNA has a protein composition different from transcriptionally silent DNA. Thus, chromatin is a dynamic structure that is constantly reorganized by corepressors and coactivators. Two types of reorganization can be distinguished: chromatin modification and chromatin remodeling.

Chromatin modification corresponds largely to post-translational modification of histones (Strahl and Allis, 2000). Histones can be phosphorylated, methylated, or acetylated at their N-terminal tails. In *Xenopus*, histone acetylation, methylation, and phosphorylation have been shown to increase TR regulation of transcription (Li *et al.*, 2002a; Sachs and Shi, 2000; Wong *et al.*, 1998). Modifications allow neutralization of positive charges and reduce the affinity of histone N-terminal tails for DNA. Chromatin is then open and accessible for transcriptional machinery. Inversely favoring positively charged histone N-terminal tails increases their affinity for DNA, causing chromatin closure and rending it inaccessible to the transcriptional machinery.

This simple model of targeted histone modifications as a major source of transcriptional control cannot easily explain the rapid repression of transcription mediated by unliganded TRs. The chromatin remodeling also affects DNA accessibility by localized alteration of nucleosomic structure (nucleosome is the unit of chromatin: 180 base pairs of DNA associated with a histone octamer). The *Xenopus* TRs have been extensively studied to analyze TR effects on chromatin remodeling. TRs and RXRs bind DNA organized around histones (Wong *et al.*, 1995). TR makes use of the chromatin assembly process to silence transcription efficiently and directs the disruption of local chromatin structure in response to TH (Li *et al.*, 1999; Wong *et al.*, 1995, 1997a,b, 1998). As for chromatin modifications, chromatin remodeling is necessary but insufficient for transcriptional activation (Wong *et al.*, 1997a).

C. MODEL OF TR ACTION

Wolffe (1997) first proposed a model to explain the mechanism of repression and activation of transcription by TR (Fig. 1). In the absence of T_3, TRs and RXRs bind T_3REs and recruit an NCoR corepressor complex that contains at least one form of HDAC activity. An exclusive preference of TR for NCoR over SMRT has been observed in mammals (Ishizuka and Lazar, 2003; Jepsen *et al.*, 2000), but the preference does not seem to be relevant in all cell types or assay systems (Yoon *et al.*, 2003). In addition to histone deacetylation, the inaccessibility of chromatin to the transcriptional machinery could result from chromatin remodeling or direct protein interaction with transcriptional machinery. Indeed, in mammals, some components of chromatin remodeling complexes copurified with NCoR (Underhill *et al.*, 2000). TRs could also directly interact with transcriptional machinery to repress transcription (Baniahmad *et al.*, 1993; Fondell *et al.*, 1993). Furthermore, NCoR has also been shown to interact with and block transcriptional machinery independently of the presence of HDAC (Muscat *et al.*, 1998; Wong and Privalsky, 1998).

T_3 binding induces a conformational change in TR that relieves its inhibitory effect through the release of the corepressor complex and the recruitment of a coactivator complex. This coactivator complex contains histone acetyltransferase (HAT) activity carried by the steroid receptor coactivator (SRC) (Li *et al.*, 2000b) and p300/CBP (Li *et al.*, 1999, 2000b). In mammals, methyltransferases, such as coactivator-associated arginine methyltransferase 1 (CARM1), are also part of the corepressor complex (Chen *et al.*, 1999). Histone modifications lead to chromatin opening that will promote transcription activation. Upon T_3 treatment, however, TRs can also recruit coactivator complexes without HAT activity, like the SWI/SNF complex involved in chromatin remodeling and the mediator complex directly involved in transcriptional activation. The diversity in coactivator complexes may reflect the diversity of TH and TR effects. Different complexes might be required for regulating distinct cis-DNA elements in a temporal and tissue-specific manner. Because these complexes have different functions, however, TR may require all of them. Recently, analysis of transcriptional activation by TR in *Xenopus* has revealed that histone acetylation by the SRC/p300 complex may be a prerequisite for recruitment and function of SWI/SNF and mediator complexes at specific T_3-response promoters (Huang *et al.*, 2003).

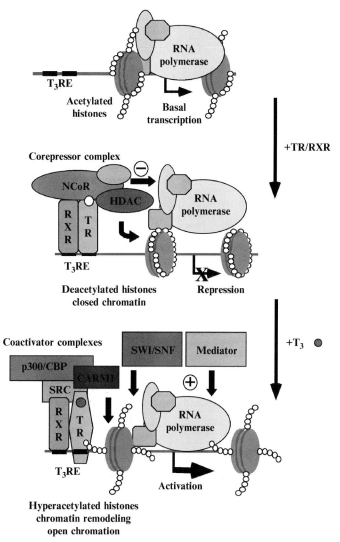

FIGURE 1. A proposed mechanism for transcriptional regulation by TRs. TRs are transcription factors that bind as heterodimers with RXRs to T_3REs in target genes. In the absence of T_3, unliganded TRs probably repress gene transcription through the recruitment of corepressor complexes containing NCoR and HDAC. Upon T_3 binding, a conformational change takes place in the heterodimer. Liganded TRs release the corepressor complex and recruit several coactivator complexes. One of them will contain coactivators such as SRC or p300/CBP, which function in part through chromatin modification, as they possess HAT activity. However, chromatin remodeling is also required because of the recruitment of the SWI/SNF complex. Finally, the mediator complex will be recruited for full activation of the target gene. In addition, RNA polymerase and some other basal transcription machinery factors are depicted in the figure.

III. DUAL ROLE OF TRs DURING AMPHIBIAN DEVELOPMENT

The physiologic effects of liganded TR gene activation have been extensively studied (Shi, 1999). However, unliganded TR active gene repression could also be implicated in specific developmental processes.

A. TRs AND T_3 RESPONSE GENES: EXPRESSION PROFILES DURING AMPHIBIAN DEVELOPMENT

In *Xenopus*, TRα and TRβ are differentially regulated during development (Fig. 2). Their mRNAs are present at very-low levels in oocytes and during embryogenesis, though their functions are unknown (Banker *et al.*, 1991; Oofusa *et al.*, 2001). High levels of TRα mRNA are present after larval hatches (stage NF35/38) well before the appearance of a functional

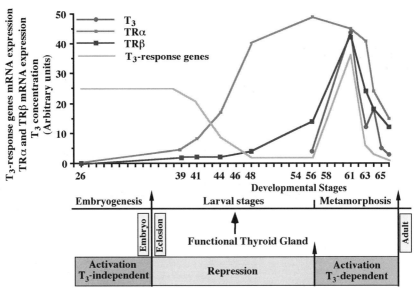

FIGURE 2. Developmental levels of TH and expression of TRα, TRβ, and T_3-response genes throughout embryogenesis, tadpole growth, and metamorphosis. Development of *Xenopus laevis* is indicated according to the Nieuwkoop and Faber (1967) developmental stages. During developmental progression, the stage of the appearance of a functional thyroid gland is shown. The plasma concentrations of T_3 are from Leloup and Buscaglia (1977). The mRNA levels for TRα and TRβ are based on data from Yaoita and Brown (1990). The mRNA levels for T_3-response genes, such as ST3 or hh, are based on published data (Patterton *et al.*, 1995; Stolow and Shi, 1995). For clarity, T_3 and mRNA levels are plotted as arbitrary units on different scales. For each phase of development, the state of T_3-response gene transcription is indicated in correlation with current knowledge on TR mechanism of action (repression versus activation).

thyroid gland at stages NF45. TRα mRNA continues to be highly expressed until the end of metamorphosis when levels decrease, remaining low in juveniles and adults (Yaoita and Brown, 1990). TRβ mRNA levels are barely detectable before stage NF52 (Yaoita and Brown, 1990). During metamorphosis, however, TRβ mRNA levels increase in parallel with endogenous TH levels (Yaoita and Brown, 1990). The *Xenopus* TRβ gene was found to be a T_3 direct-response gene (Ranjan *et al.*, 1994). Just as for TRα, TRβ mRNA levels decrease in juveniles and adults (Yaoita and Brown, 1990). Furthermore, TR and RXR genes are coordinately regulated during amphibian development. (Wong and Shi, 1995).

Analysis of TRs and RXRs binding on T_3-response gene promoters was carried out by chromatin immunoprecipitation (ChIP) assays using antibodies against TRs or RXRs on embryonic or tadpole chromatin fragments (Sachs and Shi, 2000). Immunoprecipitated DNA fragments were analyzed by PCR for the presence of T_3RE-containing regions of two known T_3-response genes, the TRβ and TH/bZip genes, respectively (Furlow and Brown, 1999; Ranjan *et al.*, 1994). The results revealed that few TRs or RXRs are present at the T_3REs in embryos with little TRs expression, but both T_3REs are bound by TR/RXR in tadpoles with TR (at least TRα) expression (Sachs and Shi, 2000) (Fig. 2). This result agrees well with the ability of TH to induce the expression of T_3-response genes in tadpoles but not embryos (Sachs and Shi, 2000; Yaoita and Brown, 1990). Moreover, during the entire larval development, TRs and RXRs bind T_3-response gene promoters regardless of the presence of T_3 (Havis *et al.*, 2003; Sachs and Shi, 2000).

Several genes have been identified as T_3-response genes during amphibian metamorphosis (Shi, 1996). Some of them also appear to participate in embryogenesis of *Xenopus* (Fig. 2). Examples include the putative morphogene Sonic hedgehog (hh) (Stolow and Shi, 1995) and the matrix metalloproteinase stromelysin 3 (ST3) (Patterton *et al.*, 1995). Both genes are first activated around stages NF16 to 17 as revealed by Northern blot analysis and reach their peak levels of expression by stages NF33 to 34 prior to tadpole hatching (Patterton *et al.*, 1995; Stolow and Shi, 1995). Their mRNA levels are subsequently down-regulated by feeding stage NF45 and up-regulated again during metamorphosis when T_3 is present (Patterton *et al.*, 1995; Stolow and Shi, 1995). The down-regulation of ST3 and hh at the beginning of larval stages coincides with the increase of TRα mRNA levels (Fig. 2). This repression was hypothesized to involve unliganded TR action.

B. EMBRYONIC DEVELOPMENT AS A MODEL

Early stages of development provide a good model to study function of TRs and RXRs because few endogenous TRs and RXRs are present at these stages. The overexpression of mRNA encoding TRα and RXRα

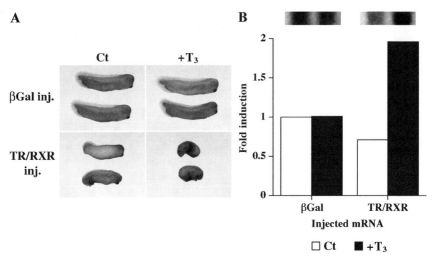

FIGURE 3. Effects of combined overexpression of TRs and RXRs on embryonic development. (A) Overexpression of TRα and RXRα results in embryonic abnormalities. Embryos were injected with 500 pg of TRα and RXRα mRNA (TR/RXR inj) and cultured for 3 days in the absence (Ct) or presence of 100-nM triiodothyronine (+T_3). Injection of 1 ng of mRNA coding for β-galactosidase (βGal inj.) was used as a negative control. Note first that βGal-injected embryos were normal regardless of T_3 presence, and note second that TRα and RXRα expression led to abnormal development, with differences in observed phenotypes between the T_3-treated and T_3-untreated groups. (B) Overexpression of TRα and RXRα results in specific regulation of T_3-response genes. The embryos shown in 3A were subject to Northern blot analysis for expression of ST3. Quantification of the hybridization signals shows that ST3 was slightly repressed by unliganded TR/RXR and that the addition of T_3 led to activation of ST3.

microinjected into fertilized eggs causes developmental abnormalities in embryos with distinct phenotypes depending on the presence of T_3 (Puzianowska-Kuznicka et al., 1997) (Fig. 3A). More importantly, overexpression of TRs and RXRs during embryogenesis efficiently and specifically regulates known T_3-response genes, such as ST3 or hh, repressing them in the absence of T_3 but activating them when T_3 is present (Puzianowska-Kuznicka et al., 1997) (Fig. 3B). These results suggest that unliganded TRs are present in T_3-response gene repression at the end of embryogenesis. Finally, use of this embryo model points out the critical corequirement of RXR for the developmental functions of TR. Without RXRs, TRs alone cannot induce teratogenic effects or affect T_3-response gene transcription (Puzianowska-Kuznicka et al., 1997).

C. HYPOTHESIS: THE DUAL ROLE OF TRs DURING *XENOPUS* DEVELOPMENT

The expression profiles together with the ability of TRs to both repress and activate T_3-response genes in the absence and presence of TH, respectively, suggest dual functions for TRs during development (Fig. 2). In premetamorphic tadpoles, TRs (primarily TRα) act to repress T_3-response genes. As many of these genes are likely to participate in metamorphosis (Shi, 1996), their repression by unliganded TRs will ensure tadpole growth. As the thyroid gland matures, T_3 is synthesized and secreted into the plasma to transform TRs from repressors to activators. This process induces the expression of T_3-response genes, including the TRβ genes, and leads to metamorphosis.

IV. COREPRESSOR FUNCTION DURING AMPHIBIAN DEVELOPMENT

Several results indicate that TRs recruit corepressors to mediate gene repression. However, the physiologic requirement of corepressors during development requires investigation as discussed in this section.

A. COREPRESSOR RECRUITMENT BY TR: REPRESSION OF T_3-RESPONSE GENES DURING LARVAL GROWTH

In *Xenopus*, several lines of evidence suggest that corepressors are recruited *in vivo* at the promoters of T_3-response genes in absence of TH. First, as Figure 4A illustrates, overexpression of a dominant negative NCoR led to increased transcription from a T_3-dependent promoter (Sachs *et al.*, 2002). The dominant negative NCoR corresponded to a CoRNR box of the NCoR TR binding domain. The transgene coding for the CoRNR peptide was introduced in premetamorphic tadpole tail muscle by *in vivo* gene transfer (de Luze *et al.*, 1993). Binding of the peptide to TRs blocked wild-type NCoR interaction and induced the loss of repression of the T_3-dependent promoter, but did not have an effect on the T_3 nonresponse gene. To further test that unliganded TRs recruit NCoR to mediate gene repression *in vivo*, association of NCoR with the promoter of T_3-response genes was analyzed by ChIP assay in premetamorphic tadpole tail and intestine, with an antibody against NCoR (Sachs *et al.*, 2002). NCoR was strongly recruited on T_3 RE-containing regions of T_3-response genes, such as TRβ and TH/bZip genes, only in the absence of T_3 (Fig. 4B). These two results indicate that unliganded TRs make use of NCoR to repress target gene transcription.

Second, using the ChIP assay with an acetylated histone H4 antibody, H4 acetylation at T_3-response gene promoters was increased by T_3

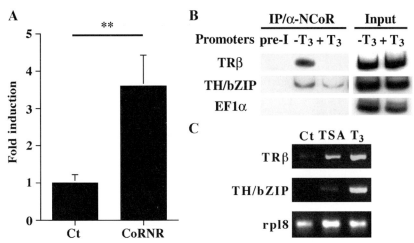

FIGURE 4. Corepressor requirements for TR-mediated gene repression. (A) Dominant negative NCoR (CoRNR) abolishes the repression of T_3RE-containing promoter by unliganded TRs. Overexpression of CoRNR, but not the control peptide (Ct), increased the expression level of firefly luciferase expression vector driven by a T_3RE. Overexpression was realized by *in vivo* gene transfer (de Luze *et al.*, 1993). *Xenopus laevis* dorsal muscles were coinjected with 0.5 μg of vector expressing CoRNR or control vector, 0.5 μg of T_3RE–tk–firefly luciferase vector, and 0.1 μg of SV40–*Renilla* luciferase vector as an internal control for transfection efficiency. After 2 days, the luciferase activities were assayed in muscle homogenates. The ratio of the firefly luciferase over *Renilla* luciferase was plotted as fold induction with standard errors. Each point represents the mean from at least seven tadpoles. Statistical signification was analyzed using Student's t-test. The symbol ** refers to $p < 0.001$. (B) Chromatin immunoprecipitation (ChIP) assays show that NCoR is released from the T_3-response gene promoter following T_3 treatment. Premetamorphic tadpole tail nuclei were isolated after a 10-nM T_3 treatment for 48 hr. Chromatin was formaldehyde cross-linked, fragmented by sonication, and immunoprecipitated with anti-NCoR antibody. Finally, the precipitated DNA was analyzed by PCR as described for the presence of T_3-response gene promoters, such as TRβ and TH/bZip promoters (Sachs and Shi, 2000). EF1α, which is not regulated by T_3, served as a negative control. (C) T_3 and the histone deacetylase inhibitor (trichostatin A, TSA) increase mRNA levels of T_3-response genes. Premetamorphic tadpoles were treated for 2 days with T_3 (10 nM) or TSA (100 nM). RNA was extracted from the tail and used for PCR after reverse transcription analysis of TRβ and TH/bZip expression. The expression of the ribosomal protein L8 (rpl8) was used as an internal control because its mRNA levels are not affected by T_3 (Shi and Liang, 1994).

treatment of premetamorphic tadpoles (Havis *et al.*, 2003; Sachs and Shi, 2000; Sachs *et al.*, 2002). Thus, in premetamorphic tadpoles, histones H4 are deacetylated, as one would predict from the unliganded TR mode of action (Fig. 1). To test the role of HDAC in the level of H4 acetylation in premetamorphic tadpoles, association of HDAC with the promoter of T_3-response genes was analyzed by ChIP assay using an antibody against HDAC1 (Havis *et al.*, 2003; Sachs and Shi, 2000). In whole tadpoles treated with high levels of T_3 (100 nM), HDAC1 was partially released from the TRβ promoter but significantly from the TH/bZip promoter (Sachs and

Shi, 2000). Similarly, following tadpole tail treatment with physiologic T_3 concentrations (10 nM), HDAC1 was not released from the TRβ promoter but clearly released from the TH/bZip promoter (Havis et al., 2003). Using a reconstituted system to analyze TR mechanism of action in oocytes, HDAC1 was also found to be constitutively present on the TRβ promoter (Li et al., 2002b). These results suggest that recruitment of the NCoR/HDAC1 complex by TRs could be promoter specific. In the case of the TRβ promoter, other HDACs contribute to TR repression, such as HDAC3, which has been shown to be recruited in the oocyte system (Li et al., 2002b). However, HDAC3 recruitment has not yet been analyzed during larval development.

To further analyze the role of HDAC in transcriptional repression of T_3-response genes, premetamorphic tadpoles were treated with trichostatin A (TSA), an HDAC inhibitor (Yoshida et al., 1995). TSA treatment of tadpoles increased H4 acetylation at T_3-response gene promoters and increased mRNA levels of T_3-response genes (Fig. 4C) (Sachs and Shi, 2000; Sachs et al., 2001a). TSA treatment was as effective as T_3 treatment on TRβ gene expression but not as effective on TH/bZip gene expression. Thus, for TRβ, TSA does not only relieve repression but also activates transcription. For TH/bZip, TSA seems to only release repression and does not lead to gene activation, as TSA does for T_3.

B. COREPRESSOR EXPRESSION DURING METAMORPHOSIS

The few corepressors studied so far show specific expression patterns from embryonic development to juvenile frogs. In whole embryos (Sachs unpublished data), NCoR mRNAs of maternal origin are present before the mid-blastula transition (MBT). After MBT, NCoR mRNA levels increase until stage NF40 and then decrease to lower levels at stage NF44. These lower levels are maintained until metamorphosis when they increase again (Sachs et al., 2002). More precisely, in the intestine, NCoR mRNA levels increase to reach maximum levels at stage NF62 when organ transformation takes place. In the tail, NCoR mRNA levels also increase until stage NF64 at tail regression. In the hind limb, levels are high just before climax when organogenesis takes place, and levels decrease at the beginning of metamorphosis. In the intestine and tail, the variations of NCoR mRNA levels were reproduced for this particular study by T_3 treatment of premetamorphic tadpoles. Results show that NCoR mRNA levels increase after 3 days of T_3 treatment, displaying a profile similar to the one observed during natural metamorphosis. Analysis of SMRT expression profiles shows similar results to those of NCoR (Sachs et al., 2002).

Further, Sin3 and HDAC1 protein levels were analyzed by Western blot analysis. The levels were coordinately regulated, which is consistent with their presence in the same corepressor complex (see section IIA). The results are as follows. In whole embryos, Sin3 and HDAC1 are present at high levels until stage NF40. Similar to NCoR mRNA profiles, Sin3 and HDAC1 protein levels decrease at stage NF44, to increase again at stage NF52, and then are maintained until metamorphosis (Sachs et al., 2001b). Two types of profiles are observed during metamorphosis (Sachs et al., 2001a,b). First, just as in the intestine, Sin3 and HDAC1 levels dramatically increase during metamorphosis to reach maximum levels at stage NF62. Second, in the tail and hind limb, the expression levels decrease as metamorphosis progresses. The increase in the intestine could also be reproduced by T_3 treatment of premetamorphic tadpoles after 3 days of treatment. In the tail and hind limb, the levels do not change during 5 days of treatment.

These few expression profiles need to be completed by spatiotemporal studies to be related with organ transformation and TR function. The present data indicate, however, that corepressor expression is high during embryogenesis and metamorphosis, which is when organ transformation occurs. These correlations suggest that corepressors are not only involved at larval stages for TR-mediated repression but are also involved in metamorphic processes.

C. HDAC INVOLVEMENT DURING METAMORPHOSIS

The evidence that corepressors are essential to TR action during larval growth and metamorphosis suggests that the alterations in histone acetylation levels may influence metamorphosis. In particular, blocking HDAC activity might be expected to precociously activate T_3-inducible genes and possibly activate metamorphosis as well. This chapter's study used TSA to block the function of HDAC. Premetamorphic tadpoles treated with 100-nM TSA were blocked at their starting developmental stage, whereas untreated tadpoles completed metamorphosis as shown in Figure 5A (Sachs et al., 2001a, b). TSA's inhibitory effect was confirmed on T_3-induced metamorphosis (Sachs et al., 2001a,b).

In mammal models, TSA affects only a restricted set of genes onto which HDACs are recruited (Van Lint et al., 1996). In tadpole intestine (Fig. 5B), TSA can induce T_3-response genes, such as TRβ, as well as the gene coding for the sodium-phosphate cotransporter (NaP, only expressed in intestinal epithelial cells). TSA has no effect on other control genes, such as TRα and ribosomal protein L8 (rpl8), (Shi and Liang, 1994). T_3 and TSA together also induce large increases in T_3-response gene expression (Fig. 5B). Thus, TSA can block metamorphosis in animals that express T_3-response genes. These observations could seem contradictory at first sight; however, HDACs could be recruited by other transcription factors involved in the cascade gene regulation induced during metamorphosis. TSA probably affects a step downstream of T_3-response induction as revealed by analysis

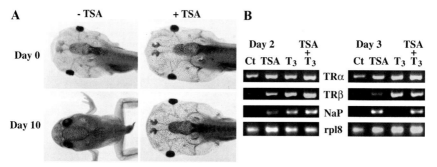

FIGURE 5. The histone deacetylase inhibitor (trichostatin A, TSA) blocks metamorphosis by affecting late gene regulation. (A) TSA inhibits natural metamorphosis. Premetamorphic tadpoles (day 0) were treated with 100 nM TSA (+TSA). After 10 days, all the untreated animals (−TSA) were juvenile frogs, but 90% of the TSA treated tadpoles showed no transformations and were morphologically identical to their day 0 state. The other 10% displayed strong delays in metamorphic progression. (B) T_3 and TSA treatments affect metamorphosis-regulated genes in premetamorphic tadpole intestines. Tadpoles were treated for 2 or 3 days with T_3 (10 nM) and/or TSA (100 nM). RNA was extracted from the intestine and used for PCR after reverse transcription analysis of TRα, TRβ, and sodium phosphate cotransporter (NaP) expression. The expression of the ribosomal protein L8 (rpl8) was used as an internal control. Note first that TRα and Rpl8 expression was not affected by T_3 and/or TSA treatment. Second, note that 2 days of treatment with TSA and/or T_3 increased TRβ and NaP mRNA levels. Finally, after 3 days of treatment, loss of NaP expression due to T_3 was not reproduced in the presence of TSA.

of late gene regulation in metamorphic responses (Sachs et al., 2001a). For example, TSA treatment abrogates the down-regulation of NaP that normally takes place after 3 days of T_3 treatment (Fig. 5B). The future challenge is to determine how early T_3-response genes regulate the TSA-sensitive step and which genes are the downstream targets of HDAC.

D. TR RELATED FUNCTION: THE FEEDBACK HYPOTHESIS

Because liganded TRs do not recruit corepressors, the proposed role for corepressors during metamorphosis is to mediate the function of transcription factors other than TRs. Several observations suggest that corepressors might also be involved in TR function during metamorphosis. The first observation came from analysis of insect metamorphosis and shows the many similarities between insect and amphibian metamorphosis. Almost 30 years ago, before the discovery of NRs, Ashburner was studying chromosomal puffing in *Drosophila* and concluded that an action of ecdysone is to up-regulate the expression of early genes. The products of early genes would in turn induce the late genes but inhibit the early genes themselves. (Ashburner et al., 1974). The validity of this model (Fig. 6A) has

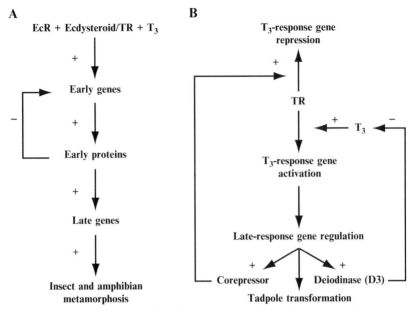

FIGURE 6. Models for control of TH and TR function during metamorphosis. (A) The Ashburner model of ecdysone shows the effects during insect metamorphosis transposed to amphibian metamorphosis. The liganded ecdysteroid receptor or TR activates early genes. The production of early genes induces the late genes but inhibits the early genes themselves. (B) This is a proposed negative feedback loop at the cellular level for stopping activation of T_3-response gene when the metamorphic program is underway. Liganded TRs activate T_3-response genes that will increase synthesis of deiodinase D3 and corepressors. Inactivation of TH by deiodinase increases unliganded TR levels, which will facilitate the recruitment of corepressors and repression of T_3-response genes. The requirement of corepressors and deiodinase establishes an important novel biologic function for active repression of TR target genes when late gene regulation is induced and tadpole transformation has started.

been confirmed at the molecular level (Thummel, 1993). This model can be transposed to the amphibian metamorphosis (Fig. 6A). Because TRβ is an early response gene, it may induce late genes and may inhibit itself and other early genes. To efficiently repress transcription, TRs will need to recruit corepressors.

The Ashburner model could be compared at the cellular level to a negative feedback loop, which will stop the first step of the gene regulation cascade when it has been induced. The action of TH is regulated at multiple levels. First, secretion of TH could be regulated by complex neuroendocrine negative feedback systems in the hypothalamus–pituitary–thyroid axis that have not been well characterized during *Xenopus* development (Shi, 1999). Second, at the cellular level, different types of deiodinase convert 3,5,3′,5′-tetraiodothyronine (T_4) to T_3 and/or inactivate both hormones. Interestingly, 5-deiodinase (D3) that inactivates TH is upregulated by T_3

treatment in premetamorphic tadpoles (St. Germain *et al.*, 1994). Thus, during metamorphosis, with the increase of D3 expression leading to the inactivation of TH and the increase of corepressor expression, all the conditions are brought together to favor T_3-response gene repression (Fig. 6B) after metamorphosis begins and the transformations have started.

V. CONCLUSION AND PERSPECTIVES

This chapter has focused mainly on the roles of NCoR and HDAC corepressor families in TR-mediated gene repression and in larval stages, including metamorphosis. The author's studies on TR function during *Xenopus* development have provided some *in vivo* evidence that supports a dual role of TRs (Fig. 7). Unliganded TRs actively repress T_3-response genes during the larval period. Then, liganded TRs activate T_3-response genes during metamorphosis. This dual role has also been demonstrated in mice. Comparing mice devoid of all TRs and mice devoid of TH (Knockout mice), the *in vivo* importance of the ligand-independent function of the TRs was highlighted (Flamant *et al.*, 2002). Moreover, mouse mutant strains reveal that during postnatal development, a period characterized by a peak in TH production resembles amphibian metamorphosis (Flamant and Samarut, 2003). In amphibians, however, the physiologic role of the repression has not been established and needs to be further analyzed. For *Xenopus laevis*, it can be proposed that the repression will prevent precocious induction of metamorphosis by inhibiting metamorphic T_3-response genes

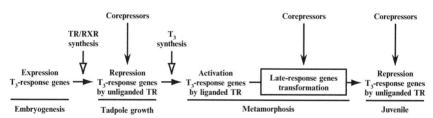

FIGURE 7. A model for dual function of TRs and for corepressor functions during amphibian development. During embryogenesis, T_3-response genes are required for organogenesis. To repress these T_3-response genes when organogenesis is completed, TRα and RXRα are expressed at the beginning of larval development (tadpole growth). In the absence of TH, TRs and RXRs recruit corepressors to repress transcription. At metamorphosis, TH is secreted and then liganded TRs release corepressors and recruit coactivators, thus activating T_3-response genes. The up-regulation of T_3-response genes corresponds to the first step in the cascade of gene regulation that will allow metamorphosis. The genetic program required for tadpole transformation involves corepressors. Finally, when this program of transformation has begun, T_3-response gene activation must be down-regulated. The decrease of TH level (local inactivation of TH by deiodinase) will lead to T_3-response gene repression (in the juvenile stage) by unliganded TRs and again their corepressors.

after embryogenesis. Hence, repression favors larval growth and the acquisition of full competence for full or correct metamorphosis. The author's studies on NCoR and HDACs have suggested that these corepressors are implicated in unliganded TR repression of T_3-response genes during the larval period and in the genetic program, causing tadpole transformation during metamorphosis (Fig. 7). Functional analyses, however, are necessary to definitively demonstrate their physiologic involvement in the larval and metamorphic periods. Finally, the demonstrated requirement of corepressors in normal development highlights how active gene repression occupies a central position that is as important as gene activation in current concepts of signaling mechanisms through transcription factors. Studies using the transgenic technology in *Xenopus* (Kroll and Amaya, 1996) will probably provide direct evidence for unliganded TR and corepressor functions during amphibian development in the same manner that this technology has shown roles for liganded TRs during metamorphosis (Das *et al.*, 2002; Schreiber *et al.*, 2001).

ACKNOWLEDGMENTS

I thank N. Becker, T. Collingwood, B. Demeneix, E. Havis, P. Jones, V. Laudet, Y. B. Shi, J. R. Tata, F. Urnov, P. Wade, and A. Wolffe for their helpful discussions on TR and corepressor function during these last 7 years. Professor B. Demeneix is also thanked for the constructive criticism and English revision of this manuscript. I also appreciate the support of the Association pour la Recherche contre le Cancer, the CNRS, and the MNHN.

REFERENCES

Alland, L., Muhle, R., Hou, Jr., H., Potes, J., Chin, L., Schreiber-Agus, N., and DePinho, R. A. (1997). Role for NCoR and histone deacetylase in Sin3-mediated transcriptional repression. *Nature* **387,** 49–55.

Ashburner, M., Chihara, C., Meltzer, P., and Richards, G. (1974). Temporal control of puffing activity in polytene chromosomes. *Cold Spring Harb. Quant. Biol.* **38,** 655–662.

Baniahmad, A., Ha, I., Reinberg, D., Tsai, S., Tsai, M. J., and O'Malley, B. W. (1993). Interaction of human thyroid hormone receptor β with transcription factor TFIIB may mediate target gene derepression and activation by thyroid hormone. *Proc. Natl. Acad. Sci. USA* **90,** 8832–8836.

Banker, D. E., Bigler, J., and Eisenman, R. N. (1991). The thyroid hormone receptor gene (c-erbAα) is expressed in advance of thyroid gland maturation during the early embryonic development of *Xenopus laevis*. *Mol. Cell. Biol.* **11,** 5079–5089.

Blumberg, B., Mangelsdorf, D. J., Dyck, J. A., Bittner, D. A., Evans, R. M., and de Robertis, E. M. (1992). Multiple retinoid-responsive receptors in a single cell: Families of retinoid X receptors and retinoic acid receptors in the *Xenopus* egg. *Proc. Natl. Acad. Sci. USA* **89,** 2321–2325.

Chen, D., Ma, H., Hong, H., Koh, S. S., Huang, S. M., Schurter, B. T., Aswad, D. W., and Stallcup, M. R. (1999). Regulation of transcription by a protein methyltransferase. *Science* **284,** 2174–2177.

Damm, K., Thompson, C. C., and Evans, R. M. (1989). Protein encoded by v-erbA functions as a thyroid hormone receptor antogonist. *Nature* **339**, 593–597.

Das, B., Schreiber, A. M., Huang, H., and Brown, D. D. (2002). Multiple thyroid hormone-induced muscle growth and death programs during metamorphosis in *Xenopus laevis*. *Proc. Natl. Acad. Sci. USA* **99**, 12230–12235.

de Luze, A., Sachs, L., and Demeneix, B. (1993). Thyroid hormone-dependent transcriptional regulation of exogenous genes transferred into *Xenopus* tadpole muscle in vivo. *Proc. Natl. Acad. Sci. USA* **90**, 7322–7326.

Flamant, F., Poguet, A. L., Plateroti, M., Chassande, O., Gauthier, K., Streichnberger, N., Mansouri, A., and Samarut, J. (2002). Congenital hypothyroid Pas8$^{-/-}$ mutant mice can be rescued by inactivating the TRα gene. *Mol. Endocrinol.* **16**, 24–32.

Flamant, F., and Samarut, J. (2003). Thyroid hormone receptors: Lessons from knockout and knock-in mutant mice. *Trends Endocrinol. Metabol.* **14**, 85–90.

Fondell, J. D., Roy, A. L., and Roeder, R. G. (1993). Unliganded thyroid hormone receptor inhibits formation of a functional preinitiation complex: Implications for active repression. *Genes Dev.* **7**, 1400–1410.

Furlow, J. D., and Brown, D. D. (1999). In vitro and in vivo analysis of the regulation of a transcription factor gene by thyroid hormone during *Xenopus laevis* metamorphosis. *Mol. Endocrinol.* **13**, 2076–2089.

Glass, C. K., and Rosenfeld, M. G. (2000). The coregulator exchange in transcriptional functions of nuclear receptors. *Genes Dev.* **14**, 121–141.

Havis, E., Sachs, L. M., and Demeneix, B. A. (2003). Metamorphic T$_3$-response genes have specific coregulator requirements. *EMBO Rep.* **4**, 883–888.

Heinzel, T., Lavinsky, R. M., Mullen, T. M., Söderström, M., Laherty, C. D., Torchia, J., Yang, W. M., Brard, G., Ngo, S. D., Ravie, J. R., Seto, E., Eisenman, R. N., Rose, D. W., Glass, C. K., and Rosenfeld, M. G. (1997). A complex containing NCoR, mSin3, and histone deacetylase mediates transcriptional repression. *Nature* **387**, 43–48.

Hörlein, A. J., Näär, A. M., Heinzel, T., Torchia, J., Gloss, B., Kurokawa, R., Ryan, A., Kamei, Y., Söderström, M., Glass, C. K., and Rosenfeld, M. G. (1995). Ligand-independent repression by the thyroid hormone receptor mediated by a nuclear receptor corepressor. *Nature* **377**, 397–404.

Hu, I., and Lazar, M. A. (1999). The CoRNR motif controls the recruitment of corepressors by nuclear hormone receptors. *Nature* **402**, 93–96.

Huang, Z. Q., Li, J., Sachs, L. M., Cole, P. A., and Wong, J. (2003). A role for cofactor–cofactor and –histone interactions in targeting p300, SWI/SNF, and mediator for transcription. *EMBO J.* **22**, 2146–2155.

Ishizuka, T., and Lazar, M. A. (2003). The NCoR/histone deacetylase 3 complex is required for repression by thyroid hormone receptor. *Mol. Cell. Biol.* **23**, 5122–5131.

Jepsen, K., Hermanson, O., Onami, T. M., Gleiberman, A. S., Lunyak, V., McEvilly, R. J., Kurokawa, R., Kumar, V., Liu, F., Seto, E., Hedrick, S. M., Mandel, G., Glass, C. K., Rose, D. W., and Rosenfeld, M. G. (2000). Combinatorial roles of the nuclear receptor corepressor in transcription and development. *Cell* **102**, 753–763.

Jepsen, K., and Rosenfeld, M. G. (2002). Biological roles and mechanistic actions of corepressor complexes. *J. Cell. Sci.* **115**, 689–698.

Jones, P. L., Sachs, L. M., Rouse, N., Wade, P. A., and Shi, Y. B. (2001). Multiple NCoR complexes contain distinct histone deacetylases. *J. Biol. Chem.* **276**, 8807–8811.

Koide, T., Downes, M., Chandraratna, R. A. S., Blumberg, B., and Umesono, K. (2001). Active repression of RAR signaling is required for head formation. *Genes Dev.* **15**, 2111–2121.

Kroll, K. L., and Amaya, E. (1996). Transgenic *Xenopus* embryos from sperm nuclear transplantations reveal FGF signaling requirements during gastrulation. *Development* **122**, 3173–3183.

Laherty, C. D., Billin, A. N., Lavinsky, R. M., Yochum, G. S., Bush, A. C., Sun, J. M., Mullen, T. M., Davie, J. R., Rose, D. W., Glass, C. K., Rosenfeld, M. G., Ayer, D. E., and Eisenman, R. N. (1998). SAP30, a component of the mSin3 corepressor complex involved in NCoR-mediated repression by specific transcription factors. *Mol. Cell* **2**, 33–42.

Leloup, J., and Buscaglia, M. (1977). La triiodothyronine: Hormone de la metamorphose des amphibiens. *C. R. Acad. Sci. Paris* **284**, 2261–2263.

Levine, M., and Tjian, R. (2003). Transcription regulation and animal diversity. *Nature* **424**, 147–151.

Li, Q., Imhof, A., Collingwood, T. N., Urnov, F. D., and Wolffe, P. A. (1999). p300 stimulates transcription instigated by ligand-bound thyroid hormone receptor at a step subsequent to chromatin disruption. *EMBO J.* **18**, 5634–5652.

Li, J., Wang, J., Wang, J., Nawaz, Z., Liu, J. M., Qin, J., and Wong, J. (2000a). Both corepressor proteins SMRT and NCoR exist in large protein complexes containing HDAC3. *EMBO J.* **19**, 4342–4350.

Li, J., O'Malley, B. W., and Wong, J. (2000b). p300 requires its histone acetyltransferase activity and SRC-1 interaction domain to facilitate thyroid hormone receptor activation in chromatin. *Mol. Cell. Biol.* **20**, 2031–2042.

Li, J., Lin, Q., Yoon, H. G., Huang, Z. Q., Strahl, B. D., Allis, C. D., and Wong, J. (2002a). Involvement of histone methylation and phosphorylation in regulation of transcription by thyroid hormone receptor. *Mol. Cell. Biol.* **22**, 5688–5697.

Li, J., Lin, Q., Wang, W., Wade, P., and Wong, J. (2002b). Specific targeting and constitutive association of histone deacetylase complexes during transcriptional repression. *Genes Dev.* **16**, 687–692.

Marklew, S., Smith, D. P., Mason, C. S., and Old, R. W. (1994). Isolation of a novel RXR from *Xenopus* that most closely resembles mammalian RXRβ and is expressed throughout early development. *Biochim. Biophys. Acta* **1218**, 267–272.

Muscat, G. E., Burke, L. J., and Downes, M. (1998). The corepressor NCoR and its variant RIP13a and RIP13Delta1 directly interact with the basal transcription factors TFIIB, TAFII32, and TAFII70. *Nucleic Acid Res.* **26**, 2899–2907.

Nieuwkoop, P. D., and Faber, J. (1967). *Normal table of* Xenopus laevis *(Daudin)* 2nd Ed. Amsterdam: North-Holland.

Oofusa, K., Tooi, O., Kashiwagi, A., Kashiwagi, K., Kondo, Y., Watanabe, Y., Sawada, T., Fujikawa, K., and Yoshizato, K. (2001). Expression of thyroid hormone receptor beta A gene assayed by transgenic *Xenopus laevis* carrying its promoter sequences. *Mol. Cell. Endocrinol.* **181**, 97–110.

Patterton, D., Pär Hayes, W., and Shi, Y. B. (1995). Transcriptional activation of the matrix metalloproteinase gene *stromelysin-3* coincides with thyroid hormone-induced cell death during frog metamorphosis. *Dev. Biol.* **167**, 252–262.

Puzianowska-Kuznicka, M., Damjanovski, S., and Shi, Y. B. (1997). Both thyroid hormone and 9-cis retinoic acid receptors are required to efficiently mediate the effects of thyroid hormone on embryonic development and specific gene regulation in *Xenopus laevis*. *Mol. Cell. Biol.* **17**, 4738–4749.

Ranjan, M., Wong, J., and Shi, Y. B. (1994). Transcriptional repression of *Xenopus* TRβ gene is mediated by a thyroid hormone response element located near the start site. *J. Biol. Chem.* **269**, 24699–24705.

Sachs, L. M., and Shi, Y. B. (2000). Targeted chromatin binding and histone acetylation *in vivo* by thyroid hormone receptor during amphibian development. *Proc. Natl. Acad. Sci. USA* **97**, 13138–13143.

Sachs, L. M., Amano, T., Rouse, N., and Shi, Y. B. (2001a). Requirement of histone deacetylase at two distinct steps in thyroid hormone receptor-mediated gene regulation during amphibian development. *Dev. Dyn.* **222**, 280–291.

Sachs, L. M., Amano, T., and Shi, Y. B. (2001b). An essential role of histone deacetylases in postembryonic organ transformations in *Xenopus laevis*. *Int. J. Mol. Med.* **8,** 595–601.
Sachs, L. M., Jones, P. L., Havis, E., Rouse, N., Demeneix, B. A., and Shi, Y. B. (2002). Nuclear receptor corepressor recruitment by unliganded thyroid hormone receptor in gene repression during *Xenopus laevis* development. *Mol. Cell. Biol.* **22,** 8527–8538.
Schreiber, A. M., Das, B., Huang, H., Marsh-Armstrong, N., and Brown, D. D. (2001). Diverse developmental programs of *Xenopus laevis* metamorphosis are inhibited by a dominant negative thyroid hormone receptor. *Proc. Natl. Acad. Sci. USA* **98,** 10739–10744.
Shi, Y. B. (1996). Thyroid hormone-regulated early and late genes during amphibian metamorphosis. In Metamorphosis: Postembryonic reprogramming of gene expression in amphibian and insect cells (L. I. Gilbert, J. R. Tata, and B. G. Atkinson, Eds.). Academic Press, New York.
Shi, Y. B. (1999). *Amphibian metamorphosis: From morphology to molecular biology.* John Wiley & Sons, New York.
Shi, Y. B., and Liang, V. C. T. (1994). Cloning and characterization of the ribosomal protein L8 gene from *Xenopus laevis*. *Biochem. Biophys. Acta* **1217,** 227–228.
Steinbac, O. C., Wolffe, A. P., and Rupp, R. A. (2000). Histone deacetylase activity is required for the induction of the MyoD muscle cell lineage in *Xenopus*. *Biol. Chem.* **381,** 1013–1016.
St. Germain, D. L., Schwartzman, R. A., Croteau, W., Kanamori, A., Wang, Z., Brown, D. D., and Galton, V. A. (1994). A thyroid hormone-regulated gene in *Xenopus laevis* encodes a type III iodothyronine 5-deiodinase. *Proc. Natl. Acad. Sci. USA* **91,** 7767–7771.
Stolow, M. A., and Shi, Y. B. (1995). *Xenopus* sonic hedgehog as a potential morphogen during embryogenesis and thyroid hormone-dependent metamorphosis. *Nucleic Acids Res.* **23,** 2555–2562.
Strahl, B. D., and Allis, D. (2000). The language of covalent histone modifications. *Nature* **403,** 41–45.
Tata, J. R. (2002). Signalling through nuclear receptors. *Nat. Rev. Mol. Cell. Biol.* **3,** 702–710.
Thummel, C. S. (1993). From embryogenesis to metamorphosis: The regulation and function of *Drosophila* nuclear receptor superfamily members. *Cell* **83,** 871–877.
Underhill, C., Qutob, M. S., Yee, S. P., and Torchia, J. (2000). A novel nuclear receptor corepressor complex, NCoR, contains components of the mammalian SWI/SNF complex and corepressor KAP-1. *J. Biol. Chem.* **275,** 40463–40470.
Urnov, F. D., Yee, J., Sachs, L., Collingwood, T. N., Bauer, A., Beug, H., Shi, Y. B., and Wolffe, A. P. (2000). Targeting of NCoR and histone deacetylase 3 by the oncoprotein v-erbA yields a chromatin infrastructure-dependent transcriptional repression pathway. *EMBO J.* **19,** 4074–4090.
Van Lint, C., Emiliani, S., and Verdin, E. (1996). The expression of a small fraction of cellular genes is changed in response to histone hyperacetylation. *Gene Expression* **5,** 245–253.
Webb, P., Anderson, C. M., Valentine, C., Nguyen, P., Marimuthu, A., West, B. L., Baxter, J. D., and Kushner, P. J. (2000). The nuclear receptor corepressor (NCoR) contains three isoleucine motifs (I/LXXII) that serve as receptor interaction domains (IDs). *Mol. Endocrinol.* **14,** 1976–1985.
Weston, A. D., Blumberg, B., and Underhill, T. M. (2003). Active repression by unliganded retinoid receptors in development: Less is sometimes more. *J. Cell. Biol.* **161,** 223–228.
Wolffe, A. P. (1997). Sinful repression. *Nature* **387,** 16–17.
Wong, J., and Shi, Y. B. (1995). Coordinated regulation of and transcriptional activation by *Xenopus* thyroid hormone and retinoid X receptors. *J. Biol. Chem.* **270,** 18479–18483.
Wong, J., Shi, Y. B., and Wolffe, A. P. (1995). A role for nucleosome assembly in both silencing and activation of the *Xenopus* TRβA gene by thyroid hormone receptor. *Genes Dev.* **9,** 2696–2711.
Wong, J., Shi, Y. B., and Wolffe, A. P. (1997a). Determinants of chromatin disruption and transcriptional regulation instigated by the thyroid hormone receptor: Hormone regulated

chromatin disruption is not sufficient for transcriptional activation. *EMBO J.* **16,** 3158–3171.

Wong, J., Li, Q., Levi, B. Z., Shi, Y. B., and Wolffe, A. P. (1997b). Structural and functional features of a specific nucleosome containing a recognition element for the thyroid hormone receptor. *EMBO J.* **16,** 7130–7145.

Wong, J., Patterton, D., Imhof, A., Guschin, D., Shi, Y. B., and Wolffe, A. (1998). Distinct requirements for chromatin assembly in transcriptional repression by thyroid hormone receptor and histone deacetylase. *EMBO J.* **17,** 520–534.

Wong, C. W., and Privalsky, M. L. (1998). Transcriptional repression by the SMRT–mSin3 corepressor: Multiple interactions, multiple mechanisms, and a potential role for TFIIB. *Mol. Cell. Biol.* **18,** 5500–5510.

Yaoita, Y., and Brown, D. D. (1990). A correlation of thyroid hormone receptor gene expression with amphibian metamorphosis. *Genes Dev.* **4,** 1917–1924.

Yaoita, Y., Shi, Y. B., and Brown, D. D. (1990). *Xenopus laevis* α and β thyroid hormone receptors. *Proc. Natl. Acad. Sci. USA* **87,** 7090–7095.

Yoon, H. G., Chan, D. W., Huang, Z. Q., Li, J., Fondell, J. D., Qin, J., and Wong, J. (2003). Purification and functional characterization of the human NCoR complex: The roles of HDAC3, TBL1, and TBLR1. *EMBO J.* **22,** 1336–1346.

Yoshida, M., Horinouchi, S., and Beppu, T. (1995). Trichostatin A and trapoxin: Novel chemical probes for the role of histone acetylation in chromatin structure and function. *Bioessays* **17,** 423–430.

Zhang, Y., Iratni, R., Erdjument-Bromage, H., Tempst, P., and Reinberger, D. (1997). Histone deacetylase and SAP18, a novel polypeptide, are components of a human Sin3 complex. *Cell* **89,** 357–364.

8

Cdc25B as a Steroid Receptor Coactivator

STEVEN S. CHUA,* ZHIQING MA,† ELLY NGAN,‡ AND SOPHIA Y. TSAI*

*Department of Molecular and Cellular Biology, Baylor College of Medicine
Houston, Texas 77030
†Lexicon Genetics, Woodlands, Texas 77381
‡Department of Zoology, The University of Hong Kong
Hong Kong, SAR, PRC

I. Introduction
   A. The Mammalian Cell Cycle
II. The Cdc25 Family of Proteins
   A. Historical Perspective
   B. Role and Regulation of Cdc25 Proteins in the Cell Cycle
   C. Cdc25 and Cancer
III. Evaluation of the Cdc25B Role in Murine Mammary Glands
   A. Phenotypic Consequences of Ectopic Cdc25B Expression
   B. Increased Expression of ER Responsive Genes in Cdc25B Transgenics
IV. Cdc25B: A Coactivator that Enhances Steroid Receptor-Dependent Transcription
   A. Properties of Cdc25B as a Coactivator
   B. Direct Interaction of Cdc25B with Steroid Receptors

    C. Recruitment of Histone Acetyltransferases
    D. Cdc25B Displaying Separable Coactivator and
       Cell Cycle Functions
    E. Cdc25B Enhancing the Transcription of Other
       Class I Steroid Receptors
    F. Enhancement of Cell-Free Transcription
       Dependent on PR-B
V. Coactivator Function of Cdc25B in the Prostate
    A. Overexpression of Cdc25B in Human
       Prostate Cancers
    B. Enhancement of Androgen Receptor-Responsive
       Transcriptional Activity
    C. Direct Interaction of Cdc25B with AR
VI. Conclusions
    References

The traditional role of the Cdc25 family of dual-specificity phosphatases is to activate cyclin-dependent kinases (CDKs) to enable progression through the cell cycle. This chapter reports that in addition to its cell cycle role, Cdc25B functions as a novel steroid receptor coactivator (SRC). When overexpressed in transgenic mammary glands, Cdc25B can up-regulate the expression of two estrogen receptor (ER)-target genes: cyclin D1 and Lactoferrin. In addition, when coexpressed with ER, Cdc25B can coactivate an ER-dependent reporter in the presence of estradiol. The coactivation of Cdc25B can be extended to the glucocorticoid receptor (GR), progesterone receptor (PR), and androgen receptor (AR). Because of the respective importance of ER and AR in breast and prostate cancer, this chapter focuses on the coactivation of both receptors by Cdc25B. We demonstrate that Cdc25B can interact directly with these nuclear receptors, recruit and enhance the activity of histone acetyltransferases (HATs), and potentiate cell-free transcription independent of its cell cycle regulatory function. Furthermore, because Cdc25B is up-regulated in highgrade and poorly differentiated prostate tumors, which are likely transiting from the hormone-dependent to hormone-independent state, we hypothesize that the coactivation of AR by Cdc25B may induce genes responsible for this progression. Taken together, it is highly conceivable that Cdc25B can promote neoplasia by its two disparate functions of (1) coactivation to induce higher levels of expression of steroid receptor target genes and (2) its role of activating CDKs to deregulate progression of the cell cycle, DNA replication, and mitosis. © 2004 Elsevier Inc.

I. INTRODUCTION

The Cdc25 family of proteins have traditionally played a role in the cell cycle by activating cyclin-dependent kinases (CDKs) to allow for the progression of the cell cycle (Draetta and Eckstein, 1997; Dunphy and Kumagai, 1991). Recent results from studies conducted by the chapter authors have suggested that in addition to their known cell cycle roles, the Cdc25 proteins can also function as coactivators to increase steroid receptor transactivation (Ma et al., 2001; Ngan et al., 2003). This finding is not surprising because other cell-cycle effectors play a role in the expression of genes. Cyclin A/CDK2 can phosphorylate estrogen receptors (ERs) to potentiate ligand-independent target gene expression (Rogatsky et al., 1999; Trowbridge et al., 1997), and free cyclin D1 has been shown to activate ER-dependent gene transcription in breast epithelial cells (Neuman et al., 1997; Zwijsen et al., 1998). Therefore, before this chapter delineates the role of Cdc25B in ER transactivation, it will briefly present the known functions of Cdc25B in the eukaryotic cell cycle.

A. THE MAMMALIAN CELL CYCLE

The eukaryotic cell cycle, which consists of four phases (G1, S, G2, & M), is regulated in a precise and orderly transition by a delicate interplay between positive and negative cell cycle effectors (Vermeulen, 2003). Progress through each of the cell cycle phases is driven by specific cyclin-dependent kinases (CDKs), whose activities require association with cyclin phosphorylation on a conserved threonine residue (Elledge, 1996; Morgan, 1996; Sherr, 1996) and dephosphorylation of inhibitory phosphates at two sites by the Cdc25 family of proteins (Draetta and Eckstein, 1997; Millar et al., 1991). Currently, there are at least nine known CDKs and five have been shown to play a role in the mammalian cell cycle (Vermeulen, 2003). These serine/threonine kinases interact with a specific subset of cyclins (A–H) during different stages of the cell cycle (Draetta, 1994; Pines, 1993) by phosphorylating important key substrates like pRB, DNA synthetic enzymes, and the transcriptional machinery to express genes required for the progression of the cell cycle. Working to impede progression through the cell cycle are two classes of cyclin-dependent kinase inhibitors (CDKIs) (Hunter and Pines, 1994; Sherr and Roberts, 1995). The CIP/KIP family, which includes p21, p27 and p57, can associate with and inhibit all known G1 cyclin-CDK complexes (Harper et al., 1995; Xiong et al., 1993). On the other hand, the INK4 (inhibitor of CDK4) family, which includes p15, p16, p18, and p19, is selective for complexes containing the D-type cyclins and CDK4 or CDK6 (Hirai et al., 1995; Serrano et al., 1993). An additional level of negative regulation of CDK action is imposed by phosphorylation on two sites near its amino-terminal by the protein kinase wee-1 (Igarashi et al., 1991; Russell and Nurse, 1986) and the related kinases mik-1 and myt-1 (Booher *et al.*, 1993; Muller *et al.*, 1995). The removal of these inhibitory phosphates is necessary for CDK function and is mediated by the Cdc25 family of dual-specificity protein phosphatases (Draetta and Eckstein, 1997; Millar *et al.*, 1991), which form the basis of this review. A more thorough description of the cell cycle is beyond the scope of this review, and the reader should consult reviews by Elledge (1996), Pines (1999), and Vermeulen (2003).

II. THE Cdc25 FAMILY OF PROTEINS

A. HISTORICAL PERSPECTIVE

The Cdc25 gene was first characterized in the fission yeast *Schizosaccharomyces pombe* to be a dose-dependent inducer of mitosis (Russell and Nurse, 1986; Russell and Nurse, 1987). Temperature-sensitive yeast mutants in Cdc25 have been shown to arrest at G2/M prior to entry into mitosis. In mammals, three different Cdc25 genes have been cloned that can complement and suppress the yeast Cdc25 mutant (Galaktionov and Beach, 1991). Each of these Cdc25 proteins share approximately 40 to 50% homology at the amino acid level and possess conserved catalytic domains at the C-terminal and varying N-terminal regions that may be subjected to differing regulation (Galaktionov and Beach, 1991; Moreno and Nurse, 1991; Sadhu *et al.*, 1990).

B. ROLE AND REGULATION OF Cdc25 PROTEINS IN THE CELL CYCLE

The regulation of the activity of a cyclin/CDK complex is an intricate process that requires a number of proteins. The wee-1 family of proteins exert their negative influence on CDKs by phosphorylating threonine 14 and tyrosine 15 of CDKs (Booher *et al.*, 1993; Igarashi *et al.*, 1991; Muller *et al.*, 1995; Russell and Nurse, 1986). The Cdc25 family of dual-specificity phosphatases function to remove these inhibitory phosphate groups (Draetta and Eckstein, 1997; Millar *et al.*, 1991). To attain full catalytic activity, however, CDKs need to be phosphorylated at threonine 160 and the CAK (CDK-activating kinase) complex, functions (Fisher and Morgan, 1994; Wu *et al.*, 1994). Conversely, a phosphatase called *KAP* (CDK-associated phosphatase) dephosphorylates threonine 160 to dampen the activity of a cyclin/CDK complex (Poon and Hunter, 1995). The precise timing and execution of these regulatory processes ensure that the cell cycle proceeds smoothly.

Unlike the single-celled organisms like *S. pombe*, which contain only one Cdc25 protein to regulate transition into mitosis, mammalian cells have evolved to possess three Cdc25 gene products to regulate a more complex cell cycle. Cdc25A regulates entry into the S phase (Hoffmann *et al.*, 1994;

Jinno *et al.*, 1994), S phase progression (Blomberg and Hoffmann, 1999; Sexl *et al.*, 1999), and mitosis while Cdc25C has a more limited role and regulates entry into mitosis (Donzelli and Draetta, 2003; Millar *et al.*, 1991). Because Cdc25B is activated earlier than Cdc25C, Cdc25B most likely plays, a role in early G2/M transition to initially activate cyclin B/Cdc2, which then activates Cdc25C by phosphorylation. This process creates an autoamplification loop to irreversibly commit the cell toward mitosis (Hoffmann *et al.*, 1993; Karlsson *et al.*, 1999).

Though the Cdc25 genes display variable patterns of expression in embryonic and adult tissues, the finding that they are also expressed abundantly in proliferating tissues suggests their roles in cell growth. The Cdc25 proteins are downstream targets of proliferation signal pathways. For example, in the ras signaling pathway, both Cdc25A and Cdc25B can interact with the raf-1 protein kinase and the 14-3-3 family of proteins (Conklin *et al.*, 1995; Galaktionov *et al.*, 1995a). Their phosphatase activities can also be induced by raf-dependent phosphorylation (Galaktionov *et al.*, 1995a). At the transcriptional level, both Cdc25A and Cdc25B are targets of the c-myc oncogene and E2F (Galaktionov *et al.*, 1996; Vigo *et al.*, 1999). At the post-translational level, the activity of the Cdc25 proteins can be regulated by phosphorylation by cyclin-CDK complexes (Hoffmann *et al.*, 1993, 1994) and polo-like kinase (plk-1) family members (Glover *et al.*, 1998; Nigg, 1998). Their stability can be regulated by proteolytic degradation (Baldin *et al.*, 1997; Donzelli *et al.*, 2002) and checkpoint-mediated degradation during DNA and UV damage (Shiloh, 2001, 2003; Zhou and Elledge, 2000). Cdc25C can also be compartmentalized in the cell by phosphorylation at ser216 by Cdc25C-associated protein kinase C-Tak1 (Peng *et al.*, 1998), Chk1, and Chk2 (Matsuoka *et al.*, 1998; Peng *et al.*, 1998; Sanchez *et al.*, 1997). Phosphorylation creates a binding site for the 14-3-3 family of proteins whose role is to sequester Cdc25C in the cytoplasm (Kumagai *et al.*, 1998; Lopez-Girona *et al.*, 1999). Taken together, this evidence suggests that the Cdc25 family of proteins play an important role in cell cycle progression by allowing diverse signaling pathways to impinge on them. For a more extensive discourse, consult the review by Donzelli and Draetta (2003).

C. Cdc25 AND CANCER

Given that the overexpression of positive regulators like cyclin D1 and E or the inactivation of negative regulators like p16 can predispose cells toward malignancy (Barnes, 1997; Gray-Bablin *et al.*, 1996; Keyomarsi and Pardee, 1993), it is reasonable to suggest that the deregulation of the Cdc25 proteins can contribute to cancer progression. Although Cdc25A, Cdc25B, or activated ras alone did not induce foci formation of rat embryonic fibroblasts, each of these proteins could elicit a robust transformation of these cells in the absence of the retinoblastoma gene product, Rb (Galaktionov et al., 1995b). In addition Cdc25A or Cdc25B can cooperate with activated ras in rat embryonic fibroblasts to induce oncogenic foci, grow on soft agar, and develop high-grade tumors when injected into nude mice (Galaktionov et al., 1995b). The observation that gene amplifications and overexpression of Cdc25A and Cdc25B have been found in human primary breast cancers and tumors further lends credence to this important family of proteins as potential oncogenes (Galaktinov et al., 1995b; Hernandez et al., 1998; Kudo et al., 1997; Wu et al., 1998) as the increased levels of these proteins may attenuate and/or overcome normal checkpoint proteins and inhibitors.

Based on the important correlation of Cdc25B with cancer, we evaluated the role of human Cdc25B by overexpressing this gene in the murine mammary glands. Our initial studies on Cdc25B-mediated mammary tumorigenesis led to the subsequent finding that Cdc25B can coactivate ER-mediated transcription of hormonally responsive genes. We will first highlight our studies of Cdc25B in the rodent mammary glands and then focus on the novel coactivation function of Cdc25B.

III. EVALUATION OF THE CDC25B ROLE IN MURINE MAMMARY GLANDS

A number of positive cell-cycle regulators (i.e., cyclin and CDKs) have been successfully targeted to the mammary gland to affect homeostatic perturbations (Bortner and Rosenberg, 1995, 1997; Wang et al., 1994). Given the association of Cdc25 with many primary tumors of the breast (Galaktionov et al., 1996), esophagus (Hu et al., 2001), head and neck (Gasparotto et al., 1997), colorectum (Takemasa et al., 2000), gastric area (Kudo et al., 1997) as well as with small cell lung cancer (Sasaki et al., 2001; Wu et al., 1998), we sought to address if the overexpression of Cdc25 proteins can lead to abnormal mammary development and/or carcinogenesis. To achieve this goal, the human Cdc25B was overexpressed in transgenic murine animals, namely mice, under the control of the mouse mammary tumor virus (MMTV) promoter which directs transgene expression predominantly to the mammary gland (Pattengale et al., 1989).

A. PHENOTYPIC CONSEQUENCES OF ECTOPIC CDC25B EXPRESSION

We obtained multiple transgenic lines that express high levels of both the Cdc25B mRNA and protein and showed that these mice displayed an increased rate of mammary epithelial cell proliferation and developmental abnormalities (Ma et al., 1999). Phenotypically, the transgenic virgin mammary gland developed precocious alveolar hyperplasia characteristic of early pregnancy and became more pronounced with increasing age. Furthermore, the transgenic mammary glands showed retarded postactational involution, a process marked by apoptosis and remodeling. We surmised that Cdc25B acts as a positive effector to deregulate cell cycle progression. Cdc25B's overexpression may relieve part of the cell cycle brakes imposed by a negative effector like TGF-β and a downstream effector like p21, which normally would prevent the activation of CDK2 by nullifying formation of Cdc25A–cyclin/E–CDK2 complexes. The finding that at the molecular level, Cdc25B induced higher cyclin E-CDK2 kinase activity serves as testimony to this observation, and this enhanced proliferation imposed by Cdc25B is reflected phenotypically as an increase in the number of side and alveolar buds (Ma *et al.*, 1999). Furthermore, by inducing the expression of cyclin D1, a growth sensor, and likely survival factor and by decreasing the expression of p53 and myc (both of which have been shown to mediate apoptosis), Cdc25B most likely attenuated the apoptotic process (Ma *et al.*, 1999).

None of the Cdc25B transgenic mice, however, developed tumors during their lifetime (Ma *et al.*, 1999). This result is not surprising and is consistent with the finding that overexpression of either Cdc25A or Cdc25B failed to elicit transformation of rat embryo fibroblasts (Galaktinov *et al.*, 1995b). This finding further emphasizes that additional events, other than the perturbation of the G1/S transition, are required for full tumorigenic transformation. Nonetheless, the overexpression of Cdc25B may provide the initiating events for tumorigenesis. Indeed, when Cdc25B transgenic mice were challenged with the carcinogen 9,10-dimethyl-1, 2-benzanthracene (DMBA), their susceptibility toward tumor formation increased (Yao *et al.*, 1999).

Furthermore, the availability of the Cdc25B transgenic mice permitted us to validate the synergy between Cdc25B and ras *in vivo*. We bred MMTV–Cdc25B transgenic mice with MMTV–v-ras transgenic mice (Sinn *et al.*, 1987) and examined the latency of mammary tumor formation. We found that the bitransgenic mice harboring both the Cdc25B and activated ras transgenes developed tumors with a much shorter latency than the monogenic MMTV–v-ras mice (Fig. 1). The average time for 50% of the monotransgenic MMTV–v-ras mice to develop mammary tumors was about 29 weeks, whereas mammary tumor formation in the MMV–v-ras/Cdc25B bitransgenic mice was much faster, with an average duration of about 21 weeks. By 29 weeks, all the bitransgenic mice developed tumors. Taken together, these results suggest that the Cdc25B and activated ras signaling pathways can synergize to more dramatically deregulate growth and precipitate a more robust and shorter latency of mammary tumor formation.

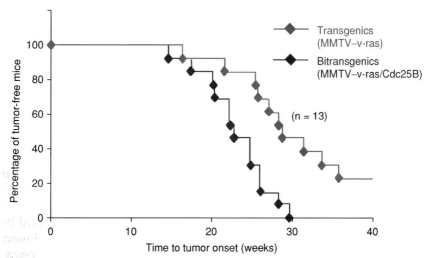

FIGURE 1. Cdc25B and activated ras synergize to shorten tumor latency. MMTV–Cdc25B transgenic mice were bred with MMTV–v-ras transgenic mice to generate bitransgenic mice. Age-matched monotransgenic MMTV–v-ras mice (lighter diamonds and lines) and bitransgenic MMTV–v-ras/Cdc25B (darker diamonds and lines) were checked weekly for tumor formation by palpation. Of the bitransgenic mice, 50% developed mammary tumors by 21 weeks, and 50% of the ras monotransgenic mice developed tumors by 29 weeks. These results indicate that the overexpression of Cdc25B could synergize with activated ras *in vivo* to promote mammary tumor formation. n = 13 signifies total number of mice used for each group.

B. INCREASED EXPRESSION OF ER RESPONSIVE GENES IN CDC25B TRANSGENICS

The intriguing finding, of (1) elevated levels of cyclin D1 protein at the molecular level in both immunohistochemical and Western blot analyses (2- to 4-fold higher than the wild type) and (2) increased cyclin D1 mRNA levels by ribonuclease protection assays (2-fold higher than the wild type) in Cdc25B transgenic versus wild-type mice led us to surmise that Cdc25B may activate cyclin D1 at the transcriptional level (Ma *et al.*, 2001). The cyclin D1 gene is directly up-regulated by ER in response to hormone stimulation; the enhanced expression of cyclin D1 can mediate estrogen-induced mitogenesis possibly via its cell cycle and transcriptional role (Altucci *et al.*, 1996; Neuman *et al.*, 1997; Prall *et al.*, 1998; Sabbah *et al.*, 1999; Zwijsen *et al.*, 1998, 1997). These facts are also significant because increased levels of cyclin D1 are frequently encountered in breast cancers overexpressing ER (Hui *et al.*, 1996; Kenny *et al.*, 1999) and because ablation of cyclin D1 dramatically affects lobuloalveolar development during pregnancy (Sicinski *et al.*, 1995). These significant correlations led us to address the potential role of Cdc25B on steroid receptor-mediated transcription by further characterizing the expression of another ER-responsive gene, Lactoferrin.

After normalization, we observed a greater than 20-fold increase in Lactoferrin mRNA expression in Cdc25B transgenics compared to expression of wild-type controls (Ma et al., 2001). Taken together, this preliminary but significant evidence suggests that the overexpression of Cdc25B can enhance the responsiveness of mammary gland to steroid stimulation, and we sought to further dissect the role of Cdc25B in mediating this response.

IV. Cdc25B: A COACTIVATOR THAT ENHANCES STEROID RECEPTOR-DEPENDENT TRANSCRIPTION

A. PROPERTIES OF Cdc25B AS A COACTIVATOR

To elucidate the mechanism by which Cdc25B enhances ER-dependent transcription, we first tested the ability of Cdc25B to potentiate ER-mediated transcription in transient transfections in HeLa cells (Ma et al., 2001). The addition of 17β-estradiol (E2) activated an ER-responsive target reporter by 10-fold, and the addition of Cdc25B further increased reporter activity by about 4-fold in a receptor- and hormone-dependent fashion as shown in Figure 2A (Ma et al., 2001). The addition of Cdc25B in the absence of ER had no effect on reporter activity, whereas the coexpression of ER with Cdc25B in the absence of hormone only weakly increased reporter activity, possibly via ligand-independent means (Ma et al., 2001). It is important to note that the coexpression of Cdc25 and ER did not affect ER expression levels. To assess if the enhancement of ER transcription by Cdc25B is agonist-specific, we tested the ability of Cdc25B to activate ER-responsive transcription in the presence of a couple of ER antagonists, such as tamoxifen and ICI 163484. We found that Cdc25B was not able to potentiate ER-dependent transcription in the presence of antagonists (Fig. 2A). Similar results were also obtained with Cdc25A (data not shown). Taken together, we showed that Cdc25B is specific in its ability to further potentiate ER-dependent transcription only in the presence of the agonist 17β-estradiol (E2) or weakly by ligand-independent means in the presence of ER without hormone treatment; we also found that Cdc25B is not able to potentiate ER-dependent transcription of reporter activity in the presence of ER antagonists (Ma et al., 2001).

To determine which of the functional domains in ER are necessary for Cdc25B coactivation (Fig. 2B), we performed transient transfections using a number of ER deletion mutants, including AF-1 (179C), AF-2 (N282G), and helix 12 (3X) (Fig. 2B). Our transfection results showed that mutations of helix 12 (3X) and the AF-2 domain (N282G) abrogated their ability to respond to hormone-dependent activation of reporter activity and

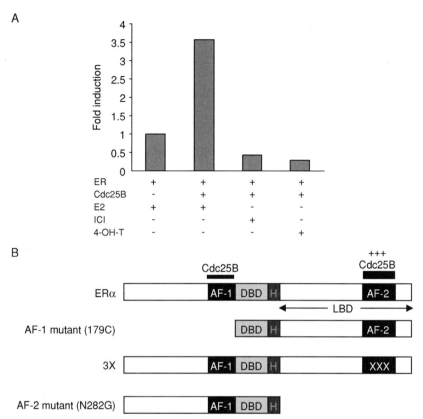

FIGURE 2. Coactivation of estrogen receptor by Cdc25B. (A) Cdc25B and ER coactivation. Cdc25B is able to coactivate ER only in the presence of (E2) 17β-estradiol but not in the presence of the ER antagonists (4-OH-T) tamoxifen or ICI. The positive (+) and negative (−) symbols indicate the presence and absence of exogenously added receptors and hormones. (B) Wild-type ERα, AF-1 (179C), helix 12 (3X), and AF-2 (N282G) as mutants of ER. Cdc25B coactivation of ER depends on the presence of both the AF-1 and AF-2 domains but depends predominantly on the AF-2 domain (+++) for full hormone-dependent coactivation. The black bars under Cdc25B indicate the regions on ER that can be mediated by Cdc25B coactivation (AF-2 and AF-1 domains). H: hinge region; DBD: DNA binding domain; AF-1 and AF-2: activation functions domains 1 and 2, respectively; LBD: ligand binding domain.

that they could only be weakly coactivated by Cdc25B (Ma et al., 2001) as compared to the ERα controls. On the other hand, mutation of the AF-1 domain (179C) did not lead to a loss of hormone-dependent coactivation by Cdc25B (Ma et al., 2001). Therefore, these results indicate that the AF-2 domains of ER play a major role in mediating hormone-dependent coactivation by Cdc25B and that the AF-1 domain plays only a minor role (Fig. 2B).

B. DIRECT INTERACTION OF Cdc25B WITH STEROID RECEPTORS

Given the ability of Cdc25B to enhance steroid receptor transactivation of reporter expression, we next ascertained if Cdc25B would interact with steroid receptors by three independent methods (Ma *et al.*, 2001). First, we used a mammalian two-hybrid system to determine the interaction of Cdc25B with ER. We coexpressed chimeric Cdc25B/VP16 with ER and assayed interaction on a reporter harboring three copies of the estrogen response element (ERE) by means of luciferase activity. We found that a robust interaction occurs only when these proteins are coexpressed in the presence of estradiol (Ma *et al.*, 2001). In contrast, coexpression of VP16 and ER does not significantly affect luciferase activity, indicating that VP16 and ER do not interact as significantly and strongly as Cdc25B/VP16 and ER, even in the presence of estradiol.

We next tested for interaction between ER and Cdc25B in cells by coimmunoprecipitation analyses of MCF-7 cell extracts. By immunoprecipitating cell extracts with a Cdc25B specific antibody followed by electrophoresis of these immunoprecipitated products on a SDS–PAGE gel, we were able to detect ER with an ER-specific antibody in immunocomplexes pulled down by the Cdc25B antibody. We were not able to detect ER, however, if the Cdc25B antibody was preadsorbed with a Cdc25B peptide (Ma *et al.*, 2001). Therefore, this result indicates that Cdc25B can interact with ER *in vivo*.

To further confirm that the interaction between Cdc25B and ER occurs directly, we performed GST pull-down assays using full-length Cdc25B and GST–ER in the presence or absence of estradiol. The results showed that Cdc25B interacts with GST–ER in a hormone-independent fashion (Ma *et al.*, 2001). Given that Cdc25B further potentiates ER-dependent transactivation of target reporters only in the presence of hormone, we surmise that Cdc25B interacts with the *in vitro* translated GST–ER because of exposed target sites normally masked in the cell. This is consistent with the notion that the addition of a hormone *in vivo* unmasks sites on ER to allow Cdc25B to interact with ER.

Since Cdc25B can interact directly with ER, we next determined the exact domain(s) on Cdc25B that allows this interaction to take place. By using a number of GST–Cdc25B deletion mutants (Fig. 3), we mapped two interaction domains in Cdc25B that interact with ER: a C-terminal region spanning amino acids 352 to 539, which make up a highly conserved region among Cdc25B proteins, and a middle region of Cdc25B spanning amino acids 81 to 273 (Ma *et al.*, 2001). Surprisingly, a putative nuclear receptor (NR) box (LXXLL motif) located at the N-terminal of Cdc25B failed to show any interaction with ER. In addition, a Cdc25B fragment containing amino acids 274 to 351 composed of multiple phosphorylation sites (P) critical for the cell cycle function of Cdc25B failed to show any

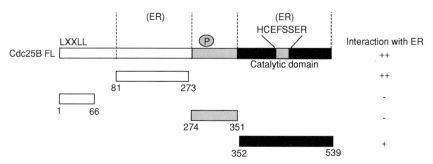

FIGURE 3. Schematic representation of domains on Cdc25B responsible for interacting with ER. The various Cdc25B deletion mutants are as indicated. The LXXLL domain located at the N-terminal of Cdc25B is a putative NR box motif necessary for interaction with a steroid receptor. The phosphorylation domain (P) is a site for phosphorylation of Cdc25B for modulation of activity; the catalytic domain, which contains the amino acid sequence HCEFSSER, possesses phosphatase activity. GST pull-down assays show that ER interacts in the middle of Cdc25B (aa81–273) and in the catalytic domain (aa352–539) indicated by (ER). Surprisingly, the LXXLL motif on Cdc25B does not interact with Cdc25B. The symbol ++ denotes detectable interaction with ER; + denotes weaker interaction with ER; and − denotes no interaction with ER. The amino acid positions are indicated.

interaction with ER. Taken together, these *in vitro* and *in vivo* studies show that Cdc25B physically interacts with NRs to enhance their ability to increase transcriptional activity of their target genes.

C. RECRUITMENT OF HISTONE ACETYLTRANSFERASES

Because steroid receptor coactivators can acetylate histones or recruit histone acetyltransferases (HATs) to modify chromatin and permit greater accessibility for transcription factors to bind and activate genes, we sought to test if Cdc25B also possesses this property. We included two known HATs, p300/CBP-associated factor (PCAF) and CREB binding protein (CBP), in transfections with Cdc25B on ER-mediated reporters. As shown in Figure 4, Cdc25B with PCAF or CBP could synergize to increase estradiol and ER-dependent transcription tremendously in comparison to PCAF or CBP alone (Cdc25B alone ∼2.8-fold; PCAF alone ∼1.6-fold; Cdc25B + PCAF ∼9.2-fold; CBP alone ∼1.4-fold; Cdc25B + CBP ∼7.8-fold). On the other hand, the coexpression of steroid receptor coactivator-1 (SRC-1) with Cdc25B showed a less than additive effect on ER-mediated transcription (Cdc25B alone ∼2.8-fold; SRC-1 alone ∼3.4-fold; Cdc25B + SRC-1 ∼4.8-fold). These results suggest that Cdc25B shows preferential recruitment of HATs and prefers PCAF or CBP over SRC-1 on the estrogen receptor.

To further substantiate that Cdc25B is able to interact with PCAF and CBP, we performed GST pull-down assays and showed that GST Cdc25B

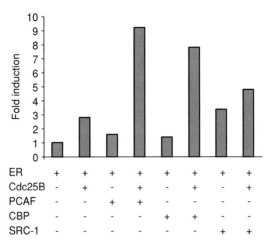

FIGURE 4. Synergy in coactivation by Cdc25B and the HATs PCAF and CBP. Cdc25B can coactivate ER transcription and synergizes with both PCAF and CBP to further potentiate reporter gene expression in the presence of a hormone. Cdc25B and SRC-1, however, do not synergize and display a less than additive effect of coactivation. The symbols + and − respectively represent the presence and absence of the exogenously added receptor and coactivators.

can directly associate with PCAF and CBP, albeit the association is a little weaker with the latter protein (Ma *et al.*, 2001). In agreement with the weak additive effects of Cdc25B and SRC-1 coactivation of an ER-dependent reporter, we found that GST Cdc25B is only able to weakly interact with SRC-1. We also ruled out the possibility that Cdc25B acts as a bridging factor to increase association of ER with PCAF by showing that no increase in PCAF coprecipitated with ER in the presence or absence of Cdc25B (Ma *et al.*, 2001). To assess if the enhancement of ER-dependent transcription is mediated by an increase in HAT activity, we tested the efficiency of acetylation of histone substrates in the presence of Cdc25B. We observed an increase in histone H3 but not H4 acetylation in the presence of Cdc25B over a nonspecific control (Ma *et al.*, 2001). According to data not shown, in Figure 4 we have ascertained that Cdc25B does not display any HAT activity. Collectively, these results suggest that Cdc25B directly interacts with both ER and HATs to enhance transcription by remodeling chromatin via acetylation of histones.

D. CDC25B DISPLAYING SEPARABLE COACTIVATOR AND CELL CYCLE FUNCTIONS

Our domain mapping studies indicate that ER interacts with two domains of Cdc25B that reside outside the cell cycle regulatory region of amino acids 274 + 0351. Since we cannot rule out the possibility that the cell cycle regulatory region of Cdc25B can mediate activation of CDK2 to enhance transcription of ER, we used roscovitine (Meijer et al., 1997) to block potential cyclin/CDK dependent pathways in a transient transfection study. We demonstrated that though the addition of roscovitine led to a general decrease in transcriptional activity of ER-dependent reporter, Cdc25B was still able to consistently coactivate ER-mediated transcription an additional 4-fold over that of ER in the presence of hormone and a similar fold induction by Cdc25B in the absence of roscovitine (Ma et al., 2001). Therefore, these results clearly show that the coactivation function of Cdc25B occurs independent of its cell cycle function.

To further ascertain if the phosphatase activity of Cdc25B is necessary for its ability to coactivate steroid receptor transactivation, we mutated critical residues at its active site from Cys446 to Ser446 and Arg452 to Ala452 to generate phosphatase inactive mutants C466S and R452C; these mutants respectively cannot bind significant amounts of cyclin E and cannot associate with cyclin A and CDK2 (Xu and Burke, 1996). We tested the ability of these mutants to coactivate ER in the presence of estradiol and found that they retained their ability to coactivate ER-dependent transactivation of reporter in HeLa cells when compared to wild-type Cdc25B (Ma et al., 2001). Overall, our studies have revealed that Cdc25B functions as an SRC to interact with ER and HATs and to coactivate ER-dependent transcription of ER target genes in the presence of estradiol in a cell cycle independent fashion.

E. CDC25B ENHANCING THE TRANSCRIPTION OF OTHER CLASS I STEROID RECEPTORS

To ascertain if the coactivation effect of Cdc25B is specific to ER, we performed transient transfections with other steroid receptors in the presence of their cognate hormones. As shown in Figure 5, Cdc25B is able to further activate GR-, PR-, and AR-dependent reporters ~3.5,6-, and 6-fold, respectively (Ma et al., 2001). In contrast, Cdc25B had only a minimal effect on RAR-mediated transactivation of its reporter under similar assay conditions. Taken together, these results indicate that Cdc25B appears to preferentially activate transcription of certain steroid receptors.

F. ENHANCEMENT OF CELL-FREE TRANSCRIPTION DEPENDENT ON PR-B

To determine if we can reconstitute the coactivation function of Cdc25B in a cell-free system, we used purified His-tagged Cdc25B and suboptimal levels of purified PR-B with an *in vitro* packaged chromatin template pPRE3-E4 (Ma et al., 2001). We found that Cdc25B was able to enhance

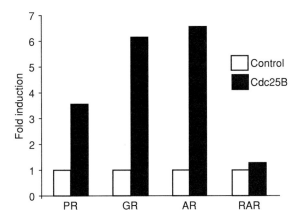

FIGURE 5. Coactivation of class I receptors by Cdc25B. Cdc25B is able to coactivate class I receptors like PR, GR, and AR but not RAR in the presence of their cognate hormones. Each receptor-dependent activation of its target reporter in the presence of its cognate hormone has been normalized to 1-fold. The relative fold coactivation by Cdc25B is then obtained by dividing the relative luciferase activity in the presence of Cdc25B, receptor, and hormone (solid boxes) over the value obtained in the presence of receptor and hormone only (white boxes).

transcription of the chromatin template in a dose-dependent fashion in the presence of progestin. At maximal levels of Cdc25B added, a more than 10-fold level of transcription was registered (Ma *et al.*, 2001).

V. COACTIVATOR FUNCTION OF CDC25B IN THE PROSTATE

Prostate cancer is an extremely common nonskin cancer-related cause of death in American men (Parker *et al.*, 1997). It initially begins in a hormone-dependent fashion, whereby ablation of androgen receptor function proves highly successful. This ablation therapy, however, becomes inefficacious with the progression of the cancer to a hormone-resistant state. To better understand the mechanism of prostate cancer progression, it is imperative to more extensively elucidate the function of AR and consequently delineate cellular factors that may precipitate the hormone-resistant condition. Given that Cdc25B can induce precocious alveolar hyperplasia (Ma *et al.*, 1999) and can coactivate ER, GR, and AR (Ma *et al.*, 2001) and that overexpression of Cdc25B is present in many cell lines and primary tumors including those of the breast (Galaktionov *et al.*, 1995b), we sought to better characterize the role of Cdc25B in the prostate (Ngan *et al.*, 2003).

A. OVEREXPRESSION OF Cdc25B IN HUMAN PROSTATE CANCERS

We first analyzed the expression levels of Cdc25B by immunohistochemistry. This analysis was followed by a semiquantitative evaluation of 30 prostate sections collected from men between the ages of 39 and 76 who underwent radical prostatectomy without any treatment prior to surgery at our institution (Baylor Prostate SPORE Program) using a protocol according to Lombardi et al. (1999). These samples included tumors at different stages and Gleason scores (Ngan et al., 2003). We observed ubiquitous staining of Cdc25B in the glandular epithelium from both neoplastic and normal samples. More robust staining was consistently observed in the tumor regions when compared to adjacent nontumor areas (29 out of 30 samples; 97%), indicating that Cdc25B expression was upregulated in these neoplastic regions (Ngan et al., 2003). When the 30 tumors were analyzed accordingly to stages (T2, T3a, and T3b) or Gleason scores (5, 6, and 7), we observed a trend of increasing Cdc25B expression when the stage increased from T2 (organ restricted) to T3b (seminal vesicle invasion) and when the Gleason score increased from 5 to 7 (Ngan et al., 2003). For example, we discovered that Cdc25B overexpression is more prominent in later stage tumors than earlier ones (comparing stage T3b: 158.3 ± 12.8 with stage T3a: 140.0 ± 17.1 or stage T2: 90.6 ± 14.0). Similarly, as the Gleason scores increased, Cdc25B staining became more intense (comparing Gleason score 7: 132.7 ± 12.5 with Gleason score 5: 94.0 ± 24.4). Collectively, the results of this study indicate a propensity of Cdc25B overexpression in prostate tumors and is consistent with the strong association of Cdc25B up-regulation, microvessel densities, and higher histological grades of breast cancers (Galaktionov et al., 1995b).

B. ENHANCEMENT OF ANDROGEN RECEPTOR-RESPONSIVE TRANSCRIPTIONAL ACTIVITY

As our initial studies have shown that Cdc25B can coactivate AR transactivation potential (Ma et al., 2001), we sought to analyze this property of Cdc25B further (Ngan et al., 2003). We performed transient transfections in lymph node cancer of the prostate (LNCaP) cells and consistently observed a 15-fold induction of a reporter with three copies of the androgen response element (ARE_3) in the presence of R1881, an AR agonist. When Cdc25B was coexpressed with AR in the presence of R1881, the target reporter expression was induced 2-fold (Ngan et al., 2003). This induction of AR-mediated transcription of a reporter is not cell specific and can also be recapitulated in HeLa cells (~6-fold higher coactivation by Cdc25B when AR is coexpressed with Cdc25B).

To rule out the possibility that Cdc25B's coactivation is not dependent on its cell cycle role and that our initial findings in ER (Ma et al., 2001) can also be extended to AR, we repeated the transfections using phosphatase inactive mutants C446S and R452-C (Ma *et al.*, 2001; Xu and Burke, 1996). We found that these mutants can still coactivate AR-dependent reporter activity to levels similar to Cdc25B wild type in the presence of R1881. This result indicates that similar to studies performed on ER (Ma *et al.*, 2001), Cdc25B is able to coactivate AR independent of Cdc25B's cell cycle regulatory function (Ngan *et al.*, 2003).

As a means to ensure that the ability of Cdc25B to coactivate AR-dependent gene transcription was possible only in the presence of R1881, we performed transient transfections with the antiandrogenic compound cyproterone acetate (CPA). Similar to studies performed on ER in the presence of tamoxifen (Ma *et al.*, 2001), we abrogated the ability of Cdc25B to coactivate AR-mediated reporter in the presence of CPA (Ngan *et al.*, 2003).

To further characterize the domain of Cdc25B responsible for the coactivation of AR-dependent reporter activity, we performed transient transfections in HeLa cells using a number of AR deletion mutants: ARΔAF1 (mutant deleted in AF-1) and ARΔAF-2 (mutant deleted in AF-2). We observed that the ablation of the AF-1 domain leads to a loss of induction by R1881 and that the ARΔAF-1 mutant can only recapitulate basal reporter activity similar to reporter activity in the absence of R1881 (Ngan *et al.*, 2003). The addition of Cdc25B also failed to coactivate the AF-1 mutant, indicating that Cdc25B appears to be acting through the AF-1 domain of AR (Ngan *et al.*, 2003). On the other hand, the AF-2 mutant significantly loses dependency of R1881 but nonetheless maintains high transcriptional activity, and the addition of Cdc25B only potentiates reporter activity about 1.5-fold compared with ~6-fold when wild-type AR was used (Ngan *et al.*, 2003). Therefore, the results of this analysis indicate that Cdc25B mediates coactivation of AR-dependent transactivation of target genes chiefly through the AF-1 domain, with the AF-2 domain conferring hormone-dependence and attainment of maximal transcriptional activity.

C. DIRECT INTERACTION OF Cdc25B WITH AR

To determine if the coactivation of AR by Cdc25B is mediated via interaction of these proteins, we performed mammalian two-hybrid assays. We ascertained that we would obtain similar results to those obtained with ER (Ma *et al.*, 2001): a significant interaction was observed only when Cdc25B–VP16 was coexpressed with ER and not when VP16 and ER were coexpressed (Ngan *et al.*, 2003). This finding suggests that Cdc25B specifically interacts with AR *in vivo*. To examine if Cdc25B is directly interacting with AR, we performed GST pull-down assays and observed that in agreement with results obtained with ER (Ma *et al.*, 2001), a significant amount of GST–Cdc25B can pull down AR either in the presence or absence of hormone and there was no interaction of AR with GST alone (Ngan et al., 2003).

Next, we dissected the interaction of Cdc25B with AR further and tested the ability of the AF-1 and AF-2 mutants of AR to associate with Cdc25B by immunoprecipitation studies using a Cdc25B antibody. We observed that both the Cdc25B antibody and GST–Cdc25B can pull down ^{35}S-methionine (labeled wild-type AR) and the two mutants AF-1 and AF-2 (Ngan et al., 2003). Consistent with the requirement of both AF-1 and AF-2 to confer full Cdc25B-mediated coactivation of AR function, these results cement the importance of the direct interaction between Cdc25B and AR.

Next, we ascertained if Cdc25B could also recruit HATs to act in synergy to potentiate an AR transactivation function by cotransfecting Cdc25B with PCAF and CBP (Ngan et al., 2003). We found that there exists a significant synergistic enhancement of reporter activity, especially with CBP over ER-dependent transcription in the presence of estradiol without exogenously added coactivators (Cdc25B alone: ∼2.3-fold; CBP alone: ∼3.5-fold; Cdc25B + CBP: ∼15-fold; P/CAF alone: ∼3-fold; Cdc25B + PCAF: ∼8.3-fold). In summary, these results provide a possible mechanism in which the increased expression of Cdc25B in prostate cancer cells can enhance the expression of AR target genes. Up-regulation of Cdc25B can facilitate the AF-2 domain of AR to recruit Cdc25B, thereby shifting the equilibrium to favor the formation of AR–Cdc25B and/or AR–Cdc25B–CBP to stabilize the transcription machinery and induce higher expression of AR target genes to precipitate the mitogenic effects of androgen. Furthermore, the finding that Cdc25B up-regulation is associated at the period of transition from androgen dependence to independence suggests that, in addition to perhaps contributing to this transition, Cdc25B can also perturb checkpoint functions, deregulate DNA synthesis, and promote cell cycle progression. These actions can induce cellular transformation by activating other positive effectors of the cell cycle like the cyclin–CDK complexes.

VI. CONCLUSIONS

The Cdc25 family of dual-specificity phosphatases has been traditionally thought to function exclusively in the cell cycle to activate CDKs by dephosphorylating two inhibitory phosphate groups on threonine 14 and tyrosine 15 (Draetta and Eckstein, 1997; Galaktionov and Beach, 1991). This chapter has presented evidence that in addition to their known cell cycle functions, Cdc25B and, to a similar extent, Cdc25A can function as SRCs to further enable the transcriptional potential of a number of steroid receptors, including ER, GR, and AR (Ma et al., 2001; Ngan et al., 2003). This is an intriguing finding as it further extends our understanding on how

Cdc25B uses its two seemingly disparate functions to promote cell cycle progression and predispose cells toward neoplasia.

The normal cell cycle is regulated by a tight balance of positive effectors promoting forward progression of the cell cycle and negative effectors restraining forward movement. The shift in equilibrium to favor the gain in function of positive effectors or the loss of function of negative effectors, which frequently function as checkpoint proteins and antagonists to the positive effectors, can lead to deregulated growth and neoplasia. It is not surprising that as positive effectors of the cell cycle, members of the Cdc25 family have been overexpressed in cell lines and primary tumors (Galaktionov and Beach, 1991). Indeed, other positive effectors like cyclin D1, cyclin E, cyclin A, and CKD2 have similarly been implicated in promoting deregulated and abnormal growth as well as the formation of tumors (Bortner and Rosenberg, 1995, 1997; Wang et al., 1994). Thus, the ability to understand how these positive effectors promote cell growth is instrumental toward finding specific and efficacious means of combating cancers caused by the deregulated expression of these effectors.

We validated the role of Cdc25B in inducing precocious alvelolar hyperplastic transgenic mammary glands reminiscent of pregnant glands. The severity of the hyperplasia progresses with age, and we also found that Cdc25B can elicit increased cellular proliferation and decreased apoptosis at involution (Ma et al., 1999). Though no tumors developed during the lifetime of the Cdc25B transgenic mice (Ma et al., 1999), Cdc25B was able to synergize with activated ras to shorten tumor latency in the bitransgenic mice expressing both these genes.

The finding that Cdc25B can up-regulate the expression of cyclin D1, an ER target gene, led us to believe that Cdc25B may play a role in the ER signaling pathway. Moreover, the observation that Cdc25B can induce the expression of Lactoferrin, an ER-mediated gene, in addition to cyclin D1, strongly hinted to us the potential coactivation function of Cdc25B (Ma et al., 2001). We have reported that Cdc25B, when coexpressed with ER, can activate an ERE-dependent reporter a further 2- to 6-fold if MCF-7 or HeLa cells respectively were used in the presence of the ER agonist ligand 17β-estradiol (Ma et al., 2001). This coactivation is not unique to ER and can be extended to PR, GR, and AR (Ma et al., 2001). There appears, however, to be preferential coactivation as Cdc25B only weakly potentiates RAR-dependent reporter transcription and may reflect the propensity of Cdc25B to coactivate the class I receptors. That Cdc25B can coactivate ER and AR, specifically in the presence of their cognate ligands, is further demonstrated as Cdc25B cannot coactivate when their respective antagonists tamoxifen and CPA were added (Ma et al., 2001; Ngan et al., 2003).

Because both ER and AR play such important roles in breast and prostate cancers respectively, we concentrated our efforts on understanding how Cdc25B can mediate both ER-and AR-mediated transactivation of their target genes. In this chapter, we presented evidence that Cdc25B acts as a steroid receptor coactivator and can coactivate both ER and AR as well as possibly potentiate the mitogenic effects of these receptors in breast and prostate cancers respectively (Ma *et al.*, 2001; Ngan *et al.*, 2003).

We demonstrated the ability of Cdc25B to interact with both ER and AR via three independent assays: mammalian two-hybrid, coimmunoprecipitation, and GST pull-down. In GST pull-down assays, Cdc25B can interact with both ER and AR in a hormone-independent fashion because the proteins were *in vitro*-translated and proteins normally associated with the receptors *in vivo* are not present. When hormone is added *in vivo*, it binds to the receptor to dissociate heat shock proteins, translocates the receptor to the nucleus, and also exposes binding sites for Cdc25B to bind and coactivate only in a hormone-dependent fashion.

We also mapped the interaction sites on both ER and AR for Cdc25B. Cdc25B interacts predominantly with the AF-2 domain of ER (AF-2 is found in the ligand binding domain) and less with AF-1 (Ma *et al.*, 2001). With AR, Cdc25B interacts predominantly with AF-1, and full ligand dependence is accorded by the presence of the AF-2 domain of AR (Ngan *et al.*, 2003). In both receptors, for full hormone-dependent coactivation by Cdc25B, AF-1 and AF-2 domains must be present. By mapping reciprocal domains on Cdc25B that interact with ER, we found that these interaction domains reside in the middle of the molecule and also in the catalytic region that is highly conserved in the Cdc25 family of proteins (Ma *et al.*, 2001). Surprisingly, a putative NR box (LXXLL motif) found in Cdc25B does not interact with steroid receptors (Ma *et al.*, 2001). Indeed, consistent with this observation, an L29A mutant lacking the LXXLL motif retained the ability to interact with steroid receptors.

In addition to interacting with steroid receptors and coactivating steroid dependent transactivation, Cdc25B as an SRC can remodel the chromatin environment either through its own HAT activity or by recruiting other HATs to facilitate access of general transcription factors (GTFs) to the target promoter (Freedman, 1999; McKenna *et al.*, 1999, 1998). Cdc25B is found to interact with HATs like PCAF and can synergize with HATs to further coactivate ER and AR transcription (Ma *et al.*, 2001; Ngan *et al.*, 2003). By performing HAT assays, we found that Cdc25B can modulate the HAT activity of PCAF to increase H3 acetylation (Ma *et al.*, 2001). In addition, Cdc25B may also stabilize the preinitiation complex through direct or indirect interactions with GTFs, and consistent with observation, Cdc25B can increase PR-dependent transcription of a chromatin template *in vitro* (Ma *et al.*, 2001). By interacting with steroid receptors and recruiting HATs, Cdc25B can remodel the chromatin environment to permit easy access for and stabilization of the general transcriptional machinery to coactivate target gene expression.

Given that Cdc25B is a cell cycle regulatory protein, we initially surmised that Cdc25B may be coactivating through its cell cycle role by activating CDK pathways through its phosphatase role since it is well known that CDK2 can stimulate ER transactivation via ligand-independent means (Rogatsky et al., 1999; Trowbridge et al., 1997). Our present report, however, provides evidence contrary to this observation. By using roscovitine, which inactivates CDK pathways, we found that Cdc25B can still coactivate ER transcription to levels similar to those in the absence of roscovitine. Second, the absence of the phosphatase domain does not affect the ability of these phosphatase domain mutants to enhance ER and GR transactivation to levels similar to those of Cdc25B wild type. Furthermore, the ability of Cdc25B to enhance cell-free transcription of a PR-chromatin template suggests that the phosphatase domain of Cdc25B is not necessary for coactivation. Finally, we ruled out the notion that Cdc25B is coactivating through cyclin D1 by performing transfections in cyclin $D1^{-/-}$ fibroblasts and found that we obtained similar levels of transactivation regardless whether cyclin $D1^{-/-}$ or wild-type fibroblasts were used. Taken together, these results indicate that Cdc25B is coactivating independent of its cell cycle regulatory role.

In hormone responsive tissues, such as in tissues of the breast and prostate, it is conceivable to suggest that cells with genetic alterations that favor hormone stimulation will attain a growth advantage and contribute to the presentation of preneoplastic lesions. This observation is consistent with the view that a majority of breast cancer cells (~70%) express ER while only about 10 to 15% of normal mammary epithelial cells express ER (Castles and Fuqua, 1996; Petersen et al., 1987). Therefore, these ER positive cells can become more vulnerable to transformation as positive effectors like Cdc25B coactivate to increase the expression of ER target genes. Indeed, Cdc25B joins the list of steroid receptor coactivators like SRC-1, SRC-2, and SRC-3, which have been found to be overexpressed in cancers like those of the prostate (Gnanapragasam et al., 2001; Gregory et al., 2001). Cdc25B may be more versatile as it can also modulate CDKs to allow progression of the cell cycle as well as initiate DNA replication and mitosis to precipitate deregulated cell cycle growth. Thus, these two seemingly diverse roles of Cdc25B can synergize to induce epithelial cell proliferation and precipitate mammary tumorigenesis. In a similar vein, it is likely that Cdc25B is promoting prostate growth through the activation of AR-mediated genes. In both cases, it is compelling to speculate that Cdc25B may potentiate the transformation of tumors from hormone-dependence to hormone-independence by up-regulating the expression of genes responsible for this process. This hypothesis requires further analyses, but evidence in favor of this hypothesis is derived from the finding that Cdc25B is predominantly overexpressed in poorly differentiated, high-grade prostate tumors that may be at the stage of transiting from hormone-dependence to hormone-independence. In conclusion, this chapter has presented a novel function of Cdc25B in addition to its traditional cell cycle role. The implication of this finding will have far-reaching consequences in the understanding of breast and prostate cancers and allow these effectors to be recognized as efficacious in combatting these cancers.

REFERENCES

Altucci, L., Addeo, R., Cicatiello, L., Dauvois, S., Parker, M. G., Truss, M., Beato, M., Sica, V., Bresciani, F., and Weisz, A. (1996). 17β-estradiol induces cyclin D1 gene transcription, p36D1–p34cdk4 complex activation, and p105Rb phosphorylation during mitogenic stimulation of G(1)-arrested human breast cancer cells. *Oncogene* **12,** 2315–2324.

Baldin, V., Cans, C., Superti-Furga, G., and Ducommun, B. (1997). Alternative splicing of the human Cdc25B tyrosine phosphatase. Possible implications for growth control? *Oncogene* **14,** 2485–2495.

Barnes, D. M. (1997). Cyclin D1 in mammary carcinoma. *J. Pathology* **181,** 267–269.

Blomberg, I., and Hoffmann, I. (1999). Ectopic expression of Cdc25A accelerates the G(1)/S transition and leads to premature activation of cyclin E- and cyclin A-dependent kinases. *Mol. Cell. Biol.* **19,** 6183–6194.

Booher, R. N., Deshaies, R. J., and Kirschner, M. W. (1993). Properties of *Saccharomyces cerevisiae* wee-1 and its differential regulation of p34Cdc28 in response to G1 and G2 cyclins. *EMBO J.* **12,** 3417–3426.

Bortner, D. M., and Rosenberg, M. P. (1995). Overexpression of cyclin A in the mammary glands of transgenic mice results in the induction of nuclear abnormalities and increased apoptosis. *Cell Growth Different.* **6,** 1579–1589.

Bortner, D. M., and Rosenberg, M. P. (1997). Induction of mammary gland hyperplasia and carcinomas in transgenic mice expressing human cyclin. E. *Mol. Cell. Biol.* **17,** 453–459.

Castles, C. G., and Fuqua, S. A. W. (1996). Alterations within the estrogen receptor in breast cancer. *In* "Hormone-Dependent Cancer" (J. R. Pasqualini and B. S. Katznellenbogen, Eds.). Marcel Dekker, New York.

Conklin, D. S., Galaktionov, K., and Beach, D. (1995). 14-3-3 proteins associate with Cdc25 phosphatases. *Proc. Natl. Acad. Sci. USA* **92,** 7892–7896.

Donzelli, M., and Draetta, G. F. (2003). Regulating mammalian checkpoints through Cdc25 inactivation. *EMBO Rep.* **4,** 671–677.

Donzelli, M., Squatrito, M., Ganoth, D., Hershko, A., Pagano, M., and Draetta, G. F. (2002). Dual mode of degradation of Cdc25 A phosphatase. *EMBO J.* **21,** 4875–4884.

Draetta, G., and Eckstein, J. (1997). Cdc25 protein phosphatases in cell proliferation. *Biochim. Biophys. Acta* **1332,** M53–M63.

Draetta, G. F. (1994). Mammalian G1 cyclins. *Curr. Opin. Cell. Biol.* **6,** 842–846.

Dunphy, W. G., and Kumagai, A. (1991). The Cdc25 protein contains an intrinsic phosphatase activity. *Cell* **67,** 189–196.

Elledge, S. J. (1996). Cell cycle checkpoints: Preventing an identity crisis. *Science* **274,** 1664–1672.

Fisher, R. P., and Morgan, D. O. (1994). A novel cycli associates with MO15/CDK7 to form the CDK-activating kinase. *Cell* **78,** 713–724.

Freedman, L. P. (1999). Increasing the complexity of coactivation in nuclear receptor signaling. *Cell* **97,** 5–8.

Galaktionov, K., and Beach, D. (1991). Specific activation of Cdc25 tyrosine phosphatases by B-type cyclins: Evidence for multiple roles of mitotic cyclins. *Cell* **67,** 1181–1194.

Galaktionov, K., Chen, X., and Beach, D. (1996). Cdc25 cell cycle phosphatase as a target of c-myc. *Nature* **382,** 511–517.

Galaktionov, K., Jessus, C., and Beach, D. (1995a). Raf-1 interaction with Cdc25 phosphatase ties mitogenic signal transduction to cell cycle activation. *Genes Develop.* **9**, 1046–1058.

Galaktionov, K., Lee, A. K., Eckstein, J., Draetta, G., Meckler, J., Loda, M., and Beach, D. (1995b). Cdc25 phosphatases as potential human oncogenes. *Science* **269**, 1575–1577.

Gasparotto, D., Maestro, R., Piccinin, S., Vukosavljevic, T., Barzan, L., Sulfaro, S., and Boiocchi, M. (1997). Overexpression of Cdc25A and Cdc25B in head and neck cancers. *Cancer Res.* **57**, 2366–2368.

Glover, D. M., Hagan, I. M., and Tavares, A. A. (1998). Polo-like kinases: A team that plays throughout mitosis. *Genes Develop.* **12**, 3777–3787.

Gnanapragasam, V. J., Leung, H. Y., Pulimood, A. S., Neal, D. E., and Robson, C. N. (2001). Expression of RAC-3, a steroid hormone receptor coactivator in prostate cancer. *Br. J. Cancer* **85**, 1928–1936.

Gray-Bablin, J., Zalvide, J., Fox, M. P., Knickerbocker, C. J., DeCaprio, J. A., and Keyomarsi, K. (1996). Cyclin E, a redundant cyclin in breast cancer. *Proc. Natl. Acad. Sci. USA* **93**, 15215–15220.

Gregory, C. W., He, B., Johnson, R. T., Ford, O. H., Mohler, J. L., French, F. S., and Wilson, E. M. (2001). A mechanism for androgen receptor-mediated prostate cancer recurrence after androgen deprivation therapy. *Cancer Res.* **61**, 4315–4319.

Harper, J. W., Elledge, S. J., Keyomarsi, K., Dynlacht, B., Tsa, i. L. H., Zhang, P., Dobrowolski, S., Bai, C., Connell-Crowley, L., and Swindell, E. E. A. (1995). Inhibition of cyclin-dependent kinases by p21. *Mol. Biol. Cell* **6**, 387–400.

Hernandez, S., Hernandez, L., Bea, S., Cazorla, M., Fernandez, P. L., Nadal, A., Muntane, J., Mallofre, C., Montserrat, E., Cardesa, A., and Campo, E. (1998). Cdc25 cell cycle-activating phosphatases and c-myc expression in human non-Hodgkin's lymphomas. *Cancer Res.* **58**, 1762–1767.

Hirai, H., Roussel, M. F., Kato, J. Y., Ashmun, R. A., and Sherr, C. J. (1995). Novel INK4 proteins, p19 and p18, are specific inhibitors of the cyclin D-dependent kinases CDK4 and CDK6. *Mol. Cell. Biol.* **15**, 2672–2681.

Hoffmann, I., Clarke, P. R., Marcote, M. J., Karsenti, E., and Draetta, G. (1993). Phosphorylation and activation of human Cdc25C by Cdc2–cyclin B and its involvement in the self-amplification of MPF at mitosis. *EMBO J.* **12**, 53–63.

Hoffmann, I., Draetta, G., and Karsenti, E. (1994). Activation of the phosphatase activity of human Cdc25A by a CDK2–cyclin E dependent phosphorylation at the G1/S transition. *EMBO J.* **13**, 4302–4310.

Hu, Y. C., Lam, K. Y., Law, S., Wong, J., and Srivastava, G. (2001). Identification of differentially expressed genes in esophageal squamous cell carcinoma (ESCC) by cDNA expression array: Overexpression of fra-1, neogenin, id-1, and Cdc25B genes in ESCC. *Clin. Cancer Res.* **7**, 2213–2221.

Hui, R., Cornish, A. L., McClelland, R. A., Robertson, J. F., Blamey, R. W., Musgrove, E. A., Nicholson, R. I., and Sutherland, R. L. (1996). Cyclin D1 and estrogen receptor messenger RNA levels are positively correlated in primary breast cancer. *Clin. Cancer Res.* **2**, 923–928.

Hunter, T., and Pines, J. (1994). Cyclins and cancer II: Cyclin D and CDK inhibitors come of age. *Cell* **79**, 573–582.

Igarashi, M., Nagata, A., Jinno, S., Suto, K., and Okayama, H. (1991). Wee-1(+)-like gene in human cells. *Nature* **353**, 80–83.

Jinno, S., Suto, K., Nagata, A., Igarashi, M., Kanaoka, Y., Nojima, H., and Okayama, H. (1994). Cdc25A is a novel phosphatase functioning early in the cell cycle. *EMBO J.* **13**, 1549–1556.

Karlsson, C., Katich, S., Hagting, A., Hoffmann, I., and Pines, J. (1999). Cdc25B and Cdc25C differ markedly in their properties as initiators of mitosis. *J. Cell. Biol.* **146**, 573–584.

Kenny, F. S., Hui, R., Musgrove, E. A., Gee, J. M., Blamey, R. W., Nicholson, R. I., Sutherland, R. L., and Robertson, J. F. (1999). Overexpression of cyclin D1 messenger

RNA predicts for poor prognosis in estrogen receptor-positive breast cancer. *Clin. Cancer Res.* **5,** 2069–2076.
Keyomarsi, K., and Pardee, A. B. (1993). Redundant cyclin overexpression and gene amplification in breast cancer cells. *Proc. Natl. Acad. Sci. USA* **90,** 1112–1116.
Kudo, Y., Yasui, W., Ue, T., Yamamoto, S., Yokozaki, H., Nikai, H., and Tahara, E. (1997). Overexpression of cyclin-dependent kinase-activating Cdc25B phosphatase in human gastric carcinomas. *J. Cancer Res.* **88,** 947–952.
Kumagai, A., Yakowec, P. S., and Dunphy, W. G. (1998). 14-3-3 proteins act as negative regulators of the mitotic inducer Cdc25 in *Xenopus* egg extracts. *Mol. Biol. Cell* **9,** 345–354.
Lombardi, D. P., Geradts, J., Foley, J. F., Chiao, C., Lamb, P. W., and Barrett, J. C. (1999). *Cancer Res.* **59,** 5724–5731.
Lopez-Girona, A., Furnari, B., Mondesert, O., and Russell, P. (1999). Nuclear localization of Cdc25 is regulated by DNA damage and a 14-3-3 protein. *Nature* **397,** 172–175.
Ma, Z. Q., Chua, S. S., DeMayo, F. J., and Tsai, S. Y. (1999). Induction of mammary gland hyperplasia in transgenic mice overexpressing human Cdc25B. *Oncogene* **18,** 4564–4576.
Ma, Z. Q., Liu, Z., Ngan, E. S., and Tsai, S. Y. (2001). Cdc25B functions as a novel coactivator for the steroid receptors. *Mol. Cell. Biol.* **21,** 8056–8067.
Matsuoka, S., Huang, M., and Elledge, S. J. (1998). Linkage of ATM to cell cycle regulation by the Chk2 protein kinase. *Science* **282,** 1893–1897.
McKenna, N. J., Lanz, R. B., and O'Malley, B. W. (1999). Nuclear receptor coregulators: Cellular and molecular biology. *Endocr. Rev.* **20,** 321–344.
McKenna, N. J., Nawaz, Z., Tsai, S. Y., Tsai, M. J., and O'Malley, B. W. (1998). Distinct steady-state nuclear receptor coregulator complexes exist *in vivo. Proc. Natl. Acad. Sci. USA* **95,** 11697–11702.
Meijer, L., Borgne, A., Mulner, O., Chong, J. P., Blow, J. J., Inagaki, N., Inagaki, M., Delcros, J. G., and Moulinoux, J. P. (1997). Biochemical and cellular effects of roscovitine, a potent and selective inhibitor of the cyclin-dependent kinases Cdc2, CDK2 and CDK5. *Eur. J. Biochem.* **243,** 527–536.
Millar, J. B., McGowan, C. H., Lenaers, G., Jones, R., and Russell, P. (1991). p80Cdc25 mitotic inducer is the tyrosine phosphatase that activates p34Cdc2 kinase in fission yeast. *EMBO J.* **10,** 4301–4309.
Moreno, S., and Nurse, P. (1991). Clues to action of Cdc25 protein. *Nature* **351,** 194.
Morgan, D. O. (1996). Under arrest at atomic resolution. *Nature* **382,** 295–296.
Muller, P. R., Coleman, T. R., Kumagai, A., and Dunphy, W. G. (1995). Myt-1: A membrane-associated inhibitory kinase that phosphorylates Cdc2 on both threonine 14 and tyrosine 15. *Science* **670,** 86–90.
Neuman, E., Ladha, M. H., Lin, N., Upton, T. M., Miller, S. J., DiRenzo, J., Pestell, R. G., Hinds, P. W., Dowdy, S. F., Brown, M., and Ewen, M. E. (1997). Cyclin D1 stimulation of estrogen receptor transcriptional activity independent of CDK4. *Mol. Cell. Biol.* **17,** 5338–5347.
Ngan, E. S., Hashimoto, Y., Ma, Z. Q., Tsai, M. J., and Tsai, S. Y. (2003). Overexpression of Cdc25B, an androgen receptor coactivator, in prostate cancer. *Oncogene* **22,** 734–739.
Nigg, E. A. (1998). Polo-like kinases: Positive regulators of cell division from start to finish. *Curr. Opin. Cell. Biol.* **10,** 776–783.
Parker, S. L., Tong, T., Bolden, S., and Wingo, P. A. (1997). Cancer statistics, 1997, CA. *Cancer J. Clin.* **47,** 5–27.
Pattengale, P. K., Stewart, T. A., Leder, A., Sinn, E., Muller, W., Tepler, I., Schmidt, E., and Leder, P. (1989). Animal models of human disease. Pathology and molecular biology of spontaneous neoplasms occurring in transgenic mice carrying and expressing activated cellular oncogenes. *Amer. J. Path.* **135,** 39–61.

Peng, C. Y., Graves, P. R., Ogg, S., Thoma, R. S., Byrnes, M. J., III, Wu, Z., Stephenson, M. T., and Piwnica-Worms, H. (1998). C-TAK1 protein kinase phosphorylates human Cdc25C on serine 216 and promotes 14-3-3 protein binding. *Cell Growth Differen.* **9**, 197–208.

Petersen, O. W., Hoyer, P. E., and van Deurs, B. (1987). Frequency and distribution of estrogen receptor-positive cells in normal, nonlactating human breast tissue. *Cancer Res.* **47**, 5748–5751.

Pines, J. (1993). Cyclins and cyclin-dependent kinases: Take your partners. *Trends Biochem. Sci.* **18**, 195–197.

Pines, J. (1999). Four-dimensional control of the cell cycle. *Nat. Cell. Biol.* **1**, E73–E79.

Poon, R. Y., and Hunter, T. (1995). Dephosphorylation of CDK2 Thr160 by the cyclin-dependent kinase-interacting phosphatase KAP in the absence of cyclin. *Science* **270**, 90–93.

Prall, O. W., Rogan, E. M., Musgrove, E. A., Watts, C. K., and Sutherland, R. L. (1998). c-Myc or cyclin D1 mimics estrogen effects on cyclin E-CDK2 activation and cell cycle re-entry. *Mol. Cell. Biol.* **18**, 4499–4508.

Rogatsky, I., Trowbridge, J. M., and Garabedian, M. J. (1999). Potentiation of human estrogen receptor alpha transcriptional activation through phosphorylation of serines 104 and 106 by the cyclin A-CDK2 complex. *J. Biol. Chem.* **274**, 22296–22302.

Russell, P., and Nurse, P. (1986). Cdc25+ functions as an inducer in the mitotic control of fission yeast. *Cell* **45**, 145–153.

Russell, P., and Nurse, P. (1987). The mitotic inducer nim-1+ functions in a regulatory network of protein kinase homologs controlling the initiation of mitosis. *Cell* **49**, 569–576.

Sabbah, M., Courilleau, D., Mester, J., and Redeuilh, G. (1999). Estrogen induction of the cyclin D1 promoter: Involvement of a cAMP response-like element. *Proc. Natl. Acad. Sci. USA* **96**, 11217–11222.

Sadhu, K., Reed, S. I., Richardson, H., and Russell, P. (1990). Human homolog of fission yeast Cdc25 mitotic inducer is predominantly expressed in G2. *Proc. Natl. Acad. Sci. USA* **87**, 5139–5143.

Sanchez, Y., Wong, C., Thoma, R. S., Richman, R., Wu, Z., Piwnica-Worms, H., and Elledge, S. J. (1997). Conservation of the Chk1 checkpoint pathway in mammals: Linkage of DNA damage to CDK regulation through Cdc25. *Science* **277**, 1497–1501.

Sasaki, H., Yukiue, H., Kobayashi, Y., Tanahashi, M., Moriyama, S., Nakashima, Y., Fukai, I., Kiriyama, M., Yamakawa, Y., and Fujii, Y. (2001). Expression of the Cdc25B gene as a prognosis marker in nonsmall cell lung cancer. *Cancer L.* **173**, 187–192.

Serrano, M., Hannon, G. J., and Beach, D. (1993). A new regulatory motif in cell cycle control causing specific inhibition of cyclin D/CDK4. *Nature* **366**, 704–707.

Sexl, V., Diehl, J. A., Sherr, C. J., Ashmun, R., Beach, D., and Roussel, M. F. (1999). A rate limiting function of Cdc25A for S phase entry inversely correlates with tyrosine dephosphorylation of CDK2. *Oncogene* **18**, 573–582.

Sherr, C. (1996). Cancer cell cycles. *Science* **274**, 1672–1677.

Sherr, C. J., and Roberts, J. M. (1995). Inhibitors and mammalia G1 cyclin-dependent kinases. *Genes Develop.* **9**, 1149–1163.

Shiloh, Y. (2001). ATM and ATR: Networking cellular responses to DNA damage. *Curr. Opin. Genet. Develop.* **11**, 71–77.

Shiloh, Y. (2003). ATM and related protein kinases: Safeguarding genome integrity. *Nature Rev. Canc.* **3**, 155–168.

Sicinski, P., Donaher, J. L., Parker, S. B., Li, T., Fazeli, A., Gardner, H., Haslam, S. Z., Bronson, R. T., Elledge, S. J., and Weinberg, R. A. (1995). Cyclin D1 provides a link between development and oncogenesis in the retina and breast. *Cell* **82**, 621–630.

Sinn, E., Muller, W., Pattengale, P., Tepler, I., Wallace, R., and Leder, P. (1987). Coexpression of MMTV–v-Ha-ras and MMTV-c-myc genes in transgenic mice: Synergistic action of oncogenes *in vivo*. *Cell* **49**, 465–475.

Takemasa, I., Yamamoto, H., Sekimoto, M., Ohue, M., Noura, S., Miyake, Y., Matsumoto, T., Aihara, T., Tomita, N., Tamaki, Y., Sakita, I., Kikkawa, N., Matsuura, N., Shiozaki, H., and Monden, M. (2000). Overexpression of Cdc25B phosphatase as a novel marker of poor prognosis of human colorectal carcinoma. *Cancer Res.* **60,** 3043–3050.

Trowbridge, J. M., Rogatsky, I., and Garabedian, M. J. (1997). Regulation of estrogen receptor transcriptional enhancement by the cyclin A/CDK2 complex. *Proc. Natl. Acad. Sci. USA* **94,** 10132–10137.

Vermeulen, K., Van Bockstaele, D. R., and Berneman, Z. N. (2003). The cell cycle: A review of regulation, deregulation, and therapeutic targets in cancer. *Cell Prolif.* **36,** 131–149.

Vigo, E., Muller, H., Prosperini, E., Hateboer, G., Cartwright, P., Moroni, M. C., and Helin, K. (1999). Cdc25A phosphatase is a target of E2F and is required for efficient E2F-induced S phase. *Mol. Cell. Biol.* **19,** 6379–6395.

Wang, T. C., Cardiff, R. D., Zukerberg, L., Lees, E., Arnold, A., and Schmidt, E. V. (1994). Mammary hyperplasia and carcinoma in MMTV–cyclin D1 transgenic mice. *Nature* **369,** 669–671.

Wu, L., Yee, A., Liu, L., Carbonaro-Hall, D., Venkatesan, N., Tolo, V. T., and Hall, F. L. (1994). Molecular cloning of the human CAK1 gene encoding a cyclin-dependent kinase-activating kinase. *Oncogene* **9,** 2089–2096.

Wu, W., Fan, Y. H., Kemp, B. L., Walsh, G., and Mao, L. (1998). Overexpression of Cdc25A and Cdc25B is frequent in primary nonsmall cell lung cancer but is not associated with overexpression of c-myc. *Cancer Res.* **58,** 4082–4085.

Xiong, Y., Hannon, G. J., Zhang, H., Casso, D., Kobayashi, R., and Beach, D. (1993). p21 is a universal inhibitor of cyclin kinases. *Nature* **366,** 701–704.

Xu, X., and Burke, S. P. (1996). Roles of active site residues and the NH2-terminal domain in the catalysis and substrate binding of human Cdc25. *J. Biol. Chem.* **271,** 5118–5124.

Yao, Y., Slosberg, E. D., Wang, L., Hibshoosh, H., Zhang, Y. J., Xing, W. Q., Santella, R. M., and Weinstein, I. B. (1999). Increased susceptibility to carcinogen-induced mammary tumors in MMTV–Cdc25B transgenic mice. *Oncogene* **18,** 5159–5166.

Zhou, B. B., and Elledge, S. J. (2000). The DNA damage response: Putting checkpoints in perspective. *Nature* **408,** 433–439.

Zwijsen, R. M., Buckle, R. S., Hijmans, E. M., Loomans, C. J., and Bernards, R. (1998). Ligand-independent recruitment of steroid receptor coactivators to estrogen receptor by cyclin D1. *Genes Dev.* **12,** 3488–3498.

Zwijsen, R. M., Wientjens, E., Klompmaker, R., van der Sman, J., Bernards, R., and Michalides, R. J. (1997). CDK-independent activation of estrogen receptor by cyclin D1. *Cell* **88,** 405–415.

9

Vitamin D Receptor–DNA Interactions

Paul L. Shaffer and Daniel T. Gewirth

Department of Biochemistry
Duke University Medical Center
Durham, North Carolina 27710

I. Introduction
II. VDR–DR3 Binding
   A. DNA Binding and Specificity
   B. DR3 Binding
   C. Response Element Discrimination
III. Alternative Response Elements
IV. RXR–VDR Formation
V. Conclusions
   References

The vitamin D receptor (VDR) is a member of the steroid and nuclear hormone receptor superfamily of eukaryotic transcription factors and binds target DNA, or response elements, as a homodimer or heterodimer with the 9-cis retinoid X receptor (RXR). In this chapter, we survey the current understanding of VDR–DNA interactions, emphasizing recent structural insights. We highlight the stereochemical interactions that dictate DNA binding and hexameric half-site sequence affinity as well as the protein–protein interactions that account for preferential binding to a direct repeat of half-sites with three base pairs of spacer DNA (DR3). In addition, we review alternative response element arrangements other than those with DR3. Finally, the chapter discusses the VDR DNA binding domain (DBD) and suggests that it violates classical canons because it does not heterodimerize with the RXR DBD. This unique behavior of VDR is considered in light of recent results demonstrating the formation of VDR DBD–DNA and DR3 DBD–DNA complexes with RXR using a mutant VDR protomer. © 2004 Elsevier Inc.

I. INTRODUCTION

The vitamin D receptor (VDR) (Baker et al., 1998) is a ligand-activated transcription factor that plays a central role in calcium homeostasis. VDR has been implicated in regulating diverse biologic functions, including cellular proliferation and differentiation (Abe et al., 1981; Bouillon et al., 1995; DeLuca and Zierold, 1998; Feldman et al., 1997). VDR belongs to the steroid and nuclear hormone receptor superfamily whose members include the thyroid hormone receptor (TR), all-trans retinoic acid receptor (RAR), estrogen receptor (ER), glucocorticoid receptor (GR), 9-cis retinoid X receptor (RXR), and more than 150 other receptors (Mangelsdorf and Evans, 1995; Mangelsdorf et al., 1995). Members of this family regulate transcription in response to hydrophobic ligands that diffuse into the cell without assistance from integral membrane proteins. Upon ligand binding, these receptors bind avidly to specific DNA sequences, known as response elements, and modulate the expression of target genes.

Members of the steroid and nuclear hormone superfamily all share a characteristic modular organization consisting of a variable amino-terminal region, a highly conserved DNA binding domain (DBD), a nonconserved hinge domain, and a highly conserved carboxy-terminal ligand binding domain (LBD) (Fig. 1A). The amino-terminal region often contains a ligand-independent transcriptional activation function, but in VDR this region is very short and has not been shown to modulate transcription. The DBD consists of a 66 residue core made up of two zinc-nucleated modules that fold into a unified globular domain (Figs. 1B and C) and bind to hexameric DNA sequences (Luisi et al., 1991, 1994; Schwabe and Rhodes, 1991). The hinge domain varies in both length and sequence between receptors and is now considered to be a carboxy-terminal extension (CTE) of the DBD, as it often imparts additional dimerization and sequence specificity to the conserved DBD core region (Khorasanizadeh and Rastinejad, 2001; Rastinejad et al., 1995a; Zhao et al., 1998). The LBD contains the binding pocket that specifies the high-affinity ligand for each receptor, which for VDR is 1,25-dihydroxyvitamin D_3. Ligand-dependent activation is mediated through the activation function-2 (AF-2) domain of the LBD that changes confirmation upon ligand binding and interacts with coactivators and corepressors (McKenna et al., 1999; Rachez and Freedman, 2001; Rochel et al., 2000, 2001).

With only a few exceptions, the members of the steroid and nuclear receptor (NR) superfamily act as either homodimers or heterodimers. In each dimer, there are two distinct dimerization interfaces. Partner selection, which does not require DNA binding, is accomplished via the association of the LBDs and is ligand dependent (Cheskis and Freedman, 1995, 1996; Cheskis et al., 1995). This primary dimerization event is not sufficient for DNA target recognition, however, and upon binding to the correct bipartite response element, a second dimer interface is also formed between the DBDs (Luisi et al., 1994). In addition to being spatially distinct, experiments with chimeric receptors have led to the conclusion that the two dimerization interfaces are functionally distinct (Mader et al., 1993; Perlmann et al., 1996). In the case of VDR, VDR DBD–TR LBD chimeras have been shown to bind to VDR response elements (VDREs) but activate transcription only in response to thyroid hormone (Miyamoto et al., 2001). These studies have led to the canonical view that the isolated LBDs retain the same dimerization and ligand specificity as the full length molecule and, similarly, that the DBDs retain the dimerization and DNA binding characteristics of the intact receptor (Mader et al., 1993; Perlmann et al., 1993; Towers et al., 1993; Zechel et al., 1994a,b; Khorasanizadeh and Rastinejad, 2001). This separation of partner selection and ligand binding from that of DNA targeting has allowed for the study of these activities in isolated domains of the receptors, which have often proven to be more tractable experimentally.

Despite the large number of receptors that belong to the superfamily, the response elements to which they bind share remarkably similar consensus sequences, which is not surprising given the high degree of conservation in the core DBDs. The steroid receptors bind to response elements whose hexameric half-sites have the consensus sequence $5' - AGAACA - 3'$, and all other receptors bind to $5' - AGGTCA - 3'$ half-sites (the distinguishing bases are in italics) (Klein-Hitpass et al., 1986; Scheidereit et al., 1986). In addition to tissue-specific expression and subcellular localization, diversity is achieved largely by binding as dimers to bipartite elements and by varying the arrangement of the half-sites relative to one another. These arrangements include inverted, everted, or direct repeats (DRs). In the DR series, which includes the response elements for VDR, RXR is a common heterodimeric partner (Mangelsdorf and Evans, 1995; Umesono et al., 1991). Further sequence specificity, in addition to half-site sequence and arrangement, is imparted by varying the number of neutral base pairs separating the repeats. This was formalized as the *1–5 rule*, which specifies the spacer requirement for high-affinity binding of RXR:RXR (DR1), RXR:RAR (DR2), RXR:VDR (DR3), RXR:TR (DR4), and RXR:RAR (DR5) heterodimers (Mangelsdorf and Evans, 1995).

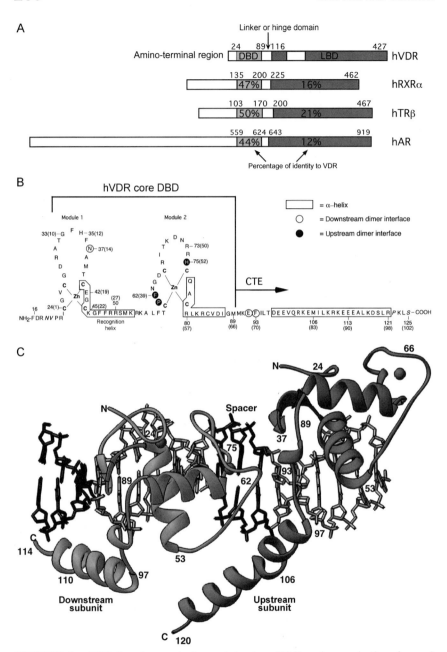

FIGURE 1. DBD domain organization and structure. (A) Domain organization of several common steroid nuclear hormone receptors. Shown are the human vitamin D receptor (hVDR), 9-cis retinoid X receptor alpha (hRXRα), thyroid hormone receptor (hTR), and androgen receptor (hAR). The DNA binding domains (DBDs) are shown in light grey and the ligand binding domains (LBDs) in dark grey, with their percentage sequence identity to VDR noted.

Structures of dimeric nuclear receptor DBDs bound to DR1–4 DNA elements have now been determined and have provided key insights into the structural basis for recognition of sequence, spacing, and orientation of half-sites (Rastinejad, 2001; Rastinejad *et al.*, 1995b; Shaffer and Gewirth, 2002; Zhao *et al.*, 1998, 2000). Because each additional neutral base pair changes the relative position of the DBDs by 3.4 Å along the DNA axis and 36° azimuthally, changes in half-site spacing cause juxtaposition of different regions of the receptor DBDs. From these studies, it was found that whereas the core DBDs all use conserved residues to bind to their consensus half-sites in an identical manner, the geometry of the bipartite response element is read out by unique protein–protein contacts that match the spacing between repeats.

Interestingly, the VDR DBD does not recapitulate the dimerization behavior of the intact full-length receptor. A recent report has shown that the VDR DBD does not form heterodimers with the RXR DBD in the presence of a DR3 response element (Shaffer and Gewirth, 2002). This finding contrasts with earlier reports that showed that RAR–RXR and TR–RXR DBDs readily formed heterodimers in the presence of cognate DNA targets. VDR is thus at odds with the canonical view that the dimerization activities of the LBD and DBD are independent. A pressing issue in the VDR field is thus to account for how the RXR:VDR heterodimer forms the intermolecular associations necessary for spacer discrimination if their DBDs do not associate.

II. VDR–DR3 BINDING

A. DNA BINDING AND SPECIFICITY

To date, all DNA-bound DBD structures, including those of VDR, have shown a nearly identical core fold and mode of binding to their hexameric DNA target. Each DBD has two structural zinc ions that are coordinated by four cysteines and buttress two perpendicular α-helices that pack together in the domain via their hydrophobic faces (Fig. 1C). One of the two helices, termed the "recognition helix," is positioned for insertion into the major groove of the DNA. Side chains of the recognition helix contact the bases of the DNA's major groove and are responsible for sequence specificity. However, response element discrimination, including half-site orientation and spacing, is accomplished through receptor-specific protein–protein

---

The VDR LBD contains an insertion region, shown in white. (B) The human VDR DBD. Sequence numbers are for the corresponding full-length receptor and those in parentheses refer to the common DBD numbering scheme. CTE: C-terminal extension. (C) Overall architecture of the VDR DBD–DR3 DNA complex. The two protomers are shown in grey, the *Zn* atoms are also grey. The hexameric half-site sequences are shown in light grey, and the 5'-flanking base pairs and the spacer are shown in black. Figure adapted from Shaffer and Gewirth (2002).

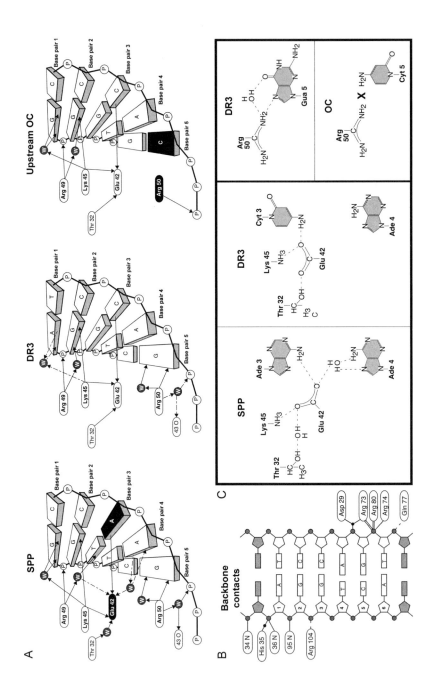

interactions that only form when the dimeric receptor is bound to a cognate response element.

The VDR DBD monomer binds to cognate DNA half-sites in the same manner as all previously characterized members of the superfamily (Shaffer and Gewirth, 2002). Residues from both the first and second zinc modules make conserved contacts with the phosphate backbone, thus positioning the receptor so that the recognition helix is placed in the major groove (Fig. 2B). In this way, specific and discriminating contacts are also made between the receptor side chains and the DNA bases. For the VDR DBD, studies have shown that the minimal DBD extends well past the end of the conserved core (residue 89), and efficient DNA binding requires a construct containing residues to Lys111 (Hsieh *et al.*, 1995, 1999). Biochemical studies have implicated the basic residues from 102 to 104 and from 109 to 111 in the C-terminal extension (CTE) of the VDR DBD as important for DNA binding and transactivation (Hsieh *et al.*, 1995). Although not all of these residues are ordered in the VDR DBD crystal structure, it is likely that they increase affinity by aiding in nonspecific interaction with the DNA backbone and minor groove.

Sequence-specific contacts between VDR and the hexameric half-site are accomplished through side chains of the recognition helix (residues 41 to 53). Biochemical studies have shown that VDR prefers hexameric sequences of 5′-RGTTCA-3′ as opposed to the consensus sequences of 5′-RGGTCA-3′ (R = A or G) (Freedman and Towers, 1991). Further evidence of this divergence from the nuclear receptor consensus sequence was seen in experiments that determined the bipartite sequence preference of RXR:VDR heterodimers. The sequences that were isolated had an upstream RXR site with a preference for an RGGTCA hexamer, while the downstream VDR-occupied site showed a preference for RGTTCA sequences (Colnot *et al.*, 1995; Nishikawa *et al.*, 1994). Similarly, in the naturally occurring osteopontin response element, both half-sites have the sequence GGTTCA, and this element has been shown to have higher affinity for VDR homodimers than the canonical DR3 element (Freedman and Towers, 1991).

The stereochemical basis for these sequence preferences was revealed in structural studies of VDR DBD homodimers bound to osteopontin, osteocalcin, and canonical DR3 response elements (Fig. 2A) (Shaffer and Gewirth, 2002). The presence of a guanine at the third position of the sense

---

FIGURE 2. Protein–DNA contacts observed in naturally occurring (SPP, OC) and canonical DR3 response elements. (A) Base-specific contacts. The DNA is drawn underwound for clarity only. Hydrogen bonds are depicted as arrows and dashed lines denote interactions observed in only one of the two half-sites. (B) Backbone contacts. Dotted lines indicate interactions seen only in the upstream half-sites. (C) Details of the Glu42 and Arg50 hydrogen bonds in selected complexes showing key specifying interactions. Figure adapted from Shaffer and Gewirth (2002).

strand (as opposed to a thymine in the consensus sequence) allows the productive rearrangement of the side chain of Glu42, enabling it to buttress additional water-mediated hydrogen bonds to the DNA bases (Fig. 2C). This altered sequence requirement for VDR versus RXR is also seen in the naturally occurring osteocalcin response element, which does not bind VDR DBD homodimers at physiologic concentrations, but binds avidly to VDR:RXR heterodimers. Studies of the VDR DBD bound to this element showed that the key to lowered binding affinity was at position 5 of the upstream half-site. The substitution of a guanine for the canonical cytosine results in a lack of hydrogen bond acceptors at this position and loss of conserved contacts with Arg50 (Figure 2C). The loss of these contacts in the context of a full-length heterodimer might be less important due to the obligatory colocalization of the RXR and VDR DBDs imposed by the strong heterodimer interface formed by the LBDs.

Together, these results showed that in naturally occurring response elements, the basis for differential affinity is due mainly to the gain or loss of specific hydrogen bonds in the recognition helix–DNA interface. In contrast, an earlier comparison of the structures of the ER–DNA (Schwabe *et al.*, 1993) and GR–DNA (Luisi *et al.*, 1991) complexes with that of a noncognate complex (Gewirth and Sigler, 1995) showed that the mechanism for the near absolute discrimination between steroid sites (AGAACA) and nonsteroid sites (AGGTCA) was a function of the DNA target geometry, which led to the unfavorable incorporation of many more water molecules in the protein–DNA interface. This may point to a more general phenomenon whereby subtle variations in affinity–such as those between VDR and the naturally occurring response elements—are modulated enthalpically via the gain or loss of hydrogen bonds, whereas absolute discrimination is achieved entropically by the capture or liberation of solvent in the interface (Gewirth and Sigler, 1995; Lundback *et al.*, 1994).

B. DR3 BINDING

Initial studies characterized vitamin D response elements (VDREs) as tandem repeats of hexameric half-sites with three intervening neutral base pairs (DR3) (Umesono *et al.*, 1991). These findings were confirmed with *in vitro* response element selection studies that showed a clear preference of VDR homodimers and VDR:RXR heterodimers for DR3 elements (Colnot *et al.*, 1995; Nishikawa *et al.*, 1994). Biochemical (Hsieh *et al.*, 1995; Miyamoto *et al.*, 2001), modeling (Quack *et al.*, 1998; Rastinejad *et al.*, 1995b; Towers *et al.*, 1993) and recent structural studies (Shaffer and Gewirth, 2002) have revealed the molecular basis for the cooperative assembly of VDR homodimers on these elements.

The three base pair spacer of the response element separates the centers of the two half-sites by a fixed rotation ($\sim 40°$) and translation (~ 30.5 Å). The separation in turn specifies the disposition of the two VDR promoters upon DNA binding and juxtaposes specific protein surfaces. Studies identified the region around Phe93 of the CTE and residues Asn37, Phe62, His75 in the core DBD as important to dimerization (Quack *et al.*, 1998; Towers and Freedman, 1998). The importance of these residues was confirmed in the structure of the VDR homodimer. The side chains of Pro61 (P61), Phe62 (F62), and His75 (H75) of the upstream protomer and residues Asn37 (N37), Glu92 (E92), and Phe93 (F93) of the downstream subunit form the inter-DBD interface (Fig. 3). These six residues are nearly invariant among all known VDRs, and the combination of these six residues is unique among hormone receptors, thus underscoring the uniqueness of the homodimer interface.

The primary mechanism of the DBD association across the interfacial gap is via *van der Waals contacts* that produce a smooth, complementary interface. All of the interfacial residues have a strong hydrophobic character, especially the phenylalanines, and removal of these residues from contact with solvent is likely to be a driving force that greatly stabilizes the dimerization. Remarkably, in contrast to the other hormone receptor DBD dimerization interfaces that have been studied structurally (Rastinejad *et al.*, 1995b, 2000; Zhao *et al.*, 1998, 2000), none of the VDR residues involved in dimerization are supported by buttressing or simultaneous contacts with the DNA. Instead, in each protomer, the dimer interface appears to be preformed by internal van der Waals contacts or, in the case of Pro61/Phe62, restricted backbone conformations. In the structures of the

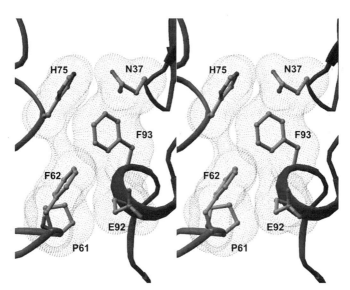

FIGURE 3. Stereo view of the VDR homodimerization interface in a van der Waals surface representation. Figure adapted from Shaffer and Gewirth (2002).

DNA bound RevErb, RXR–TR, and RAR–RXR complexes, those residues participating in dimerization were buttressed by simultaneous DNA contacts. Thus, these dimer interfaces require DNA support to form. The VDR homodimer interface instead resembles the interfaces seen in GR (Luisi *et al.*, 1991) and ER (Schwabe *et al.*, 1993), in which DNA contacts stabilize the overall confirmation of the subunits but do not directly brace any of the interacting residues. This lack of direct DNA support for the VDR homodimer interface is consistent with the likelihood that the tertiary conformation of the interfacial residues is insensitive to DNA binding. This would amount to a partial "prepayment" of the entropic cost of dimerization and may explain the stability of the VDR DBD homodimer relative to the RXR:VDR DBD heterodimer.

C. RESPONSE ELEMENT DISCRIMINATION

With a few exceptions to be discussed subsequently in this chapter, the known naturally occurring VDREs have a DR3 arrangement of half-sites. The basis for high-affinity cooperative assembly on these elements has been discussed. Modeling studies based on the known VDR–DNA structure, however, have also now led to an understanding of the unfavorable interactions that occur when VDR is bound to response elements with disfavored spacers such as DR2 and DR4 (Figs. 4A and 4B) (Shaffer and Gewirth, 2002). When the VDR DBD is modeled on a response element with a spacer length of less than three base pairs, the CTE helix of the downstream protomer clashes with the backbone of the upstream partner core. This steric clash is likely to occur also in the RXR:VDR heterodimer, since RXR replaces VDR as the upstream partner, and its core backbone is likely to be nearly identical to that of the VDR DBD. On the other hand, for response elements longer than three base pairs, modeling predicts that the VDR subunits are too far away to make the direct contacts necessary to form a stable dimer interface. Without the cooperativity that results from protein–protein interactions across the dimer interface, the affinity of VDR for these elements would resemble the much weaker affinity of a monomeric VDR for a single isolated half-site.

III. ALTERNATIVE RESPONSE ELEMENTS

Although most naturally occurring VDREs display a DR3 arrangement of half-sites, other arrangements activate transcription in heterologous reporter constructs. The VDRE in the rat Pit-1 gene consists of what appears to be a DR4 arrangement of half-sites (Rhodes *et al.*, 1993). This element and canonical DR4 sequences have been shown in various studies to modulate transcription in a vitamin D-dependent manner (Quack and Carlberg, 2000).

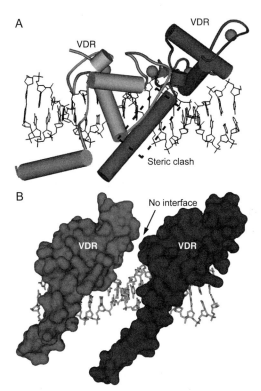

FIGURE 4. Modeling studies based on the VDR DBD structure. (A) Model of the VDR homodimer bound to a DR2 element. A likely steric clash is boxed. (B) Model of the VDR homodimer bound to a DR4 element. Molecular surfaces of the proteins are shown. Figure adapted from Shaffer and Gewirth (2002).

The mechanism for this apparent affinity for a DR4 element is puzzling because modeling of a VDR homodimer on DR4 elements revealed that they could not form a cooperative protein–protein interface. One possible explanation for this conundrum is that perhaps with RXR as an upstream partner, some productive RXR–VDR DBD interactions could be formed that were not anticipated from the VDR structural studies.

The presence of a DR4–VDRE is also puzzling because these elements are classically responsive to the thyroid hormone and the formation of RXR:TR heterodimers. The Pit-1 element, however, appears to be unresponsive to 3,5,3′-triiodothyronine (T_3) ligand in heterologous constructs containing the promoter region of Pit-1 fused to a reporter gene. One proposed reason for this violation of the *1–5* rule is the fact that the half-site sequences in the Pit-1 promoter are 5′-AG*T*TCA-3′, which are the high affinity VDR (but not TR) half-site sequences. Perhaps a DR4 composed of these half-sites is a VDRE, while a DR4 element containing the canonical 5′-AG*G*TCA- 3′-half-sites is a TRE.

A third alternative interpretation of the Pit-1 puzzle is that the VDR DBD is binding to the DR4 Pit-1 element in a DR3 arrangement, with one of the subunits interacting with a suboptimal downstream half-site. This would leave a cognate 5′ AGTTCA-3′ upstream site with a 5′-*GA*G*T*TC-3′ downstream site. As noted by the italizing, positions 1, 3, and 4 match the high affinity VDR consensus sequence. Furthermore, an adenine at the second position does not preclude protein–DNA interactions, since the A could still accept a hydrogen bond at the N7 position as seen in many structures that contain the consensus guanine at this position. Similarly, an adenine in the antisense strand at position 5 could still accept hydrogen bonds to its N7 atom, maintaining many of the contacts seen with the consensus guanine. The sixth base pair forms no direct contacts with the protein in any of the DBD and DNA structures seen to date and in this position shows variability in the downstream half-site of naturally occurring DR3 elements. Therefore, it might be the case that despite the apparent DR4 arrangement of the half-sites, the VDR DBD is still binding in a DR3 geometry to form the cooperative protein–protein interactions inherent in this mode of binding.

Several VDREs with DR6 and IP9 (inverted palindromic hexamers with nine base pairs of neutral DNA) arrangements of half-sites have also been described in the literature (Carlberg *et al.*, 1993; Polly *et al.*, 1996; Schrader *et al.*, 1995, 1997; Xie and Bikle, 1997). These elements would, based on modeling, place the DBDs too far apart to make direct contacts with one another. This raises the question of how VDR would bind cooperatively to these targets. One possible molecular explanation for cooperative heterodimerization is to invoke yet-unknown quaternary intramolecular and intermolecular interactions between the LBD, DBD, and perhaps the linker when bound to these response elements. However, as described for the DR4 Pit-1 element, an alternative interpretation is that the DBDs are binding in a DR3 arrangement with a one high-affinity half-site and one nonconsensus hexamer. In the known DR3 response elements, there is considerable degeneracy in the half-site sequences, with the vast majority differing from the consensus. This shows that the VDR DBD has the ability to bind to and modulate transcription from suboptimal binding sites *in vivo*. An excellent example of this pliability in other members of the superfamily was seen in the original GR DBD–IR4 structure, in which one DBD protomer bound to a noncognate site to maintain the preferred IR3 protein arrangement (Luisi *et al.*, 1991). Similar maintenance of dimerization contacts at the cost of nonspecific or suboptimal protein–DNA interactions has also been observed recently for the AR DBD bound to a pair of steroid half-sites arranged as a direct repeat (Shaffer *et al.*, 2004). These examples highlight the fact that in the absence of direct visualization, it can be difficult to determine the exact arrangement of two proteins on a bipartite DNA target.

IV. RXR–VDR FORMATION

As mentioned previously, the VDR DBD and RXR DBD do not form heterodimers in the presence of DR3 DNA (Shaffer and Gewirth, 2002). This observation places them squarely at odds with the classical canon that the dimerization and DNA target discrimination behavior of the full-length receptors is recapitulated in the isolated DBDs. To test whether this resulted from a failure of the VDR DBD and the RXR DBD to associate or competition from VDR DBD homodimers, a triple mutation (Pro61Ala, Phe62Ala, and His75Ala) was made in the VDR DBD that was predicted to destabilize the homodimer association but not any potential heterodimer interactions (Fig. 5). Indeed, once the homodimer was destabilized, the

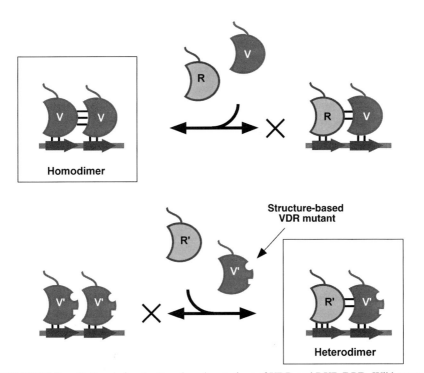

FIGURE 5. Rationale for structure-based mutations of VDR and RXR DBD. Wild-types VDR and RXR DBDs are labeled as V and R, mutant proteins are labeled V' and R', and the half-site DNA is represented as an arrow. The favored dimeric species are boxed, and the disfavored assembly pathway is indicated with an X. Figure adapted from Shaffer and Gewirth (2002).

RXR DBD:VDR DBD heterodimer was observed, showing that competition from the homodimer for the VDRE was responsible for the absence of the DBD heterodimer. The extent and nature of the RXR:VDR DBD interactions, however, have yet to be elucidated.

V. CONCLUSIONS

The crystal structure of the VDR DBD–DNA complex has provided stereochemical details of the protein–protein and protein–DNA interactions that dictate high-affinity response element binding and corroborated mutational and biochemical analyses done previously. Visualization of the VDR DBD with nonconsensus, naturally occurring response elements has led to the emerging view that coarse target discrimination (such as steroid versus nonsteroid hexameric sequences) is achieved entropically, whereas fine-tuning of affinity is accomplished by subtle enthalpic rearrangements. Moreover, the structure has allowed targeted mutagenesis of the VDR DBD and formation of RXR:VDR DBD heterodimers for the first time. When the structure of this complex is solved, it should allow further insight into the mode of response element discrimination employed by the full-length receptors.

The results described here represent a significant, but still expanding, understanding of VDR–DNA interactions, and many aspects are still uncharacterized. For example, if the interaction between the RXR DBD and VDR DBD is weak, how do RXR:VDR heterodimers discriminate between correct and incorrect response element spacers? Although modeling has shown that it is likely that DR2 binding is blocked sterically, as modeled for the VDR homodimer, and that DR4 binding fails to produce a cooperative interface, what intramolecular interactions, if any, on a DR3 element specify high-affinity binding? Two scenarios at this stage seem to be possible: (1) there are few but sufficient contacts between the RXR and VDR DBDs to allow cooperative and preferential binding, or (2) there are no DBD interactions and discrimination is achieved through interactions in other domains, such as the linker, or through quaternary interactions between the domains, such as the LBD and DBD. Systematic studies of the length and sequence requirements of the linker region should be performed to obtain needed results. What is the minimal length for a functional hinge domain, and are there specific sequences in the linker that are important? Another interesting area that bears further investigation is alternative response element arrangements. Does VDR bind cooperatively to DR4, DR6, and IP9 arranged half-sites, or does it bind to one high-affinity half-site and one non-consensus hexamer to maintain the cooperative interface available to a DR3 arrangement of promoters? These and other questions about VDR–DNA binding are fundamental to a detailed understanding of VDR action and may be answered by further structural and functional studies.

REFERENCES

Abe, E., Miyaura, C., Sakagami, H., Takeda, M., Konno, K., Yamazaki, T., Yoshiki, S., and Suda, T. (1981). Differentiation of mouse myeloid leukemia cells induced by 1 alpha, 25-dihydroxyvitamin D3. *Proc. Natl. Acad. Sci. USA* **78**, 4990–4994.

Baker, A. R., McDonnell, D. P., Hughes, M., Crisp, T. M., Mangelsdorf, D. J., Haussler, M. R., Pike, J. W., Shine, J., and O'Malley, B. W. (1998). Cloning and expression of full-length cDNA encoding human vitamin D receptor. *Proc. Natl. Acad. Sci. USA* **85**, 3294–3298.

Bouillon, R., Okamura, W. H., and Norman, A. W. (1995). Structure–function relationships in the vitamin D endo0crine system. *Endocr. Rev.* **16**, 200–257.

Carlberg, C., Bendik, I., Wyss, A., Meier, E., Sturzenbecker, L. J., Grippo, J. F., and Hunziker, W. (1993). Two nuclear signalling pathways for vitamin D. *Nature* **361**, 657–660.

Cheskis, B., and Freedman, L. P. (1994). Ligand modulates the conversion of DNA-bound vitamin D_3 receptor (VDR) homodimers into VDR:retinoid X receptor heterodimers. *Mol. Cell. Biol.* **14**, 3329–3338.

Cheskis, B., and Freedman, L. P. (1996). Modulation of nuclear receptor interactions by ligands: Kinetic analysis using surface plasmon resonance. *Biochemistry* **35**, 3309–3318.

Cheskis, B., Lemon, B. D., Uskokovic, M., Lomedico, P. T., and Freedman, L. P. (1995). Vitamin D_3:retinoid X receptor dimerization, DNA binding, and transactivation are differentially affected by analogs of 1,25-dihydroxyvitamin D_3. *Mol. Endocrinol.* **9**, 1814–1824.

Colnot, S., Lambert, M., Blin, C., Thomasset, M., and Perret, C. (1995). Identification of DNA sequences that bind retinoid X receptor: 1,25(OH)2D3 receptor heterodimers with high affinity. *Mol. Cell. Endocrinol.* **113**, 89–98.

DeLuca, H. F., and Zierold, C. (1998). Mechanisms and functions of vitamin D. *Nutr. Rev.* **56**, S4–S10; discussion S54–S75.

Feldman, D., Glorieux, F. H., and Pike, J. W., (Eds.) (1997). *Vitamin D*. Academic Press, San Diego, CA.

Freedman, L. P., and Towers, T. L. (1991). DNA binding properties of the vitamin D_3 receptor zinc finger region. *Mol. Endocrinol.* **5**, 1815–1826.

Gewirth, D. T., and Sigler, P. B. (1995). The basis for half-site specificity explored through a noncognate steroid receptor–DNA complex. *Nat. Struct. Biol.* **2**, 386–394.

Hsieh, J. C., Jurutka, P. W., Selznick, S. H., Reeder, M. C., Haussler, C. A., Whitfield, G. K., and Haussler, M. R. (1995). The T-box near the zinc fingers of the human vitamin D receptor is required for heterodimeric DNA binding and transactivation. *Biochem. Biophys. Res. Commun.* **215**, 1–7.

Hsieh, J. C., Whitfield, G. K., Oza, A. K., Dang, H. T., Price, J. N., Galligan, M. A., Jurutka, P. W., Thompson, P. D., Haussler, C. A., and Haussler, M. R. (1999). Characterization of unique DNA-binding and transcriptional-activation functions in the carboxyl-terminal extension of the zinc finger region in the human vitamin D receptor. *Biochemistry* **38**, 16347–16358.

Khorasanizadeh, S., and Rastinejad, F. (2001). Nuclear-receptor interactions on DNA-response elements. *Trends Biochem. Sci.* **26**, 384–390.

Klein-Hitpass, L., Schorpp, M., Wagner, U., and Ryffel, G. U. (1986). An estrogen-responsive element derived from the 5′ flanking region of the *Xenopus* vitellogenin A2 gene functions in transfected human cells. *Cell* **46**, 1053–1061.

Luisi, B. F., Schwabe, J. W., and Freedman, L. P. (1994). The steroid/nuclear receptors: From three-dimensional structure to complex function. *Vitam. Horm.* **49**, 1–47.

Luisi, B. F., Xu, W. X., Otwinowski, Z., Freedman, L. P., Yamamoto, K. R., and Sigler, P. B. (1991). Crystallographic analysis of the interaction of the glucocorticoid receptor with DNA. *Nature* **352**, 497–505.

Lundback, T., Zilliacus, J., Gustafsson, J. A., Carlstedt-Duke, J., and Hard, T. (1994). Thermodynamics of sequence-specific glucocorticoid receptor–DNA interactions. *Biochemistry* **33**, 5955–5965.

Mader, S., Chen, J. Y., Chen, Z., White, J., Chambon, P., and Gronemeyer, H. (1993). The patterns of binding of RAR, RXR, and TR homo- and heterodimers to direct repeats are dictated by the binding specificites of the DNA binding domains. *EMBO J.* **12**, 5029–5041.

Mangelsdorf, D. J., and Evans, R. M. (1995). The RXR heterodimers and orphan receptors. *Cell* **83**, 841–850.

Mangelsdorf, D. J., Thummel, C., Beato, M., Herrlich, P., Schutz, G., Umesono, K., Blumberg, B., Kastner, P., Mark, M., Chambon, P., and Evans, R. M. (1995). The nuclear receptor superfamily: The second decade. *Cell* **83**, 835–839.

McKenna, N. J., Xu, J., Nawaz, Z., Tsai, S. Y., Tsai, M. J., and O'Malley, B. W. (1999). Nuclear receptor coactivators: Multiple enzymes, multiple complexes, multiple functions. *J. Steroid Biochem. Mol. Biol.* **69**, 3–12.

Miyamoto, T., Kakizawa, T., Ichikawa, K., Nishio, S., Takeda, T., Suzuki, S., Kaneko, A., Kumagai, M., Mori, J., Yamashita, K., Sakuma, T., and Hashizume, K. (2001). The role of hinge domain in heterodimerization and specific DNA recognition by nuclear receptors. *Mol. Cell. Endocrinol.* **181**, 229–238.

Nishikawa, J., Kitaura, M., Matsumoto, M., Imagawa, M., and Nishihara, T. (1994). Difference and similarity of DNA sequence recognized by VDR homodimer and VDR:RXR heterodimer. *Nucleic Acids Res.* **22**, 2902–2907.

Perlmann, T., Rangarajan, P. N., Umesono, K., and Evans, R. M. (1993). Determinants for selective RAR and TR recognition of direct-repeat HREs. *Genes Dev.* **7**, 1411–1422.

Perlmann, T., Umesono, K., Rangarajan, P. N., Forman, B. M., and Evans, R. M. (1996). Two distinct dimerization interfaces differentially modulate target gene specificity of nuclear hormone receptors. *Mol. Endocrinol.* **10**, 958–966.

Polly, P., Carlberg, C., Eisman, J. A., and Morrison, N. A. (1996). Identification of a vitamin D_3 response element in the fibronectin gene that is bound by a vitamin D_3 receptor homodimer. *J. Cell. Biochem.* **60**, 322–333.

Quack, M., and Carlberg, C. (2000). Ligand-triggered stabilization of vitamin D receptor:retinoid X receptor heterodimer conformations on DR4-type response elements. *J. Mol. Biol.* **296**, 743–756.

Quack, M., Szafranski, K., Rouvinen, J., and Carlberg, C. (1998). The role of the T-box for the function of the vitamin D receptor on different types of response elements. *Nucleic Acids Res.* **26**, 5372–5378.

Rachez, C., and Freedman, L. P. (2001). Mediator complexes and transcription. *Curr. Opin. Cell. Biol.* **13**, 274–280.

Rastinejad, F. (2001). Retinoid X receptor and its partners in the nuclear receptor family. *Curr. Opin. Struct. Biol.* **11**, 33–38.

Rastinejad, F., Evilia, C., and Lu, P. (1995a). Studies of nucleic acids and their protein interactions by 19F NMR. *Methods Enzymol.* **261**, 560–575.

Rastinejad, F., Perlmann, T., Evans, R. M., and Sigler, P. B. (1995b). Structural determinants of nuclear receptor assembly on DNA direct repeats. *Nature* **375**, 203–211.

Rastinejad, F., Wagner, T., Zhao, Q., and Khorasanizadeh, S. (2000). Structure of the RXR–RAR DNA-binding complex on the retinoic acid response element DR1. *EMBO J.* **19**, 1045–1054.

Rhodes, S. J., Chen, R., DiMattia, G. E., Scully, K. M., Kalla, K. A., Lin, S. C., Yu, V. C., and Rosenfeld, M. G. (1993). A tissue-specific enhancer confers Pit-1-dependent morphogen inducibility and autoregulation on the pit-1 gene. *Genes Dev.* **7**, 913–932.

Rochel, N., Tocchini-Valentini, G., Egea, P. F., Juntunen, K., Garnier, J. M., Vihko, P., and Moras, D. (2001). Functional and structural characterization of the insertion region in the ligand binding domain of the vitamin D nuclear receptor. *Eur. J. Biochem.* **268**, 971–979.

Rochel, N., Wurtz, J. M., Mitschler, A., Klaholz, B., and Moras, D. (2000). The crystal structure of the nuclear receptor for vitamin D bound to its natural ligand. *Mol. Cell* **5**, 173–179.

Scheidereit, C., Westphal, H. M., Carlson, C., Bosshard, H., and Beato, M. (1986). Molecular model of the interaction between the glucocorticoid receptor and the regulatory elements of inducible genes. *DNA* **5**, 383–391.

Schrader, M., Kahlen, J. P., and Carlberg, C. (1997). Functional characterization of a novel type of 1 alpha, 25-dihydroxyvitamin D_3 response element identified in the mouse c-fos promoter. *Biochem. Biophys. Res. Commun.* **230**, 646–651.

Schrader, M., Nayeri, S., Kahlen, J. P., Muller, K. M., and Carlberg, C. (1995). Natural vitamin D_3 response elements formed by inverted palindromes: Polarity-directed ligand sensitivity of vitamin D_3 receptor: retinoid X receptor heterodimer-mediated transactivation. *Mol. Cell. Biol.* **15**, 1154–1161.

Schwabe, J. W., Chapman, L., Finch, J. T., and Rhodes, D. (1993). The crystal structure of the estrogen receptor DNA binding domain bound to DNA: How receptors discriminate between their response elements. *Cell* **75**, 567–578.

Schwabe, J. W., and Rhodes, D. (1991). Beyond zinc fingers: Steroid hormone receptors have a novel structural motif for DNA recognition. *Trends Biochem. Sci.* **16**, 291–296.

Shaffer, P. L., and Gewirth, D. T. (2002). Structural basis of VDR–DNA interactions on direct-repeat response elements. *EMBO J.* **21**, 2242–2252.

Shaffer, P. L., Jivan, A., Dollins, D. E., Claessens, F., and Gewirth, D. T. (2004). Structural basis of androgen receptor binding to selective androgen response elements. *Proc. Natl. Acad. Sci. USA* **101**, 4758–4763.

Towers, T. L., and Freedman, L. P. (1998). Granulocyte-macrophage colony-stimulating factor gene transcription is directly repressed by the vitamin D_3 receptor. Implications for allosteric influences on nuclear receptor structure and function by a DNA element. *J. Biol. Chem.* **273**, 10338–10348.

Towers, T. L., Luisi, B. F., Asianov, A., and Freedman, L. P. (1993). DNA target selectivity by the vitamin D_3 receptor: Mechanism of dimer binding to an asymmetric repeat element. *Proc. Natl. Acad. Sci. USA* **90**, 6310–6314.

Umesono, K., Murakami, K. K., Thompson, C. C., and Evans, R. M. (1991). Direct repeats as selective response elements for the thyroid hormone, retinoic acid, and vitamin D_3 receptors. *Cell* **65**, 1255–1266.

Xie, Z., and Bikle, D. D. (1997). Cloning of the human phospholipase C-gamma1 promoter and identification of a DR6-type vitamin D-responsive element. *J. Biol. Chem.* **272**, 6573–6577.

Zechel, C., Shen, X. Q., Chambon, P., and Gronemeyer, H. (1994a). Dimerization interfaces formed between the DNA binding domains determine the cooperative binding of RXR:RAR and RXR:TR heterodimers to DR5 and DR4 elements. *EMBO J.* **13**, 1414–1424.

Zechel, C., Shen, X. Q., Chen, J. Y., Chen, Z. P., Chambon, P., and Gronemeyer, H. (1994b). The dimerization interfaces formed between the DNA binding domains of RXR, RAR, and TR determine the binding specificity and polarity of the fulllength receptors to direct repeats. *EMBO J.* **13**, 1425–1433.

Zhao, Q., Chasse, S. A., Devarakonda, S., Sierk, M. L., Ahvazi, B., and Rastinejad, F. (2000). Structural basis of RXR–DNA interactions. *J. Mol. Biol.* **296**, 509–520.

Zhao, Q., Khorasanizadeh, S., Miyoshi, Y., Lazar, M. A., and Rastinejad, F. (1998). Structural elements of an orphan nuclear receptor–DNA complex. *Mol. Cell* **1**, 849–861.

INDEX

Page numbers followed by f and t indicate figures and tables, respectively.

A

A/B domain. *See* N-terminal region
Acetylation
  histone, 106
  of lysine residues, 98–99
  SRCs and, 242
Activating signal cointegrator-2 (ASC-2), 108–109
Activation domain
  transcriptional, 159
Activation functions
  of NRs, 146–147
Active repression domain
  in LBD, 11
AF-2 helix
  in GR LBD, 59–63
AGGTCA core motif
  direct repeats of, 8
AI-regulatory protein-1 (ARP-1), 7
Alpha-helices, 5
Amino acid sequence alignment
  of NR boxes, 152f
Amino-terminal residues, 167
Amphibian development
  embryonic, 217–218
  gene expression profiles during, 216–217
  larval, 219–221
  metamorphosis in, 221–222
Androgen receptors (AR), 56
  Cdc25 coactivation of, 246–247

Cdc25 direct interaction with, 247–248
DAX-1 and, 19
interaction sites on, 250
role of, in breast cancer, 249–250
ApoAI. *See* Apolipoprotein AI
Apo-B gene, 13
Apolipoprotein AI (ApoAI), 7
AR. *See* Androgen receptors
ARC. *See* SREBP-interacting complex
A-ring polar clamp
  of GR, 64
ARP-1. *See* AI-regulatory protein-1
ASC-2. *See* Activating signal cointegrator-2
ATP-binding casette transporter 1 (ABCA)
  SHP regulation of, 29
ATP-dependent nucleosome remodeling
  factors, 160–161

B

Baculovirus
  expression system of, 57–58
Basic helix-loop-helix domain, 103

C

cAMP response element binding protein (CREB), 5–6, 14
Cancer and
  Cdc25 proteins, 235–236

CARM1. *See* Coactivator arginine
   methyltransferase 1
CBP associated factor, 154
CBP binding proteins, 153
CBP coactivators, 104–106
CBP. *See* CREB-binding protein
Cdc25 proteins
   activated ras and, 238*f*
   AR coactivation by, 246–247
   AR direct interaction with, 247–248
   cancer and, 235–236
   coactivation of, 243*f*
   domains of, 239–240, 242*f*
   ectopic expression of, 236–238
   ER interaction with, 241
   ER responsive genes in, 238–239
   expression of, 235
   historical perspective on, 234
   in murine mammary glands, 236–239
   overexpression of, 246
   in prostate, 245–248
   regulation of, in cell cycles, 234–235
   role of, 233, 248–249
   role of, in alveolar hyperplastic transgenic
      mammary glands, 249
   separable functions of, 243–244
   steroid receptor interaction with,
      241–242
   steroid receptor transcription and, 244
Cdc25B transgenic mice
   MMTV, 237, 238*f*
CDK. *See* Cyclin-dependent kinases
CDKI. *See* Cyclin-dependent
   kinase inhibitors
C/EBP, 13
Cell-free transcription
   PR-B dependent, 244–245
Chaperone protein association, 74–76
   LBD in, 74–76
Charge clamp motif, 165
Chicken ovalbumin upstream promoter-
   transcription factor (COUP-TF), 7
   functional domains of, 8*f*
   gene expression by, 10*f*
   gene silencing induced by, 9
   gene transrepression of, 13
   HDAC and, 12
   LHR gene and, 12
   target gene expression and, 11
   as transcription repressors, 8
Chromatin, 5
   ATP-dependent, 149, 160
   modifying, 102–103, 213–214

nucleosomes, 96–97
structure of, 96–97, 95*f*
Chromatin immunoprecipitation
   (ChIP), 129
   assays, 217
Chromosome condensation
   mitotic, 99
Ciliary neurotrophic factor (CNTF), 33
Class I receptors
   coactivation of, 245*f*
Coactivator activator (CoAA), 158–159
Coactivator arginine methyltransferase 1
   (CARM1), 106
   C-terminal of, 155
   functional domains of, 107*f*
   in protein function, 154–155
Coactivator modulator (COAM),
   158–159
Coactivator proteins
   indenitying, 147–148
Coactivators
   activators, 158–159
   Cdc25 proteins as, 239–240
   corepressors and, 148
   defining, 128
   dynamic cycling of, 129
   histone acetyltransferase and, 128–129
   mechanisms of action of, 148–150
   modulators, 158–159
   overexpression of, 172
   p160, 150–151
   pharmacology of, 170–174
   recognition of, 164–169
   steroid receptor, 150
   structural organization of, 151*f*
   transcription amplifying cofactors and,
      188–190
COAM. *See* Coactivator modulator
Coregulators
   two-step mechanism of, 149*f*
Coregulatory proteins
   in transcription regulation, 127–132
Corepressor NR recognition motif (CoRNR
   box), 164
Corepressors, 162–163
   coactivators and, 148
   dual function of, 225*f*
   in larval growth, 222
   mechanisms of action of, 148–150
   overexpression of, 172
   pharmacology of, 170–174
   recognition of, 169–170
   recruitment, 219–221, 223

INDEX

structural organization of, 163f
transcription silencing cofactors and,
  187–188
TR-mediated gene repression and, 220
in *Xenopus,* 212–213
CoRNR box. *See* Corepressor NR
  recognition motif
Corticosterone
  role of, 51
Cortisol
  role of, 51
COUP-TF. *See* Chicken ovalbumin upstream
  promoter-transcription factor
CREB. *See* cAMP response element
  binding protein
CREB-binding protein (CBP), 242
  coactivation of, 243f
C-ring polar interaction region
  of GR, 66
Crystallization
  of GR, 58
CtBP. *See* C-terminal binding protein
C-terminal binding protein (CtBP)
  LCoR/RIP140 recruitment of, 135
C-terminal region, 5
  of CARM1, 155
  of GR, 53
  of LBD, 16
  of TRBP, 156
Cyclin D1
  overexpression of, 235–236
  transfections in, 251
  upregulation of, 249
Cyclin D1 protein
  levels of, 238
Cyclin E
  overexpression of, 235–236
Cyclin/CDK complex, 234
Cyclin-dependent kinase inhibitors (CDKIs)
  in cell cycles, 233
Cyclin-dependent kinases (CDKs)
  in cell cycles, 233
Cyclooxygenase-2, 82–83
CYP7A1, 27
  feedback regulation of, 28
CYP11B2, 12
CYP17, 12

D

DAD. *See* Deacetyalse activating domain
DAX-1
  AHC mutants, 19

expression of, 16
gene expression and, 15f
gene silencing and, 14–20
N-terminal of, 15
as nuclear orphan receptor, 19–20
regions of, 7
SF-1 and, 17
DBD. *See* DNA binding domain
DBP. *See* DNA binding proteins
Deacetyalse activating domain
  (DAD), 163
Dex. *See* Dexamethasone
Dexamethasone (Dex), 51, 58
  crystal structure of, 67
  electron density of, 72–73
  GR binding with, 65
  mesylate, 172
  oxetanone, 172
  RU486 and, 66
DHT. *See* Dihydrotestosterone
Dihydrotestosterone (DHT), 68–69
  AR ligand, 33
Direct repeats
  of AGGTCA core motif, 8
  domains, 31
DNA binding domain (DBD), 3
  carboxy-terminal extension of, 258
  domain organization of, 260f
  of GCNF, 20
  of GR, 51
  modeling studies based on, 267f
  mutations of, 269f
  receptor dimerization and, 4
  of SHP, 25–26
  of VDRs, 261, 263
DNA binding proteins (DBP), 211
Domain C. *See* DNA binding domain
Domain D. *See* Ligand binding domain
DR0 domain
  LHR gene and, 12
DR3
  response elements of, 262f
  VDR binding, 264–266
DR6
  VDREs with, 268
D-ring interaction region
  binding pocket, 66–67
  of GR, 66
DRIP. *See* Vitamin D receptor-interacting
  protein
Drosophila
  chromosomal puffing in, 223
  nuclear proteins in, 110

E

E. coli
  expression system of, 57–58
EAR2
  LHR gene and, 12
EAR3, 7
Effector proteins
  binding sites for, 102f
Epigenetic markers
  histone modifications as, 101–102
ER. See Estrogen receptors
ERα. See Estrogen receptor α
ERR2. See Estrogen-related receptor 2
ERRα. See Estrogen-related receptor α
Estradiol, 68
Estrogen receptor α (ERα), 18
  crystal structure of, 127f, 171
Estrogen receptors (ERs)
  Cdc25 interaction with, 241
  Cdc25 transgenics and, 238–239
  coactivation of, 240f
  coexpression of, 239–240
  interaction sites on, 250
  role of, in breast cancer, 249–250
Estrogen signaling pathways
  SHP interaction with, 26
Estrogen-related receptor 2 (ERR2), 18
Estrogen-related receptor α (ERRα)
  in gene activation, 24
Eukaryotic cells
  genomic DNA in, 213

F

Farnesoid X receptor (FXR), 2
  SHP activation of, 29
Feedback hypothesis
  of TR function, 223–225
  in Xenopus, 224
FTZ-F1, 18
FXR. See Farnesoid X receptor

G

Gain of function mutations
  in GR, 56
GAL4, 19
GCNF. See Germ cell nuclear factor
Gene silencing
  COUP-TF induced, 9
General transcription factors (GTF), 188
Genestein
  structure of, 171

Germ cell nuclear factor (GCNF), 7
  DBD of, 20
  DNA binding properties of, 21f
  expression of, 21
  gene expression transrepression and, 24
  LBD of, 20
  role of, 20
  zygotic development and, 22
Glucocorticoid receptors (GR)
  activated, 55
  AF-2 helix in, 59–63
  A-ring polar clamp, 64
  C-ring polar interaction region of, 66
  crystal structure of, 60f, 61f, 62f
  crystallization of, 58
  C-terminal of, 53
  DBD of, 53
  defining, 51
  Dex binding with, 65
  domain-stabilizing mutations of, 63–68
  D-ring interaction region of, 66
  electron density in, 72–73
  endogenous ligands for, 52f
  gain of function mutations in, 56
  general features of, 59
  GREs and, 54
  helix 5 region of, 57f
  homodimer interface, 75–76
  LBD expression in, 56–57
  LBD of, 53
  ligand binding pocket of, 65f
  ligand-recognition by, 64–68
  ligand-regulated functions of, 55
  linear schematic of, 53f
  M604 interaction region, 64
  mutations in disease, 77–82
  N546 in, 71
  N-terminal of, 53
  selected mutations, 78t–80t
  selective modulators, 82–83
  site-directed mutagenesis studies, 76–77
  602 position of, 63f
  TIF2 coactivator pocket in, 73f
  transrepression activities of, 54f
Glucocorticoid response elements (GREs)
  GRs and, 54
Glucocorticoids
  endogenous, 51
  N546 in, 71
  Q642 in, 70
  selectivity for, 70–74
Glutamic acid residue, 165, 167
GR. See Glucocorticoid receptors

GRE. *See* Glucocorticoid response elements
GRIK5, 9
GR/RU486 antagonists, 60–61
GTF. *See* General transcription factors
Guanidino arginine residues, 154

H
HaCaT
  keratinocytes, 32
HAT. *See* Histone acetyltransferase
HDAC. *See* Histone deacetylase
Hepatocyte nuclear factor (HNF), 9, 33
Heterogenous nuclear ribonucleoproteins (hnRNPs), 108
Histone acetyltransferase (HAT), 149–150
  recruitment of, 242–243, 248
  SRC carrying, 214
Histone deacetylase (HDAC), 99
  class II 6, 133–134
  class II 10, 133–134
  COUP-TF and, 12
  domain organization of, 133f
  in gene expression, 132–135
  human classes, 132–133
  LCoR/RIP140 recruitment of, 134–135
  role of, in metamorphosis, 221–223
  role of, in transcriptional repression, 221
  Western blot analysis of, 221–222
Histone modifications
  as binding sites, 102f
  codes, 100–103
  differential transcription states and, 101
  as epigenetic markers, 101–102
  models of, 213–214
  NR coactivators, 103–109
  patterns of, 99f
  in regulated transcription, 98–103
HNF. *See* Hepatocyte nuclear factor
HNF3α, 13
HNF4, 13
hnRNP. *See* Heterogenous nuclear ribonucleoproteins
Hormone response elements (HREs), 3
HRE. *See* Hormone response elements
HSP90, 74–75
Hydroxylase, 16
Hyperthyroidism, 200
Hypothyroidism, 192

I
Interleukin-1, 82–83
Interleukin-8, 82–83
IP9
  VDREs with, 268

L
LBD. *See* Ligand binding domain
LCoR. *See* Ligand-dependent corepressors
Leucine, 18
LHR
  COUP-TF/EAR2 and, 12
  DR0 domain and, 12
  TATA-less, 13
Ligand binding domain (LBD), 5
  active repression domain in, 11
  AF-2 in, 59–63
  carboxy-terminal, 258
  in chaperone protein association, 74–76
  crystal structure of, 60f, 61f, 62f, 166f
  C-terminal of, 16
  electron density of, 72–73
  expression of, in GR, 56–57
  of GCNF, 20
  general features of, 59
  of GR, 53
  homodimer interface, 75–76
  mutations in disease, 77–82
  of NRs, 131–132
  in receptor dimerization, 74–76
  in region E, 126
  schematic structure of, 4f
  selected mutations, 78t–81t
  of SHP, 25–26
  site-directed mutagenesis studies, 76–77
  of VDRs, 261
Ligand-dependent corepressors (LCoR), 113–114
  CtBP recruitment of, 135
  differential binding of, 131–132
  HDAC recruitment of, 135–136
  in hormone-dependent receptor function, 136–137
  primary structure of, 134f
Liver X receptor, 2
Lobuloalveolar development, 238–239
LXR:RXR dimer, 28
Lysine residues, 165
  acetylation of, 98–99

M

M604
  interaction region, 64
Mammalian cell cycle
  phases of, 233–234
Mammary tumor virus (MMTV), 33
Metamorphosis
  corepressor expression during, 221–222
  HDAC involvement in, 222–223
  T3 initiation of, 210–211
  thyroid hormone control in, 224f
  TSA in, 223f
Methylation
  of core histones, 100
Mineralocorticoid receptors, 70
  N770 in, 71
Mineralocorticoids
  N546 in, 71
  Q642 in, 70
  selectivity for, 70–74
MIS. *See* Mllerian-inhibiting substance
MMTV. *See* Mammary tumor virus
Mllerian-inhibiting substance (MIS)
  expression of, 16
Murine mammary glands
  Cdc25 proteins in, 236–239
Mutagenesis studies
  GR LBD, 76–77
MyoD, 13

N

NCNF. *See* Neuronal cell nuclear factor
NCoA. *See* Nuclear receptor coactivators
NCoR. *See* Nuclear receptor corepressors
Nerve growth factor inducible-B (NGF1-B), 4
Neuronal cell nuclear factor (NCNF), 20
NGF1-B. *See* Nerve growth factor inducible-B
NLS. *See* Nuclear localization signals
Northern blot analysis
  of NCoAs, 194, 196
NRF-1. *See* Nuclear respiratory factor-1
NRs. *See* Nuclear receptors
NSD1. *See* Nuclear receptor SET-domain containing protein 1
N-terminal
  of DAX-1, 15
  of GR, 51
N-terminal region, 3
Nuclear localization signals (NLS), 19
Nuclear orphan receptors
  DAX-1 as, 19–20

Nuclear receptor boxes
  amino acid sequence alignment of, 152f
  classes of, 168
  crystal structures of, 165
Nuclear receptor coactivators (NCoA)
  histone modifying, 103–109
  identifying, 188
  Northern blot analysis of, 194, 196
  pharmacology of, 170–174
  recognition of, 164–171
  TR interaction with, 191
Nuclear receptor corepressors (NCoR), 11, 129–130
  box-containing, 130–131
  histone modifying, 107–114
  pharmacology of, 170–174
  recognition of, 169–170
  silencing mechanisms and, 18
  structural organization of, 163
Nuclear receptor SET-domain containing protein 1 (NSD1), 130
Nuclear receptors (NRs)
  activation functions of, 146–147
  coactivators/corepressors in, 150–163
  conserved domains of, 95f
  crystal structure of, 166f
  defining, 94
  domain organization of, 125–127, 260f
  family of, 2, 49–50
  functions of, 5
  histone modifying, 105t
  historical perspective on, 146
  LBDs of, 131–132
  ligands, 125, 173
  orphan, 95
  pharmacology of, 170–174
  roles of, 259
  schematic structure of, 4f, 126f
  subgroups of, 6
  transcriptional repression by, 132–133
Nuclear respiratory factor-1 (NRF-1), 160
Nucleosomes,
  separation of, 97–98
  structure of, 97f

O

Oct4, 9, 23
Oxosteroid receptors, 68–71
Oxytocin, 9

INDEX 281

P

P23, 74–75
P160
  coactivators, 150–151
  recruitment of, 173
  SRC-1 and, 199–200
P160/SRC family, 103–104
  differential binding of, 131–132
  functional domains of, 104f
P300 associated factor, 154
P300 binding proteins, 153–154
P300 coactivators, 104–106
P300/CBP-associated factor (PCAF), 242
  coactivation of, 243f
P625, 76
PCAF. *See* P300/CBP-associated factor
PCR. *See* Polymerase chain reaction
Per/Arnt/Sim (PAS) homology
  domain, 103
Peroxisome proliferator-activated receptor
  (PPAR), 2, 9, 10
  coactivator 1, 159–160
  SHP interaction with, 27
  signaling pathways, 32–33
Peroxisome proliferator-activated receptorγ
  coactivator 1 (PGC-1), 159–160
PGC-1. *See* Peroxisome proliferator-activated
  receptorγ coactivator 1
Phosphorylation
  histone, 99–100
Pituitary
  TR/SRC interaction in, 196–200
Pituitary thyroid axis
  physiology of, 190f
Polymerase chain reaction (PCR), 199
POU domain
  transcription factors of, 23
PPAR. *See* Peroxisome proliferator-activated
  receptor
PR-B
  cell-free transcription and, 244–245
Prednisolone, 51
Pregnane X receptor (PXR), 2
PRMT1. *See* Protein arginine
  methyltransferases
Progesterone, 69–70
Progesterone receptors (PR), 56
Prostate
  Cdc25B in, 245–248
Prostate-specific antigen (PSA), 33
Protamine 1/2, 23
Protein arginine methyltransferases
  (PRMT1), 106

  functional domains of, 107f
  in protein function, 154–155
PXR. *See* Pregnane X receptor

R

Raloxifene
  crystal structure of, 171
RAR. *See* Retinoid acid receptor
RARE. *See* Retinoic acid response elements
Receptor cDNAs
  cloning, 124–125
  identifying, 124–125
Receptor dimerization
  LBD in, 74–76
Region E
  receptor LBDs in, 126
Reproductive tissues, 14–15
Retinoic acid induced differentiation, 23
Retinoic acid receptor-related testis-
  associated receptor (RTR), 20
Retinoic acid response elements (RARE), 9
Retinoid acid receptor (RAR)
  binding of, 26
  heterodimerization of, 23
Retinoid X receptor (RXR), 2
  binding of, 26
  domain organization of, 260f
  heterodimerization of, 23
  mutations of, 269f
  overexpression of, 218
  titration of, 9
  VDR formation, 269–270
Retinoid-related orphan receptor (ROR), 4
RIP140, 113–114
  as corepressor, 130
  CtBP recruitment of, 135
  differential binding of, 131–132
  HDAC recruitment of, 135–136
  in hormone-dependent receptor function,
    136–137
  primary structure of, 134f
RNA polymerase II, 147
RNA recognition motifs (RRMs), 158
RRM. *See* RNA recognition motifs
RTR. *See* Retinoic acid receptor-related
  testis-associated receptor
RU486
  crystal structure of, 67
  Dex and, 66
  pharmacological character
    of, 172
RXR. *See* Retinoid X receptor

S

SANT domains, 109–111
Schizosaccharomyces pombe, 234
SF-1. *See* Steroidogenic factor 1
SHP. *See* Small heterodimer partners
Silencing mediator for retinoid acid and thyroid hormone receptors (SMRT), 11–12
    as corepressor, 129–130
    histone modifying, 109–113
    length of, 162
    schematic diagram of, 111*f*
    structural organization of, 163*f*
Sin3
    Western blot analysis of, 221–222
Small heterodimer partners (SHP)
    ABCA1 regulation and, 29
    binding inhibition of, 26
    cloning of, 24–25
    domains of, 25–26
    estrogen signaling pathway interaction with, 26
    gene promoter activty of, 29
    human, 25–26
    in mice, 28–29
    physiological distribution of, 27
    PPAR interaction with, 27
SMRT. *See* Silencing mediator for retinoid acid and thyroid hormone receptors
Sp1, 12
Sp3, 12
SRA. *See* Steroid receptor RNA activators
SRC. *See* Steroid receptors coactivators
SRC-2 knockout mice, 194
SREBP-interacting complex (ARC), 157
StAR. *See* Steroidogenic acute regulatory protein
Steroid receptor RNA activators (SRA), 161–162
Steroid receptors coactivators (SRCs), 150
    acetylation and, 242–243
    HAT carried by, 214
    mRNA expression levels, 200
    p160 coactivators and, 199
    in peripheral tissues, 200–202
    physiology of, 194–196
    TRβ and, 196–199
    TRs and, 188, 189*f*
Steroidogenic acute regulatory protein (StAR), 16
Steroidogenic factor 1 (SF-1), 4
    DAX-1 and, 17
Steroidogenic tissues, 14–15

T

T3. *See* 3,4,3′-triiodothyronine
Tamoxifen, 239
Target gene expression
    COUP-TF and, 10*f*
    silencing, 11
TATA-binding proteins, 147
TATA-box genes
    LHR and, 13
Testicular receptor 2 (TR2), 7
    binding properties of, 31–32
    cloning of, 30–31
    functional domains of, 31*f*
    gene expression and, 33
    regulatory functions of, 32–33
    role of, 30
    role of, in down-regulation, 32
Testicular receptor 4 (TR4), 7
    binding properties of, 31–32
    cloning of, 30–31
    functional domains of, 31*f*
    gene expression and, 33
    regulatory functions of, 32–33
    role of, 30
    role of, in down-regulation, 32
3,4,3′-triiodothyronine (T3)
    binding, 214
    expression of, 216*f*
    metamorphosis and, 210–211
    repression of, 219–221
    response genes, 211–212
    role of, 186
    treatment with, 197
Thyroglobulin, 190
Thyroid axis
    regulation of, 191–192
Thyroid function tests
    in male mice, 195*t*
Thyroid hormone
    control of, in metamorphosis, 224*f*
    deprivation of, 197*f*
    developmental levels, 216*f*
Thyroid hormone receptor alpha (TRα)
    expression of, 216*f*
    heart rate and, 201
    TRβ and, 187
Thyroid hormone receptor beta (TRβ)
    expression of, 216*f*
    SRC-1 and, 196–199
    TRα and, 187
Thyroid hormone receptor-associated protein (TRAP)

complex, 156–158
recruitment of, 173
Thyroid hormone receptor-binding protein (TRBP), 155–156
C-terminal region of, 156
Thyroid hormone receptors (TR), 9, 10, 23
alpha, 186
beta, 186
binding proteins, 155–156
dimerization of, 188
dual function of, 225f
feedback hypothesis of, 223–225
gene repression mediated by, 220f
in larval growth, 221
models of, 214–215
NCoA interaction with, 189
organization of, 186
overexpression of, 218f
in peripheral tissues, 200–202
response genes, 216–217
SRCs and, 188, 189f
transcriptional regulation by, 215f
in *Xenopus*, 212–213
Thyroid hormone regulation, 190–191
Thyroid response elements (TREs)
gene regulation and, 187
Thyroid stimulation hormone (TSH)
gene expression of, 196–199
role of, 200
TIF2. *See* Transcriptional intermediary factor 2
TR. *See* Thyroid hormone receptors
TR2. *See* Testicular receptor 2
TR4. *See* Testicular receptor 4
TRα knockout mice
Lyon, 192–193
Stockholm, 192–193
TRβ knockout mice and, 193–194
TRα. *See* Thyroid hormone receptor alpha
Transcription amplifying cofactors
coactivators and, 188–189
Transcription factors
eukaryotic, 2
Transcription receptors
COUP-TFs as, 8
Transcription regulation
coregulatory proteins in, 127–132
Transcription silencing cofactors
corepressors and, 187–188
Transcriptional intermediary factor 2 (TIF2), 52
coactivator pocket, 73f

TRAP. *See* Thyroid hormone receptor-associated protein
TRβ knockout mice
Lyon, 191–192
Nutley, 191–192
TRα knockout mice and, 193–194
TRβ. *See* Thyroid hormone receptor beta
TRBP. *See* Thyroid hormone receptor-binding protein
TRE. *See* Thyroid hormone receptors
Trichostatin A (TSA), 12
in metamorphosis, 223f
TSA. *See* Trichostatin A
TSH. *See* Thyroid stimulation hormone

V

VDR. *See* Vitamin D receptors
VDRE. *See* Vitamin D response elements
Vitamin D receptor-interacting protein (DRIP)
complex, 156–158
Vitamin D receptors (VDR), 9, 10
alternative response elements, 266–269
DBD of, 261, 263
defining, 258
DR3 binding, 261–266
homodimerization of, 265f
LBD of, 261
modeling studies based on, 267f
mutations of, 269f
RXR formation, 269–270
Vitamin D response elements (VDREs), 264–265
discrimination, 266
with DR6/IP9, 268

W

Western blot analysis
of HDAC1, 221–222
of Sin3, 221–222

X

Xenopus
development of, 219
feedback hypothesis in, 224–225
larval, 219–221
metamorphosis in, 221–222
TR corepressor in, 212–213

Z

Zygotic development
GCNF and, 22